Advances in Experimental Medicine and Biology

Volume 976

More information about this series at http://www.springer.com/series/5584

Contents

Contributors

Xia Ding Mouse Cancer Genetics Program, National Cancer Institute, NIH, MD, USA

Shengjie Feng Department of Physiology, University of California, San Francisco, CA, USA

Karina Griesi-Oliveira Hospital Albert Einstein, São Paulo, SP, Brazil

Qian He Institute of Neuroscience, Shanghai Institutes for Biological Sciences, Chinese Academy of Sciences, Shanghai, China

Zhuohao He Department of Pathology and Laboratory Medicine, Institute on Aging and Center for Neurodegenerative Disease Research, University of Pennsylvania School of Medicine, Philadelphia, PA, USA

Junbo Huang Toronto Western Hospital Research Institute, Toronto, ON, Canada

Yichang Jia School of Medicine, Tsinghua University, Beijing, China

Peking-Tsinghua Joint Center for Life Sciences, Beijing, China

IDG/McGovern Institute for Brain Research at Tsinghua, Beijing, China

Hongyu Li Columbia University Medical Center, New York, NY, USA

Shanshan Li Department of Molecular and Cellular Biology, Baylor College of Medicine, Houston, TX, USA

Yong Liu Center of Cognition and Brain Science, Institute of Basic Medical Science, Beijing, China

Rui Lu Institute of Neuroscience, Shanghai Institutes for Biological Sciences, Chinese Academy of Sciences, Shanghai, China

Yoshito Mizoguchi Department of Psychiatry, Faculty of Medicine, Saga University, Saga, Japan

Akira Monji Department of Psychiatry, Faculty of Medicine, Saga University, Saga, Japan

Alysson Renato Muotri Department of Pediatrics and Department of Cellular & Molecular Medicine, University of California San Diego School of Medicine, La Jolla, CA, USA

Rady Children's Hospital San Diego, San Diego, CA, USA

UCSD Stem Cell Program, Institute for Genomic Medicine, New York, NY, USA

Anne Schaar Department of Biomedical Sciences, School of Medicine and Health Sciences, University of North Dakota, Grand Forks, ND, USA

Senthil Selvaraj Department of Biomedical Sciences, School of Medicine and Health Sciences, University of North Dakota, Grand Forks, ND, USA

Wei Shao Center of Cognition and Brain Science, Institute of Basic Medical Science, Beijing, China

Yu Shi Department of Thoracic Surgery, Zhongshan Hospital, Fudan University, Shanghai, People's Republic of China

Brij B. Singh Department of Biomedical Sciences, School of Medicine and Health Sciences, University of North Dakota, Grand Forks, ND, USA

Pramod Sukumaran Department of Biomedical Sciences, School of Medicine and Health Sciences, University of North Dakota, Grand Forks, ND, USA

Yuyang Sun Department of Biomedical Sciences, School of Medicine and Health Sciences, University of North Dakota, Grand Forks, ND, USA

Angela May Suzuki Department of Genetics and Evolutionary Biology, Bioscience Institute, University of São Paulo, São Paulo, SP, Brazil

Yilin Tai Cold Spring Harbor Laboratory, Cold Spring Harbor, NY, USA

Junfeng Wang Research and Development of Biogen, Cambridge, MA, USA

Shenglian Yang Center of Cognition and Brain Science, Institute of Basic Medical Science, Beijing, China

Cheng Zhan Department of Thoracic Surgery, Zhongshan Hospital, Fudan University, Shanghai, People's Republic of China

Fang Zheng Department of Pharmacology and Toxicology, College of Medicine, University of Arkansas for Medical Science, Little Rock, AR, USA

Jian Zhou Laboratory of Neural Signal Transduction, Institute of Neuroscience, Shanghai Institutes for Biological Sciences, Chinese Academy of Sciences, Shanghai, China

TRP Channel Classification

1

Hongyu Li

Abstract

The transient receptor potential (TRP) ion channels are named after the discovery of the photo-transducted channels in *Drosophila*. TRPs, activated by various extracellular and intracellular stimuli, play a plethora of physiological and pathological roles. There are seven families of TRPs including TRPC (canonical), TRPV (vanilloid), TRPM (melastatin), TRPA (ankyrin), TRPP (polycystin), TRPML (mucolipin), and TRPN (*Drosophila* NOMPC) in mammals. In yeast, the eighth TRP family was recently identified and named as TRPY. We here briefly summarize the classification and function of TRP cation channel superfamily.

Keywords

Transient receptor potential (TRP) protein • *trp* gene • Ion channel

Transient receptor potential (TRP) channels are wildly expressed on the plasma membrane in numerous types of cells, including neurons. The *trp* gene was initially identified in *Drosophila melanogaster* in the late 1960s. The first human homolog was reported in 1995. Since then about 30 *trp* genes and more than 100 TRP channels have been identified. There are seven families of TRP (a phylogenetic tree of human TRP channels is shown in Fig. 1.1): TRPC (canonical), TRPV (vanilloid), TRPM (melastatin), TRPA (ankyrin), TRPP (polycystin), TRPML (mucolipin), and TRPN (*Drosophila* NOMPC). In yeast, the eighth TRP family was recently identified and named as TRPY in which Y stands for yeast. TRPC, TRPV, TRPM, and TRPA are classified as Group1 TRP channels which have the strongest similarity with the *Drosophila* TRP. The Group2 TRP channels, including TRPP and TRPML, have distal relevance to *Drosophila* TRP. The classification of TRP superfamily is based on the differences in their amino acid sequences and topological structures, while it is difficult to differentiate the function of individual family and member simply according to the classification. Actually, TRPs play a plethora of physiological and pathological roles in response to various extracellular and

H. Li (✉)
Columbia University Medical Center,
New York, NY 10032, USA
e-mail: hl2880@cumc.columbia.edu;
hongyu1982@gmail.com

Y. Wang (ed.), *Transient Receptor Potential Canonical Channels and Brain Diseases*,
Advances in Experimental Medicine and Biology 976, DOI 10.1007/978-94-024-1088-4_1

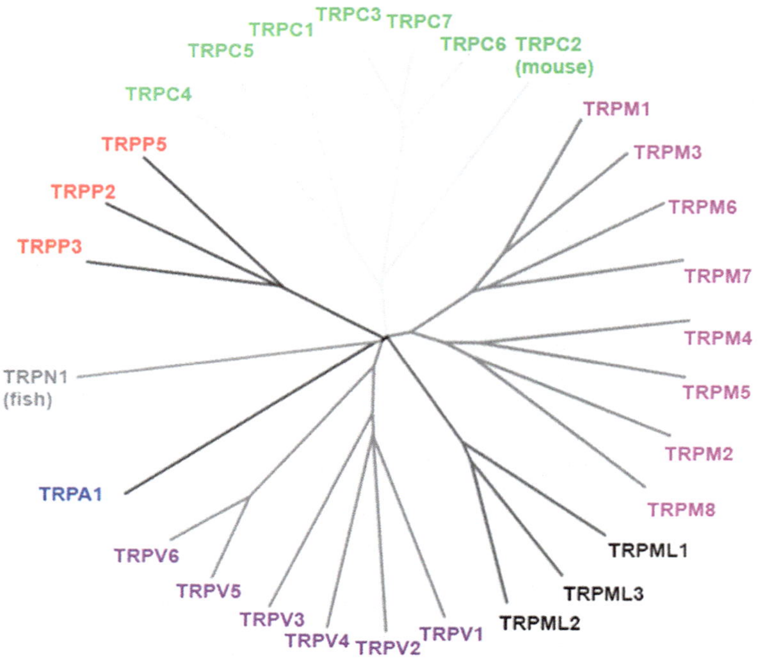

Fig. 1.1 A phylogenetic tree of human TRP channels (Modified from [40]). TRP channels fall into seven families, based on full-length sequence comparisons of human TRP channel proteins. TRPNs are not present in mammals and trpc2 is a pseudogene in human.

intracellular stimuli, such as changes of temperature, pH, or osmolarity, injury, depletion of calcium stores, as well as volatile chemicals and cytokines. Once activated, TRP channels, with homo- or hetero- tetrameric configurations, function as an integrator of several signaling pathways to elicit a serial of responses. TRPs share common structure features, including six putative transmembrane spanning domains with intracellular C and N termini and a pore lining between the fifth and sixth transmembrane domains [17, 39, 43, 51].

1.1 TRPC Channels

The TRPC family is the closest homolog to *Drosophila* TRP channels. The TRPC family consists of seven members (TRPC1–7) with trpc2 being a pseudogene in human beings [38]. The seven mammalian homologs share ≥30% amino acid identity within the N-terminal 750–900 amino acids. Based on sequence alignments and functional comparisons, the mammalian TRPCs fall into four subsets: TRPC1, TRPC2, TRPC3/6/7, and TRPC4/5.

TRPC1 protein shows a broad expression across different cell types [9]. It can interact with all other human TRPC proteins (TRPC3–7) to form a heterotetrameric channel [16, 23, 35, 55–57, 68]. TRPC1 is also able to interact with members in other families of TRP channel, such as TRPP2 [65] and TRPV4/6 [36, 54]. It is reported that TRPC1 and other TRP channels form a receptor-activated channel. For example, TRPC1/TRPP2 channel is activated by G-protein-coupled receptors (GPCRs). TRPC1 is also activated by hormone, orexin A, which is associated with the regulation of sleep/wake-up states, alertness, and appetite [27]. As expressed broadly, TRPC1 channel is related with many physiological functions.

The *trpc2* is a pseudogene in humans; however, in rodents it plays an important role in pheromone detection via the vomeronasal sensory neurons (VSN) [66]. In these cells, TRPC2 protein is mainly localized to the sensory microvilli, which are specialized for chemical signal detection.

Outside the vomeronasal organ, TRPC2 has been detected in the main olfactory epithelium [34], testes, heart, brain, liver, spleen, and erythrocytes [21, 22]. In the acrosome region of sperm, TRPC2 was also detected, which indicated its possible role in fertilization [25].

TRPC3 is most prominently expressed in specific regions of the brain, in the heart, and lungs [7, 50]. Its proteins could assemble homotetrameric and heterotetrameric channels with some other TRPC proteins. In brain synaptosomes, TRPC3 has been shown to interact with TRPC6 and 7 but not with other members of the TRPC family [16]. Many evidences show that TRPC3 is a constitutively active receptor-operated channel that can be further stimulated by DAG [29]. TRPC3 is a multifunctional cellular sensor with a wide range of physiological/pathological functions.

TRPC4 exists in many brain regions [69], endothelium and smooth muscle cells [2, 59], intestinal pacemaker cells (ICC) [26], adrenal glands [46], and kidneys [14]. Multiple signalings downstream of receptors mediate activity of TRPC4 homo-/heterotrimeric channels. TRPC4 can interact with TRPC1 and TRPC5 but not with TRPC3, TRPC6, or TRPC7 [16, 23]. TRPC4 has been shown to be involved in response to neural injury, the regulation of neurite outgrowth and neuronal exocytosis [41]. In human kidney epithelial cells, silencing TRPC4 impairs secretion of thrombopsondin-1 (TSP1) [61].

TRPC5 is mainly expressed in brain tissue [22], especially in fetal brain, indicating that TRPC5 may play an important role during brain development [56, 57]. TRPC5 participates in the formation of transient working memory in the entorhinal cortex [71]. TRPC5 is important for amygdala function and fear-related behavior [48]. In the central nervous system (CNS), TRPC5 can form heteromeric cation channels with TRPC1, and these heteromultimers are involved in store-operated Ca^{2+} entry (SOCE). The homotetrameric TRPC5 channel functions as a receptor-activated channel. Some studies report that TRPC5 is also activated by nitric oxide (NO) [67].

TRPC6 expression is the lowest in the brain compared to other TRPCs while still widely expressed in the cardiac neurons [4], retinal ganglion cells [64], olfactory epithelium neurons [12], and some areas of the brain, such as the cortex, substantia nigra, hippocampus, and cerebellum [6, 15]. TRPC6 has been reported to play important roles in brain development and diseases. Its physiological function is involved with excitatory synapse formation [72], dendritic outgrowth [58], and BDNF-mediated survival of granule cells in the cerebellum [24]. TRPC6 also participates in many pathological processes, including neuronal damage in stroke [11, 31], Aβ-production in Alzheimer's Disease (AD) [63], and human glioma cell proliferation [10, 32].

TRPC7 is mainly expressed in the kidney and pituitary gland [22, 37, 49] and closely related with TRPC6. TRPC7 has high sequence homology with TRPC3 and TRPC6 and always forms heteromultimers with TRPC3 and TRPC6. The functions of the heterotetrameric TRPC7 channel with TRPC3 and TRPC6 in kidney and heart disease were reported previously [1, 53]. TRPC7 is a receptor-activated channel activated by PLC-mediated metabolism of PIP2 and production of DAG [42, 60].

1.2 TRPV Channels

TRPVs, activated by vanillin, vanillic acid, and capsaicin in the plants, are thermo-TRP channels, as TRPV1, TRPV2, TRPV3, and TRPV4 can be activated by heat. There are six different members (TRPV1–6) in this family. Compared to TRPV1–4 which are referred to as nocireceptor that sense the damaging signals, TRPV5 and TRPV6 are epithelial calcium ion channels [52]. TRPV1 was first identified and cloned in the late 1990s. It can be activated by multiple stimuli, such as moderate heat (≥43 °C), low pH, capsaicin, ethanol, endogenous lipids, black pepper, and garlic as well as inflammatory mediators. TRPV1 is expressed in most neurons and widely investigated. In nerve injury associated with neuropathic pain and chronic pain, TRPV1 plays a central role. TRPV2 shares 50% of sequence homology with TRPV1 and is activated by higher temperature (52 °C). Other stimuli responsible for TRPV2 activation include osmotic stress and mechanical stretch. TRPV2 is expressed both in neuronal and nonneuronal cells. However, the exact physiological function of this TRPV member remains to be clarified. TRPV3 is expressed in the brain, spinal cord, trigeminal ganglia, and DRG neurons. It acts as a thermo-sensor in the skin and is activated by temperature higher than 34 °C. Other activators of TRPV3 include endogenous ligands such as PGE2, ATP, bradykinin, and histamine. TRPV4 is a nonselective cation channel and activated by temperature higher than 27 °C, mechanical stimuli, hypotonicity, and metabolites of arachidonic acid. It serves as a sensor of osmolality and mechanical stretch. TRPV5 is mainly expressed in kidney epithelial cells and plays an important role in the reabsorption of Ca^{2+}. It can be regulated by various factors, including 1,25-dihydroxyvitamin D3, parathyroid hormone, dietary Ca^{2+}, and acid-base status change of pH [30]. TRPV6 share some common features with TRPV5. They are co-expressed in several tissues, such as intestine, kidney, prostate, and testis. Unlike TRPV1–4, they are selective channels for Ca^{2+}. The gene expression of TRPV6 is upregulated in most common cancers, including prostate and breast cancers. TRPV6 is also an important channel for male fertility [13, 19].

1.3 TRPM Channels

There are eight members (TRPM1–8) of TRPM family in mammals. Based on sequence homology, TRPMs fall into three subgroups: TRPM1–3, TRPM4–5, and TRPM6–7. Some of TRPMs are located on intracellular membranes. TRPM1 is a prognostic marker for metastasis of melanoma [44]. TRPM2 is widely expressed in various tissues, including brain, heart, hematopoietic, vascular smooth muscle, and endothelial cells. It functions as a Ca^{2+} permeable channel in plasma membrane and a lysosomal calcium release channel in pancreatic beta-cell and dendritic cells [70]. TRPM3 is activated by moderate heat and steroid pregnenolone sulfate. It is expressed in islet cells of the pancreas, regulating insulin secretion, and in the brain, both in neurons and oligodendrocytes, as well as in the peripheral nervous system [3]. TRPM4 channels are Ca^{2+}-activated nonselective cation channels permeable only to monovalent ions (K^+ and Na^+). TRPM4 forms a functional channel as a tetramer which is expressed in a wide range of human tissues and involved in various physiological processes such as T cell activation, myogenic vasoconstriction, allergic reactions, and neurotoxicity [5]. TRPM5 shows 40% identity of the amino acid sequence with TRPM4 and is a Ca^{2+}-activated cation channel that mediates signaling in taste and other chemosensory cells. TRPM6 is a bifunctional protein comprising a TRP cation channel segment covalently linked to α-type serine/threonine protein kinase. TRPM6 is expressed in the intestinal and renal epithelial cells. TRPM7 has been found in mammalian tissues and plays important roles in cellular and systemic magnesium homeostasis. TRPM8 plays a critical role in the detection of environmental cold temperatures [33].

1.4 TRPA Channels

TRPA was firstly named as ANKTM1, because it has many N-terminal ankyrin repeats. TRPA1 is the only member of TRPA family in mammals, while there are two and four TRPA members in *C. elegans* and *Drosophila*, respectively [62]. TRPA1 is activated by painful cold with the temperature that is lower than 17 °C. TRPA1 is found in the plasma membrane of pain-detecting sensory nerves and activates pain pathways that trigger avoidance behaviors and pathways that promote long-lasting responses, such as inflammation. Blocking TRPA1 function is therefore a promising strategy to reduce pain. Pungent agents from wasabi and other TRPA1 triggers, known to be electrophiles, activate TRPA1 by forming covalent bonds with specific cysteine or lysine amino-acid residues [8]. TRPA1 antagonists, such as mustard oil and garlic, have potential for improving neurogenic inflammatory conditions provoked or exacerbated by irritant exposure [45].

1.5 TRPN Channels

Mammals do not encode any TRPN members. They are found in worms, flies, and zebrafish. The TRPN proteins are strong candidates for mechano-transducing channel subunits in both vertebrates and invertebrates. TRPN homologs are present in insects, nematodes, fish, and amphibians and required for tactile and proprioceptive behavior in insects and nematodes and for transduction of vibratory stimuli in zebrafish hair cells [28].

1.6 TRPP Channels

As mentioned above, TRPP and TRPML belong to Group2 of TRP channels and have limited similarity to the Group1 in homology. Furthermore, they have a large loop between transmembrane domain one and two, which discriminates them from Group1 [62]. The three human genes encoding for TRPP protein family are polycystic kidney disease 2 (PKD2, TRPP2), PKD2-like 1 (PKD2L1, TRPP3), and PKD2-like2 (PKD2L2, TRPP5). TRPP2 is a 110 kDa protein and involved in autosomal dominant polycystic kidney disease (ADPKD). TRPP2 functions as a Ca^{2+}-permeable nonselective cation channel. The highest sequence similarity between TRPP channels is in transmembrane segments S1–S6, with rather little sequence homology in their predicted amino and carboxyl terminals. All mammalian TRPP orthologues are highly conserved, with ~90% identities for TRPP2 and TRPP3 and ~80% for TRPP5. TRPP ion channels are regulated by and assemble with Polycystin-1 family proteins into receptor-channel complexes to form the core of a signaling pathway where Ca^{2+} is a second messenger [20].

1.7 TRPML Channels

TRPML includes three members TRPML1, TRPML2, and TRPML3, among which they share about 75% amino acid similarity. TRPML1 has 580 amino acids and a molecular weight of 65 kDa. It is also referred to as mucolipin-1 or MCOLN1. Mutations in this protein are responsible for lysosomal storage disorder mucolipidosis IV. TRPML1 is responsible for iron ions across the endosome/lysosome membrane into the cell [18, 47]. TRPML2 is a 566-amino acid protein with a predicted weight of 65 kDa. No evidence suggests that its mutations associated with any disorder in either human or animal models to date. TRPML3 has 533 amino acids and predicted weight of about 64 kDa. It is mainly expressed intracellularly in the inner ear. Previous reports suggested that TRPML3 is an inwardly rectifying monovalent cation channel that is permeable to Ca^{2+} and suppressed by H^+ [47].

1.8 Conclusion and Perspectives

Now there are at least 28 genes encoding mammalian TRP channel subunit proteins. Seven families of TRP are grouped, including TRPC (1–7), TRPV (1–6), TRPM (1–8), TRPA (1), TRPP (2/3/5), TRPML (1–3), and TRPN (1), respectively. Based on the similarity to *Drosophila* TRP, TRPC, TRPV, TRPM, and TRPA are classified as Group1 TRP channels, which have the strongest similarity. The Group2 TRP channels, including TRPP and TRPML, have distal relevance to *Drosophila* TRP. It has been demonstrated that these TRP family members are involved in cell physiological functions and also in many hereditary and acquired diseases. Some TRPs are involved in sensory functions such as smell, taste, pain, and pheromone sensing. Some are responsive to temperature and may help to avoid tissue-damaging noxious temperatures. Some TRPs are sensitive to natural compounds or their ingredients that have been used in medical practice.

Abnormalities in TRP channel function, as the result of alteration of protein expression levels, changes in channel properties, or changes in their myriad regulators, have been associated with numerous diseases ranging from chronic pain to cardiovascular disease, skeletal abnormalities, kidney diseases, brain diseases, and cancer, which provides numerous opportunities for therapeutic intervention.

References

1. Alvarez J, Coulombe A, Cazorla O, Ugur M, Rauzier JM, Magyar J, Mathieu EL, Boulay G, Souto R, Bideaux P, Salazar G, Rassendren F, Lacampagne A, Fauconnier J, Vassort G (2008) ATP/UTP activate cation-permeable channels with TRPC3/7 properties in rat cardiomyocytes. Am J Phys Heart Circ Phys 295(1):H21–H28. doi:10.1152/ajpheart.00135.2008
2. Beech DJ (2005) Emerging functions of 10 types of TRP cationic channel in vascular smooth muscle. Clin Exp Pharmacol Physiol 32(8):597–603. doi:10.1111/j.1440-1681.2005.04251.x
3. Brown RL, Xiong WH, Peters JH, Tekmen-Clark M, Strycharska-Orczyk I, Reed BT, Morgans CW, Duvoisin RM (2015) TRPM3 expression in mouse retina. PLoS One 10(2):e0117615. doi:10.1371/journal.pone.0117615
4. Calupca MA, Locknar SA, Parsons RL (2002) TRPC6 immunoreactivity is colocalized with neuronal nitric oxide synthase in extrinsic fibers innervating guinea pig intrinsic cardiac ganglia. J Comp Neurol 450(3):283–291. doi:10.1002/cne.10322
5. Cho CH, Lee YS, Kim E, Hwang EM, Park JY (2015) Physiological functions of the TRPM4 channels via protein interactions. BMB Rep 48(1):1–5
6. Chung YH, Sun Ahn H, Kim D, Hoon Shin D, Su Kim S, Yong Kim K, Bok Lee W, Ik Cha C (2006) Immunohistochemical study on the distribution of TRPC channels in the rat hippocampus. Brain Res 1085(1):132–137. doi:10.1016/j.brainres.2006.02.087
7. Clapham DE (2003) TRP channels as cellular sensors. Nature 426(6966):517–524. doi:10.1038/nature02196
8. Clapham DE (2015) Structural biology: pain-sensing TRPA1 channel resolved. Nature 520(7548):439–441. doi:10.1038/nature14383
9. Dietrich A, Chubanov V, Kalwa H, Rost BR, Gudermann T (2006) Cation channels of the transient receptor potential superfamily: their role in physiological and pathophysiological processes of smooth muscle cells. Pharmacol Ther 112(3):744–760. doi:10.1016/j.pharmthera.2006.05.013
10. Ding X, He Z, Zhou K, Cheng J, Yao H, Lu D, Cai R, Jin Y, Dong B, Xu Y, Wang Y (2010) Essential role of TRPC6 channels in G2/M phase transition and development of human glioma. J Natl Cancer Inst 102(14):1052–1068. doi:10.1093/jnci/djq217
11. Du W, Huang J, Yao H, Zhou K, Duan B, Wang Y (2010) Inhibition of TRPC6 degradation suppresses ischemic brain damage in rats. J Clin Invest 120(10):3480–3492. doi:10.1172/JCI43165
12. Elsaesser R, Montani G, Tirindelli R, Paysan J (2005) Phosphatidyl-inositide signalling proteins in a novel class of sensory cells in the mammalian olfactory epithelium. Eur J Neurosci 21(10):2692–2700. doi:10.1111/j.1460-9568.2005.04108.x
13. Fecher-Trost C, Weissgerber P, Wissenbach U (2014) TRPV6 channels. Handb Exp Pharmacol 222:359–384. doi:10.1007/978-3-642-54215-2_14
14. Freichel M, Vennekens R, Olausson J, Stolz S, Philipp SE, Weissgerber P, Flockerzi V (2005) Functional role of TRPC proteins in native systems: implications from knockout and knock-down studies. J Physiol 567(Pt 1):59–66. doi:10.1113/jphysiol.2005.092999
15. Giampa C, DeMarch Z, Patassini S, Bernardi G, Fusco FR (2007) Immunohistochemical localization of TRPC6 in the rat substantia nigra. Neurosci Lett 424(3):170–174. doi:10.1016/j.neulet.2007.07.049
16. Goel M, Sinkins WG, Schilling WP (2002) Selective association of TRPC channel subunits in rat brain synaptosomes. J Biol Chem 277(50):48303–48310. doi:10.1074/jbc.M207882200
17. Hardie RC (2011) A brief history of trp: commentary and personal perspective. Arch Eur J Physiol 461(5):493–498. doi:10.1007/s00424-011-0922-9
18. Harteneck C (2005) Function and pharmacology of TRPM cation channels. Naunyn Schmiedeberg's Arch Pharmacol 371(4):307–314. doi:10.1007/s00210-005-1034-x
19. Hoenderop JG, Voets T, Hoefs S, Weidema F, Prenen J, Nilius B, Bindels RJ (2003) Homo- and heterotetrameric architecture of the epithelial Ca2+ channels TRPV5 and TRPV6. EMBO J 22(4):776–785. doi:10.1093/emboj/cdg080
20. Hofherr A, Kottgen M (2011) TRPP channels and polycystins. Adv Exp Med Biol 704:287–313. doi:10.1007/978-94-007-0265-3_16
21. Hofmann T, Schaefer M, Schultz G, Gudermann T (2000) Cloning, expression and subcellular localization of two novel splice variants of mouse transient receptor potential channel 2. Biochem J 351(Pt 1):115–122
22. Hofmann T, Schaefer M, Schultz G, Gudermann T (2000) Transient receptor potential channels as molecular substrates of receptor-mediated cation entry. J Mol Med 78(1):14–25
23. Hofmann T, Schaefer M, Schultz G, Gudermann T (2002) Subunit composition of mammalian transient receptor potential channels in living cells. Proc Natl Acad Sci U S A 99(11):7461–7466. doi:10.1073/pnas.102596199
24. Jia Y, Zhou J, Tai Y, Wang Y (2007) TRPC channels promote cerebellar granule neuron survival. Nat Neurosci 10(5):559–567. doi:10.1038/nn1870
25. Jungnickel MK, Marrero H, Birnbaumer L, Lemos JR, Florman HM (2001) Trp2 regulates entry of Ca2+ into mouse sperm triggered by egg ZP3. Nat Cell Biol 3(5):499–502. doi:10.1038/35074570
26. Kim BJ, So I, Kim KW (2006) The relationship of TRP channels to the pacemaker activity of interstitial cells of Cajal in the gastrointestinal tract. J Smooth Muscle Res = Nihon Heikatsukin Gakkai kikanshi 42(1):1–7
27. Larsson KP, Peltonen HM, Bart G, Louhivuori LM, Penttonen A, Antikainen M, Kukkonen JP, Akerman KE (2005) Orexin-A-induced Ca2+ entry: evidence for involvement of trpc channels and protein kinase C regulation. J Biol Chem 280(3):1771–1781. doi:10.1074/jbc.M406073200

28. Lee J, Moon S, Cha Y, Chung YD (2010) Drosophila TRPN(=NOMPC) channel localizes to the distal end of mechanosensory cilia. PLoS One 5(6):e11012. doi:10.1371/journal.pone.0011012
29. Lemonnier L, Trebak M, Putney JW Jr (2008) Complex regulation of the TRPC3, 6 and 7 channel subfamily by diacylglycerol and phosphatidylinositol-4,5-bisphosphate. Cell Calcium 43(5):506–514. doi:10.1016/j.ceca.2007.09.001
30. Leunissen EH, Blanchard MG, Sheedfar F, Lavrijsen M, van der Wijst J, Bindels RJ, Hoenderop JG (2016) Urinary beta-galactosidase stimulates Ca2+ transport by stabilizing TRPV5 at the plasma membrane. Glycobiology. doi:10.1093/glycob/cwv172
31. Li H, Huang J, Du W, Jia C, Yao H, Wang Y (2012) TRPC6 inhibited NMDA receptor activities and protected neurons from ischemic excitotoxicity. J Neurochem 123(6):1010–1018. doi:10.1111/jnc.12045
32. Li S, Wang J, Wei Y, Liu Y, Ding X, Dong B, Xu Y, Wang Y (2015) Crucial role of TRPC6 in maintaining the stability of HIF-1alpha in glioma cells under hypoxia. J Cell Sci 128(17):3317–3329. doi:10.1242/jcs.173161
33. Liman ER (2014) Trpm5. Handb Exp Pharmacol 222:489–502. doi:10.1007/978-3-642-54215-2_19
34. Liman ER, Corey DP, Dulac C (1999) TRP2: a candidate transduction channel for mammalian pheromone sensory signaling. Proc Natl Acad Sci U S A 96(10):5791–5796
35. Lintschinger B, Balzer-Geldsetzer M, Baskaran T, Graier WF, Romanin C, Zhu MX, Groschner K (2000) Coassembly of Trp1 and Trp3 proteins generates diacylglycerol- and Ca2 + -sensitive cation channels. J Biol Chem 275(36):27799–27805. doi:10.1074/jbc.M002705200
36. Ma X, Cheng KT, Wong CO, O'Neil RG, Birnbaumer L, Ambudkar IS, Yao X (2011) Heteromeric TRPV4-C1 channels contribute to store-operated Ca(2+) entry in vascular endothelial cells. Cell Calcium 50(6):502–509. doi:10.1016/j.ceca.2011.08.006
37. Montell C (2005) The TRP superfamily of cation channels. Science's STKE: Sig Transduct Knowl Environ 2005(272):re3. doi:10.1126/stke.2722005re3
38. Montell C, Birnbaumer L, Flockerzi V, Bindels RJ, Bruford EA, Caterina MJ, Clapham DE, Harteneck C, Heller S, Julius D, Kojima I, Mori Y, Penner R, Prawitt D, Scharenberg AM, Schultz G, Shimizu N, Zhu MX (2002) A unified nomenclature for the superfamily of TRP cation channels. Mol Cell 9(2):229–231
39. Morelli MB, Amantini C, Liberati S, Santoni M, Nabissi M (2013) TRP channels: new potential therapeutic approaches in CNS neuropathies. CNS Neurol Disord Drug Targets 12(2):274–293
40. Nilius B, Owsianik G (2011) The transient receptor potential family of ion channels. Genome Biol 12(3):218. doi:10.1186/gb-2011-12-3-218
41. Obukhov AG, Nowycky MC (2002) TRPC4 can be activated by G-protein-coupled receptors and provides sufficient Ca(2+) to trigger exocytosis in neuroendocrine cells. J Biol Chem 277(18):16172–16178. doi:10.1074/jbc.M111664200
42. Okada T, Inoue R, Yamazaki K, Maeda A, Kurosaki T, Yamakuni T, Tanaka I, Shimizu S, Ikenaka K, Imoto K, Mori Y (1999) Molecular and functional characterization of a novel mouse transient receptor potential protein homologue TRP7. Ca(2+)-permeable cation channel that is constitutively activated and enhanced by stimulation of G protein-coupled receptor. J Biol Chem 274(39):27359–27370
43. Pan Z, Yang H, Reinach PS (2011) Transient receptor potential (TRP) gene superfamily encoding cation channels. Human Genomics 5(2):108–116
44. Patel S, Docampo R (2009) In with the TRP channels: intracellular functions for TRPM1 and TRPM2. Sci Signal 2(95):pe69. doi:10.1126/scisignal.295pe69
45. Paulsen CE, Armache JP, Gao Y, Cheng Y, Julius D (2015) Structure of the TRPA1 ion channel suggests regulatory mechanisms. Nature 520(7548):511–517. doi:10.1038/nature14367
46. Philipp S, Trost C, Warnat J, Rautmann J, Himmerkus N, Schroth G, Kretz O, Nastainczyk W, Cavalie A, Hoth M, Flockerzi V (2000) TRP4 (CCE1) protein is part of native calcium release-activated Ca2 + –like channels in adrenal cells. J Biol Chem 275(31):23965–23972. doi:10.1074/jbc.M003408200
47. Puertollano R, Kiselyov K (2009) TRPMLs: in sickness and in health. Am J Physiol Ren Physiol 296(6):F1245–F1254. doi:10.1152/ajprenal.90522.2008
48. Riccio A, Li Y, Moon J, Kim KS, Smith KS, Rudolph U, Gapon S, Yao GL, Tsvetkov E, Rodig SJ, Van't Veer A, Meloni EG, Carlezon WA Jr, Bolshakov VY, Clapham DE (2009) Essential role for TRPC5 in amygdala function and fear-related behavior. Cell 137(4):761–772. doi:10.1016/j.cell.2009.03.039
49. Riccio A, Mattei C, Kelsell RE, Medhurst AD, Calver AR, Randall AD, Davis JB, Benham CD, Pangalos MN (2002) Cloning and functional expression of human short TRP7, a candidate protein for store-operated Ca2+ influx. J Biol Chem 277(14):12302–12309. doi:10.1074/jbc.M112313200
50. Riccio A, Medhurst AD, Mattei C, Kelsell RE, Calver AR, Randall AD, Benham CD, Pangalos MN (2002) mRNA distribution analysis of human TRPC family in CNS and peripheral tissues. Brain Res Mol Brain Res 109(1–2):95–104
51. Santoni G, Farfariello V (2011) TRP channels and cancer: new targets for diagnosis and chemotherapy. Endocr Metab Immune Disord Drug Targets 11(1):54–67
52. Satheesh NJ, Uehara Y, Fedotova J, Pohanka M, Busselberg D, Kruzliak P (2016) TRPV currents and their role in the nociception and neuroplasticity. Neuropeptides. doi:10.1016/j.npep.2016.01.003

53. Satoh S, Tanaka H, Ueda Y, Oyama J, Sugano M, Sumimoto H, Mori Y, Makino N (2007) Transient receptor potential (TRP) protein 7 acts as a G protein-activated Ca2+ channel mediating angiotensin II-induced myocardial apoptosis. Mol Cell Biochem 294(1–2):205–215. doi:10.1007/s11010-006-9261-0
54. Schindl R, Fritsch R, Jardin I, Frischauf I, Kahr H, Muik M, Riedl MC, Groschner K, Romanin C (2012) Canonical transient receptor potential (TRPC) 1 acts as a negative regulator for vanilloid TRPV6-mediated Ca2+ influx. J Biol Chem 287(42):35612–35620. doi:10.1074/jbc.M112.400952
55. Storch U, Forst AL, Philipp M, Gudermann T, Mederos y Schnitzler M (2012) Transient receptor potential channel 1 (TRPC1) reduces calcium permeability in heteromeric channel complexes. J Biol Chem 287(5):3530–3540. doi:10.1074/jbc.M111.283218
56. Strubing C, Krapivinsky G, Krapivinsky L, Clapham DE (2001) TRPC1 and TRPC5 form a novel cation channel in mammalian brain. Neuron 29(3):645–655
57. Strubing C, Krapivinsky G, Krapivinsky L, Clapham DE (2003) Formation of novel TRPC channels by complex subunit interactions in embryonic brain. J Biol Chem 278(40):39014–39019. doi:10.1074/jbc.M306705200
58. Tai Y, Feng S, Ge R, Du W, Zhang X, He Z, Wang Y (2008) TRPC6 channels promote dendritic growth via the CaMKIV-CREB pathway. J Cell Sci 121(Pt 14):2301–2307. doi:10.1242/jcs.026906
59. Tiruppathi C, Ahmmed GU, Vogel SM, Malik AB (2006) Ca2+ signaling, TRP channels, and endothelial permeability. Microcirculation 13(8):693–708. doi:10.1080/10739680600930347
60. Trebak M, Vazquez G, Bird GS, Putney JW Jr (2003) The TRPC3/6/7 subfamily of cation channels. Cell Calcium 33(5–6):451–461
61. Veliceasa D, Ivanovic M, Hoepfner FT, Thumbikat P, Volpert OV, Smith ND (2007) Transient potential receptor channel 4 controls thrombospondin-1 secretion and angiogenesis in renal cell carcinoma. FEBS J 274(24):6365–6377. doi:10.1111/j.1742-4658.2007.06159.x
62. Venkatachalam K, Montell C (2007) TRP channels. Annu Rev Biochem 76:387–417. doi:10.1146/annurev.biochem.75.103004.142819
63. Wang J, Lu R, Yang J, Li H, He Z, Jing N, Wang X, Wang Y (2015) TRPC6 specifically interacts with APP to inhibit its cleavage by gamma-secretase and reduce Abeta production. Nat Commun 6:8876. doi:10.1038/ncomms9876
64. Warren EJ, Allen CN, Brown RL, Robinson DW (2006) The light-activated signaling pathway in SCN-projecting rat retinal ganglion cells. Eur J Neurosci 23(9):2477–2487. doi:10.1111/j.1460-9568.2006.04777.x
65. Wes PD, Chevesich J, Jeromin A, Rosenberg C, Stetten G, Montell C (1995) TRPC1, a human homolog of a Drosophila store-operated channel. Proc Natl Acad Sci U S A 92(21):9652–9656
66. Yildirim E, Birnbaumer L (2007) TRPC2: molecular biology and functional importance. Handb Exp Pharmacol 179:53–75. doi:10.1007/978-3-540-34891-7_3
67. Yoshida T, Inoue R, Morii T, Takahashi N, Yamamoto S, Hara Y, Tominaga M, Shimizu S, Sato Y, Mori Y (2006) Nitric oxide activates TRP channels by cysteine S-nitrosylation. Nat Chem Biol 2(11):596–607. doi:10.1038/nchembio821
68. Zagranichnaya TK, Wu X, Villereal ML (2005) Endogenous TRPC1, TRPC3, and TRPC7 proteins combine to form native store-operated channels in HEK-293 cells. J Biol Chem 280(33):29559–29569. doi:10.1074/jbc.M505842200
69. Zechel S, Werner S, von Bohlen Und Halbach O (2007) Distribution of TRPC4 in developing and adult murine brain. Cell Tissue Res 328(3):651–656. doi:10.1007/s00441-007-0388-4
70. Zhan KY, Yu PL, Liu CH, Luo JH, Yang W (2016) Detrimental or beneficial: the role of TRPM2 in ischemia/reperfusion injury. Acta Pharmacol Sin 37(1):4–12. doi:10.1038/aps.2015.141
71. Zhang Z, Reboreda A, Alonso A, Barker PA, Seguela P (2011) TRPC channels underlie cholinergic plateau potentials and persistent activity in entorhinal cortex. Hippocampus 21(4):386–397. doi:10.1002/hipo.20755
72. Zhou J, Du W, Zhou K, Tai Y, Yao H, Jia Y, Ding Y, Wang Y (2008) Critical role of TRPC6 channels in the formation of excitatory synapses. Nat Neurosci 11(7):741–743. doi:10.1038/nn.2127

TRPC Channel Structure and Properties

2

Shengjie Feng

Abstract

TRPC channels are the first identified members in the TRP family. They function as either homo- or heterotetramers regulating intracellular Ca^{2+} concentration in response to numerous physiological or pathological stimuli. TRPC channels are nonselective cation channels permeable to Ca^{2+}. The properties and the functional domains of TRPC channels have been identified by electrophysiological and biochemical methods. However, due to the large size, instability, and flexibility of their complexes, the structures of the members in TRPC family remain unrevealed. More efforts should be made on structure analysis and generating good tools, including specific antibodies, agonist, and antagonist.

Keywords

TRPC • Structure • Property

TRP channel subunits are rather large, ranging from 70 kD to more than 200 kD [40]. Transmembrane (TM) segment prediction suggests TRP channels resemble voltage-gated K^+ or Ca^{2+} channels [74]. A consensus has been reached by researchers that putative organization of TRP channels consists of six transmembrane (TM) domains with the carboxyl (C-) and amino (N-) terminals facing the intracellular side of the membrane [20, 73, 74]. The features of TRP channels change from family to family. However, no matter how diverse these subunits are, they are conserved in their pore region, a hydrophobic region between fifth and sixth segment and their TRP domain on the proximal C-terminal region [40, 67]. The results obtained by biochemical and optical methods strongly suggest that TRP channels are formed by four subunits [2, 29, 35],

S. Feng (✉)
Department of Physiology, University of California, San Francisco, CA, USA
e-mail: sjfeng100@gmail.com

Y. Wang (ed.), *Transient Receptor Potential Canonical Channels and Brain Diseases*, Advances in Experimental Medicine and Biology 976, DOI 10.1007/978-94-024-1088-4_2

assembling as homo- or heterotetramers [22, 25, 40, 77].

2.1 TRPC Channel Structure

2.1.1 Structure of TRPC Channels

Three factors make structural studies of TRP channels a great challenge. First, structural biology techniques require an ample supply of highly pure and stable protein samples. TRP channels, however, are not endogenously expressed in bacteria. This is perhaps a major limiting factor for the lack of high-resolution structures of full-length TRP channel. Moreover, membrane proteins are extremely difficult to be produced in large quantities and be purified in a stable native state [18]. Second, TRP channels are very large tetramers with multiple domains in each subunit. Such large protein complexes cannot be accessible to high-resolution nuclear magnetic resonance (NMR) techniques of molecules in an identical conformation [16]. Third, flexibility also complicates single-particle electron microscopy (EM) studies. The flexibility afforded by multiple domains, which probably has functional significance, often hinders crystallization [16]. For TRP channels, especially, they respond to diverse chemical or physical stimuli and are therefore believed to be conformationally dynamic. An additional obstacle to coaxing these proteins is forming well-ordered crystal lattices required for X-ray and electron crystallographic analysis [46].

Three major techniques are commonly used to obtain structural information on macromolecules: X-ray crystallography, NMR spectroscopy, and electron cryomicroscopy (cryo-EM) – either single-particle EM or electron crystallography [45]. Currently, three approaches have yielded information on the structure of TRP channels: (a) X-ray crystallography and NMR spectroscopy have been employed effectively to obtain high-resolution structures of functionally important cytosolic domains of six TRP channels. Biophysical measurements of isolated cytosolic domains of several TRP channels also gain insights into their biological function [16]. (b) Single-particle EM studies provide six low-resolution structures (TRPA1, TRPC3, TRPV1, TRPV2, TRPV4, and TRPM2) (13.6–35 Å) for full-length TRP channels [16, 21]. (c) Recently, a breakthrough in electron cryomicroscopy makes it possible to obtain a high-resolution view (~3–4 Å) of an entire TRP channel. Till now, three high-resolution structures (TRPV1, TRPV2, and TRPA1) have been determined [46, 68, 106].

Similar to most TRP channels, TRPC channel subunits have six transmembrane domains and a putative pore region between fifth and sixth transmembrane domains and assemble into tetramers to form functional channels (Fig. 2.1a, b). Though no high-resolution structures of TRPC members have been obtained yet, in the six existing low-resolution structures, mouse TRPC3 has the second highest resolution (15 Å) and is the most unique [61]. TRPC3 has a bell-shaped structure containing ample water-filled spaces within the molecule (Fig. 2.1c). The structure can be further divided into two components, a dense globular inner core and a sparse outer shell with a mesh-like structure. Viewing from the top, the antenna-like outer columns radiate from the inner chamber like terminals in the airport. Four small extracellular segments are held away from the membrane surface by slender arms. These structures are assumed to function as signal sensing modules for agonists and regulators. The overall height of the TRPC3 molecule is 240 Å, the side length at the widest position is 200 Å, and a diagonal line yields 210 Å. Therefore, the reconstituted structure of TRPC3 is much larger than that of the other five TRP structures, even though the calculated mass of a TRPC3 tetramer is the smallest in the six molecules. One explanation for this disparity is the presence of many large water-filled cavities in the TRPC3 structure [45]. Another reason is its enormous volume mostly conformed by the cytoplasmic domain, which is made of a sparse external shell [40]. Though it is still not clear why the structures of TRPC3 and TRPM2 are so different from that of TRPV1/TRPV2/TRPV4 and TRPA1, it is hypothesized that the simultaneous association of TRPC3 with other protein complexes and its multimodal activation and modulation mechanisms may underlie its expanded structure. However, the use of an auto-

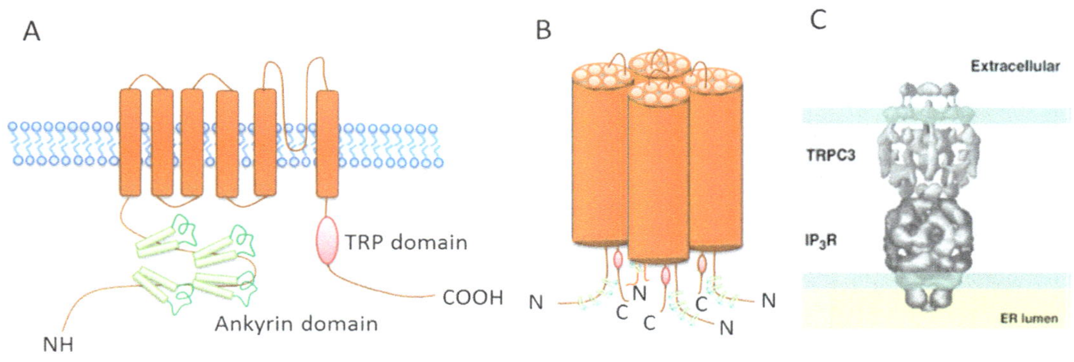

Fig. 2.1 Structures of TRPC channels. (**a**) Transmembrane topology of TRPC channels. The TRPC protein has six putative transmembrane domains, a pore region between the fifth and sixth transmembrane domains, four ankyrin domains (predicted), and a TRP domain in the proximal C-terminal region. (**b**) Structure of TRPC tetramer. The TRPC protein assembles into homotetramers or heterotetramers to form functional channels. (**c**) The large dimensions of TRPC3 [61]

mated particle-selection algorithm might have caused distortions in the data analysis and structural reconstruction of TRPC3 and TRPM2 [45].

2.1.2 Functional Domains of TRPC Channels

Structural biologists usually define a protein "domain" as an independently folding segment that can take on its native conformation even when isolated from the rest of the protein [16]. Though it has not been confirmed by experimental structure study, all members of TRPC channels were predicted to have 3–4 ankyrin repeat domains (ARD) on their N-terminal region and a conserved TRP domain at the beginning of the cytosolic C-terminal region [90] (Fig. 2.1a).

2.1.3 Ankyrin Repeat Domain

Ankyrin repeat (AR) sequences span ~33 residues and fold into a structural motif consisting of two α-helices folding back onto each other to form a helical hairpin, followed by a long hairpin loop that extends roughly perpendicular to the helical axes [63]. Several repeats of these structure motifs, ranging from 3 to over 30, are stacked side by side with their helices nearly parallel to each other, forming a modular, highly efficient, and specific protein-binding surface [15, 16]. The AR is one of the most common protein-protein interaction motifs [16, 40]. They function in various cellular processes, including regulation of transcription, cell cycle, development, cell-cell signaling, and transport.

Though little is known about TRPC ARs based on structure studies, the structure of several TRPV ARs has been published, which might give some hints to the understanding of TRPCs. When hundreds of chemicals were screened to optimize the TRPV1-ARD crystallization conditions, it was observed that the presence of ATP altered the crystal shape, likely by changing the packing interactions between protein molecules. The electron density map and biochemical assays further demonstrated that both ATP and calmodulin bind to the TRPV1-ARD [52] on the concave surface in a competitive manner. The TRPV2-ARD structure predicts that phosphorylation sites in its N-terminus (S116, Y200) should be on the surface of interaction of ARD with other proteins [31]. Furthermore, it is reported that ARDs play a role in promoting tetrameric assembly of TRPV5 and TRPV6 [6, 11]. Besides TRPVs, studies have shown that ARs of TRPA1 dictate sensitivity to thermal and chemical stimuli and ARs of TRPN (NompC) conveys force to gate the NOMPC mechanotransduction channel [8, 102].

TRPC channels likely have four ARs, which have weak similarity to the AR consensus [40]. The structure of AR in TRPC channel is therefore likely to have some unusual kinks and loops, as is observed in TRPV channels. Even there is still lack of structural evidence, biochemical assays uncover the function of TRPC ARD. TRPC1 negatively regulates TRPV6 by interaction with TRPV6 on its N-terminal ankyrin-like repeat domain [80]. The heteromerization of TRPC3 with TRPC1 was shown by GST pull-down assay of TRPC3 portions with TRPC1. The portion containing the AR region of TRPC3 was bound to TRPC1. The heteromeric TRPC3/TRPC1 is shown to participate in regulating the resting cytosolic Ca^{2+} levels in skeletal muscle [96].

The first ankyrin-like repeat is the minimum indispensable key structure for functional assembly of homo- and heteromeric TRPC4/TRPC5 channels assayed by confocal Förster resonance energy transfer (FRET) and total internal reflection fluorescence (TIRF) microscopy [79]. Consistently, by using GST pull-down, yeast two-hybrid, circular dichroism approaches and chimeras, studies show that the N-terminus of TRPC4 self-associates via the AR domain and the coiled-coil domain (CCD) to assemble the tetrameric channel of TRPC4 [43, 44]. These two domains are responsible for the association between TRPC4 and TRPC6 (members of distinct subgroups of TRPCs) [43]. TRPC4 can form complexes with TRPC6 subunits containing the N-terminal ankyrin and coil-coiled domain (residues 1-304) of TRPC4 [43]. Using both GST pull-down assay and immunoprecipitation, researchers show that MxA and RNF24 interacted with the ankyrin-like repeat domain of all TRPCs and regulated their activity or trafficking [55, 56]. The cGK-Iα interacts with the AR domain in the N-terminus of TRPC7 and phosphorylated TRPC7 at threonine 15, which contributes to the quick and accurate regulation of calcium influx and CREB phosphorylation [99]. In conclusion, ARs of TRPCs may play roles in the regulation of channel assembling, activity, and trafficking.

2.1.4 TRP Domain

Most TRP families display a conserved sequence on their proximal C-terminal region which has been regarded as a signature sequence for TRP channels. This ~25-amino acid intracellular region, just after the sixth transmembrane domain, is called the TRP domain [62]. This domain contains the TRP box, a conserved motif defined by the consensus sequence "WKFQR." All members or relatives of the mammalian TRPC family contain this highly conserved TRP box. It is possible to identify a second conserved region on the carboxy-terminal end of the TRP domain with proline-rich sequence of LPPPF (leucine on the first position is highly conserveative) [40]. However, the second TRP box is only conserved in TRPC and TRPM subfamilies. Although the main function of the TRP domain remains elusive, it may be required for PIP2 binding and regulation of channel gating [75].

TRPC3 and TRPC6 TRP domains differ in seven amino acids. Assayed by chimera and mutation experiments, the TRP domain of TRPC3, but not that of TRPC6, is found essential for association with cytoskeleton and the increased channel translocation to cell surface in response to Epo stimulation [24].

2.1.5 Pore Region

Few studies have been aimed at the identification of the pore region and the description of the pore properties of TRPC channels [67]. A theoretical prediction of pore elements seems to be ineffective to TRPC channels. Unlike TRPV channels, the segment between TM5 and TM6 of TRPC members does not share significant homology to the sequence of the pore region of bacterial K^+ channels [67]. This loss of homology to the bacterial archetype pore may signify that TRPCs are phylogenetically younger than TRPV channels [67].

The ultimate proof that a region contributes to the pore of an ion channel is demonstrating that pore properties, e.g., ion selectivity, can be altered by mutations to the putative pore-forming region [94]. To this extent, the location and structure of the pore region and selectivity filter of most TRP proteins, including all members of the TRPC subfamilies, are currently unknown. Till now, three studies have addressed the role of several amino acids in the TM5-TM6 linker of three TRPC members, which gives some hints about the pore region.

To determine the degree of cooperativity within a TRPC channel pore complex, researchers generated a dominant-negative construct of TRPC6 (TRPC6DN) by exchanging three highly conserved residues, L678, F679, and W680, in the putative pore region for alanine residues. The TRPC6DN protein is correctly inserted into the plasma membrane and was functionally silent. Transient expression of TRPC6DN nearly abrogates TRPC3- and TRPC6-dependent currents, but does not compromise TRPC4 or TRPC5 activity [27]. It is shown that substitution by positively charged lysines at Glu576 and Asp581 in TRPC1, both located just outside the putative pore region, reduced the Ca^{2+} permeability [53]. For TRPC5, mutating glutamate residues (Glu543, Glu595, and Glu598), which are located close to either TM5 or TM6, affect the potentiation and inhibitory effects of extracellular La^{3+} on channel activity [32].

A conserved glycine residue within the cytosolic S4–S5 linker of both TRPC4 and TRPC5 proteins is important for their mysterious pore function. Mutating the glycine residue by a serine forces the channels into an open conformation. Expression of the TRPC4G503S and TRPC5G504S mutants causes cell death, which could be prevented by decreasing extracellular Ca^{2+} concentration in the culture medium. Current-voltage relationships of the TRPC4G503S and TRPC5G504S mutant channels resemble that of wild-type TRPC4 and TRPC5 channels [4]. Introduction of a second mutation (S623A) into TRPC4G503S suppressed the constitutive activation. Therefore, it is likely that the S4-S5 linker is a critical constituent of TRPC4/TRPC5 channel gating and that disturbance of its sequence allows channel opening independent of stimulation [4].

Given that these different residues are not conserved within the TRPC subfamily, it is unlikely that they form part of the actual selectivity filter, but rather contribute to the extracellular mouth of the pore [94].

2.1.6 Coiled-Coil Domain

Coiled coil is a protein structure in which α-helices wrap around each other in a helical coil conformation [45]. Sequences with a propensity to assume coiled-coil structures are characterized by recurring pattern of aliphatic residues alternating every third and then fourth residue to form seven residue repeats [45]. Coiled-coil structures, functioning as oligomerization domains, are found in a variety of proteins including transcription factors, motor proteins, structural proteins, cellular and viral membrane fusion proteins, and ion channels [45, 64]. Coiled-coil domains have been implicated in subunit interaction and assembly of ion channels, including TRPV1 [14], TRPM2 [58], TRPM7 [13], and TRPM8 [10, 87].

Coiled-coils are predicted in TRPC channels at either or both the N-terminal intracellular linker between the AR and the transmembrane domain and the C-terminal domain [42, 81]. Though these TRPC coiled-coil regions still need to be confirmed through biochemical and/or structural experiments, the Orai1-activating region of STIM1 interacts with the TRPC channel coiled-coil domains (CCDs). This interaction is essential for opening the channels by STIM1 [41]. Disruption of the N-terminal CCDs by mutations eliminated TRPC surface localization and reduced binding of STIM1 to TRPC1 and TRPC5 while increasing binding to TRPC3 and TRPC6 [41]. Using a yeast two-hybrid assay, the coiled-coil domain is found to facilitate homodimerization of the N-terminus of mTRPC1 and is required for structural organization, thus forming functional channels [9]. The CCD is one of the two domains responsible for the association between TRPC4 and TRPC6 [43, 44]. Thus, TRPC channel CCDs can participate in channel gating and assembling.

2.1.7 Other Functional Domains

Binding domains for various signaling molecules exist in the N-termini and C-termini of certain TRPC proteins. In addition to these structurally defined domains mentioned above, functional domains, including PIP2- and calmodulin (CaM)-binding domains, CRIB domain, and PDZ domain, also yield insights into the biological function of TRPC channels [73].

Because of their common interaction with ion channels, PIP2 and calmodulin may be considered as channel components [23, 76]. TRPC proteins contain multiple putative CaM-binding sites in their N- and C-termini, some of which overlap with an IP3 receptor (IP3R)-binding site (CRIB domain) [85]. The GST fusion proteins of all TRPC channels interact with CaM in vitro. By using CaM and IP3R peptides, in vitro Ca^{2+}-dependent competition experiments demonstrated that Ca^{2+}/CaM and IP3Rs may dynamically regulate TRPC through competitive interactions [85]. However, the function of TRPC and IP3R interactions has been challenged when receptor-activated TRPC3 functions have been found totally normal in avian DT40 cells lacking all three IP3R isoforms [89].

In *Drosophila* photoreceptors, TRP channels are tethered into signaling complexes via PDZ interactions with the scaffolding protein INAD [19]. Both TRPC4 and TRPC5 contain a carboxyl terminal PDZ-binding motif (VTTRL) which is absent in other TRPCs. This motif in both channels mediates interactions with NHERF/EBP-50 and PLCβ1 [86]. The PDZ domain of TRPC4 controls its localization and surface expression in HEK293 cells [59]. Although PDZ-binding motifs of TRPC channels have been demonstrated to participate in plasma membrane localization [59, 82], few data so far directly implicate PDZ proteins in the control of TRPC channel activity.

A finding shows that 437–508 aa of TRPC6 is important for its inhibition of Aβ production [95]. This domain contains the first and second transmembrane regions and the first extracellular loop. When the second transmembrane (TM2) region is mutated, by point mutation, replacement, or reversal, TRPC6 is not able to reduce Aβ levels, indicating that the TM2 domain is essential for TRPC6 to regulate Aβ production.

2.2 TRPC Channel Properties

2.2.1 Expression Pattern of TRPC Channels

2.2.1.1 Tissue Distribution

TRPC channel proteins are expressed in both excitable and non-excitable cells. While mRNA and protein of TRPC1 are widely expressed in mammalian tissues, those of TRPC3 and TRPC5 are predominantly detected in the brain. A relatively weak signal of TRPC3 mRNA is present in the ovary, colon, small intestine, lung, prostate, placenta, and testis [103]. TRPC5 can also be detected in the liver, kidney, testis, and uterus in much lower levels [66]. TRPC6 mRNA is detected in the lung and at a lower level in the brain, muscle, placenta, and ovary [5, 90]. TRPC7 mRNA expression is in the heart, lung, and eye and moderate expression in the brain, spleen, and testis [65]. TRPC2 is a pseudogene in human, and its protein localizes to neuronal microvilli in rat vomeronasal organ and in the head of mouse sperm [33, 50, 54] (Table 2.1).

2.2.1.2 Expression Pattern During Development

During fetal development, TRPC1 is expressed at the highest level in the brain and at lower levels in the liver and kidneys. In the adult, TRPC1 is expressed at the highest levels in the heart, testes, ovary, and many regions of the brain [105]. In rat hippocampus, TRPC1 and TRPC3 proteins are detectable at postnatal day 14 and 7 [84], and their expression levels remain high into adulthood. By contrast, the peak expression of TRPC4, TRPC5, and TRPC6 is between postnatal days 7 and 14 [84].

Table 2.1 Properties of TRPC Channels Homotetramer

Name	Selectivity	Conductance (pS)	Effects of trivalent cations	Activation mechanism	Expression
TRPC1	Non	16	La^{3+}, Gd^{3+}	Store operated/STIM1 Mechanical stimuli	Widely expressed
TRPC2	2.7	42	La^{3+}, Gd^{3+}	Store operated /STIM1, receptor operated	VNO, sperm, testis, heart, brain
TRPC3	1.6	23/66	La^{3+}, Gd^{3+}	Store operated/IP3, Receptor operated, Mechanical stimuli	Brain (mainly), heart, placenta, muscle, lung, ovary, colon, prostate, small intestine, testis
TRPC4	1.05/7	41	La^{3+}	Store operated/STIM1, receptor operated	Brain, testis, placenta, adrenal gland, retina endothelia, testis
TRPC5	9.5/1.79	63	La^{3+}, Gd^{3+} (GTPγS) La^{3+}, Gd^{3+s} (ATP)	Store operated/STIM1, receptor operated	Brain (mainly), liver, kidney, testis, uterus [66]
TRPC6	5	35	La^{3+}, Gd^{3+}	Store operated, Receptor operated Mechanical stimuli	Lung, brain, muscle, placenta, ovary
TRPC7	1.9/5	25-50	La^{3+}, Gd^{3+s}	Store operated, Receptor operated	Heart, lung, eye, brain, spleen, testis

Non nonselective, *s* slightly, *red* inhibition, *green* activation

2.2.2 Electrophysiological Properties of TRPC Channels

Interpretation of the functional data of TRPC channels is often difficult and complex due to a noisy background caused by endogenous cation-selective, Ca^{2+}-permeable channels regulated by store depletion and/or products of PLC-dependent pathways [67]. This has led to conflicting descriptions of the pore properties of TRPC channels. For example, TRPC4 and TRPC5 have been described as either Ca^{2+} selective or nonselective between mono- and divalent cations.

2.2.2.1 Activation Properties of TRPC Channels

There are two major mechanisms proposed for the activation of TRPC channels, either through receptor-operated or store-operated Ca^{2+} channel [7]. However, the activation mechanism of mammalian TRPC channels remains controversial. The results conflict with each other when the systems are varied. TRPC homologs have been found to be gated by a direct interaction with inositol 1,4,5-trisphosphate (IP3) receptors, ryanodine receptors, or diacylglycerol [26, 38, 39, 65, 66, 78]. But studies of *Drosophila* mutants lacking its only IP3 receptor indicate that TRP activation is independent of Ca^{2+} store depletion in its native environment in photoreceptor cells [1, 60, 72]. TRPC2 can function as a store-operated channel in transfected cells and is likely involved in sperm function [33, 88]. However, within its native environment in vomeronasal neuron, TRPC2 acts as a diacylglycerol-gated cation channel participating in the chemoelectrical transduction [54]. TRPC3 was used as a representative of a store-operated Ca^{2+} channel (SOC) to study the property and machineries of SOC [70]. But abundant evidence has been shown that

TRPC3 can be activated in a receptor-operated way [7, 70].

Yet, there are similarities among TRPC channels, as all members are activated through pathways coupled to stimulation of phospholipase C (PLC) [90]. Consistently, PLC-deficient mutants also show that *Drosophila* TRP is gated in some manner downstream of PLC. Notable findings are the identification of STIM1, a Ca^{2+} sensor of endoplasmic reticulum essential for SOC and Icrac, and Orai1 or CRACM1, functioning as Icrac channels or as an Icrac subunit [12, 92, 101]. STIM1 selectively binds to TRPC1, TRPC2, and TRPC4, but not to TRPC3, TRPC6, or TRPC7, providing evidence for TRPC1, TRPC2, and TRPC4 being SOC [28]. It is worth mentioning that although the Orai and TRPC channels can function independently of each other to mediate the CRAC current [100] and nonselective Ca^{2+}-permeable current, it was shown that all Orai channels interact with TRPC channels to complex with STIM1 and enhance TRPC channels' store dependence [47–49]. In most cells, both Orai and TRPC channels appear to be required for SOC by affecting the activity of each other. Deletion of Orai1 inhibits all forms of Ca^{2+} influx in these cells. For the TRPC3/TRPC6/TRPC7 channels, the majority of published results suggest that endogenous DAG, produced upon PLC activation, or exogenous DAG (usually oleyl acetyl glycerol, or OAG) can be the signal activating these channels [7, 70, 90]. It turned out that TRPC3 can function as STIM1-dependent SOC channel only by assembling with TRPC1. TRPC6 functions as STIM1-dependent channel only in the presence of TRPC4 [98]. Notably, in the absence of TRPC1 and TRPC4, activation of TRPC3 and TRPC6 is STIM1 independent. Since TRPC4 and TRPC5 channels are highly sensitive to protein kinase C (PKC), it is difficult to obtain evidence that DAG can activate TRPC. Indeed, OAG inhibits the activation of TRPC4 or TRPC5, and this inhibition is blocked by inhibitors of PKC [70, 91]. The Gαi/o proteins, including Gαi2 and Gαi3, are proposed to be important for the activation of TRPC4 and TRPC5 [30, 37].

It is reported that TRPC1 can be activated with a latency of a few milliseconds by stretch in patches from *Xenopus* oocytes [57]. In another study, TRPC1 is found forming mechanosensitive channels in growth cones of *Xenopus* spinal neurons [36]. Recently, it has been shown that the sensitivity of podocyte cells to stretch is reduced by TRPC6 RNAi or by the expression of podocin, a protein that interacts with TRPC6 [3]. In a study, TRPC3 and TRPC6 are proposed to be essential for normal mechanotransduction in subsets of sensory neurons and cochlear hair cells [71]. These findings suggest that some members of TRPC channels can be activated in response to mechanical stimuli.

The activation mechanism of TRPC channels is summarized in Fig. 2.2 and Table 2.1.

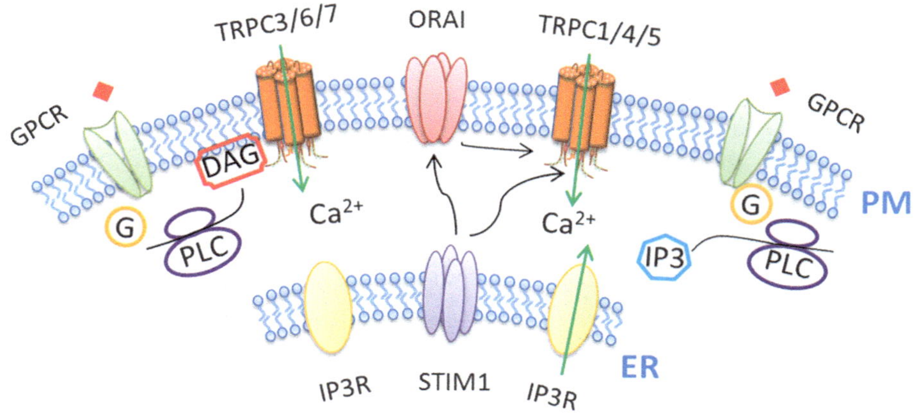

Fig. 2.2 Activation mechanisms for TRPC channels

2.2.2.2 Properties of TRPC Channel Homotetramer

TRPC1

TRPC1 has no bias of Ca^{2+} and Na^{+} selectivity. By performing noise analysis of the currents, the relation between the currents and their variances is roughly linear, and the slope of this relation indicates a single-channel amplitude of 1.1 pA at −70 mV. Assuming a linear amplitude-voltage relation, the predicted single-channel conductance is 16 pS [105].

TRPC2

The current-voltage (I-V) relationship of the SAG-activated TRPC2 conductance was nearly linear, with a reversal potential of 1.72 ± 2.4 mV. SAG activates a nonselective cation conductance that is permeable for Na^{+}, Cs^{+}, and Ca^{2+} but not for $NMDG^{+}$. Under bi-ionic conditions, the relative permeabilities PCa/PNa and PCs/PNa were 2.7 ± 0.7 and 1.5 ± 0.3, respectively. SAG-induced single-channel currents with a mean unitary current amplitude of −3.3 ± 0.5 pA at a holding potential of −80 mV, exhibiting a nearly linear I-V relationship with a slope conductance of 42 pS in symmetrical 150 mM Na^{+} solution [54].

TRPC3

The permeability for monovalent cations is PNa > PCs ≈ PK >> PNMDG, and the relative permeability PCa/PNa is 1.62 ± 0.27. The trivalent cations La^{3+} and $Gd^{3+}$are potent blockers of TRPC3 current (the IC50 for La^{3+} was 24.4 ± 0.7 μM). The single-channel conductance of bTRPC3 (cloned from bovine retina) activated by ATP, assessed by noise analysis, is 23 pS [34]. Stimulated by intracellular Ca^{2+}, hTRPC3 in inside-out patches shows cation-selective channels with 66-pS conductance and short (<2 ms) mean open times [104].

TRPC4

Following GTPγS stimulation, the PCa/PNa value calculated using the Goldman equation for divalent and monovalent cations is 1.05 for mTRPC4 [78]. According to the constant-field theory, the permeability ratio PCs/PNa of bTRPC4 is about 0.9 at nominal extracellular free Ca^{2+}. Extrapolating the ratio to higher divalent ion concentrations, the lower limits for the relative permeabilities PCs/PNa/PCa/PBa of bTRPC4 is 1:1.1:7.7:12.3 [69]. In contrast to TRPC3, La^{3+} enhances mTRPC4 current [78]. Under symmetrical buffer conditions, the single-channel I-V relation for mTRPC4 closely resembled those for whole-cell currents, showing a doubly rectifying shape and a reversal potential close to 0 mV. The single-channel chord conductances at 260 mV are 41 ± 1 picosiemens for mTRPC4 [78].

TRPC5

The PCa/PNa value calculated using the Goldman equation for divalent and monovalent cations is 1.79 for mTRPC5 following GTPγS stimulation. However, in response to ATP, on the assumption that activity coefficients are 0.3 for Ca^{2+} and 0.75 for both Na^{+} and Cs^{+}, the reversal potentials of 8 mV in the $0Ca^{2+}$ external solution and 17 mV in the 10 mM Ca^{2+} solution lead to permeability ratios PCa/PNa/PCs = 14.3:1.5:1 [66]. Similar to TRPC4, the GTPγS-induced TRPC5 current at both positive and negative membrane potentials is potentiated by La^{3+} without changing the reversal potential. Carbachol-induced Mn^{2+} entry through mTRPC5 in fura-2-loaded HEK cells was not impaired by La^{3+} at concentrations up to 300 mM [78]. However, La^{3+}, but not Gd^{3+}, significantly suppresses Ca^{2+} influx induced by ATP in TRPC5-transfected cells [66]. Like mTRPC4, the single-channel I-V relation for mTRPC5 shows a doubly rectifying shape and a reversal potential close to 0 mV. The single-channel chord conductance at 260 mV was 63 ± 1 pS for mTRPC5 [78].

TRPC6

The hTRPC6 is a nonselective cation channel that is permeable for Ca^{2+}, Cs^{+}, Na^{+}, and K^{+}, but not NMDG. The relative permeabilities PCa/PNa and PNa/PCs are 5 and 0.7, respectively, under bi-ionic conditions with Cs^{+} as the main cation in the pipette. The current-voltage (I-V) relation reveals dual inward and outward rectification.

The reversal potential of the hTRPC6 current is −3.6 ± 0.8 mV. In inside-out patches, single-channel currents with a mean unitary current amplitude of −1.7 ± 0.1 pA are detected at a holding potential of −60 mV. The hTRPC6 exhibited a linear I-V relationship with a calculated slope conductance of 35 pS and 37.5 pS in symmetrical 120 mM Cs^+ and in the bath solution with 120 mM Na^+. The open probability for hTRPC6 is higher at positive (+60 mV) than at negative (−60 mV) holding potentials, and the mean open time is within 1 ms [26].

TRPC7

The current-voltage relationship of TRPC7 current induced by ATP was almost linear, showing a slight flattening approximately between 0 and 20 mV and outward rectification at more positive potentials. Divalent cations, Ca^{2+}, Ba^{2+}, and Mn^{2+}, permeate the TRPC7 channels responsible for the spontaneous and ATP-enhanced inward currents. The calculated relative permeabilities (PCs/PNa/PCa/PBa) are 1:1.0:1.9:3.5 for the spontaneous current and 1:1.1:5.9:5.0 for the ATP-enhanced current. In the presence of 2 mM Ca^{2+}, further addition of 100 μM Gd^{3+} only slightly inhibits the currents induced by ATP [65].

2.2.2.3 Properties of TRPC Channel Heterotetramer

The combinatorial rules within the TRPC subfamily put forward by Hofmann et al. appear to apply in vivo, since the complex formed by TRPC1/TRPC4/TRPC5 is found in the embryonic brain [27, 83]. Although some ruled out heteromerization between distant relatives in the TRPC family, the nature of the heterotretramer is not clear so far. The evidence provided by several groups indicates that distant TRPC members can form heteromeric channels. For example, TRPC3 can assemble with TRPC1 [51], and a TRPC3/TRPC4 complex is able to form redox-sensitive channels in endothelial cells.

Moreover, TRPC heteromultimers with functional properties are distinct from homodmeric channels. Co-expression of TRPC1 and TRPL in *Xenopus* oocytes produced a thapsigargin-stimulated current that was not observed in oocytes expressing either TRPC1 or TRPL alone [17]. The co-expression of TRPC1 and TRPL in 293T cells produces a novel current not present in either TRPC1-expressing or TRPL-expressing cells [97, 93]). Co-expression of TRPC1 and TRPC3 results in a constitutively active cation conductance higher than that in TRPC1 or TRPC3 expression [51]. Coincidently, co-expression of TRPC1 and TRPC4 or TRPC5 results in outwardly rectifying nonselective cation conductances [83]. Homomeric TRPC5 is inwardly rectifying and has a conductance of 38 pS; TRPC1/TRPC5 is outwardly rectifying and displays an eightfold smaller conductance. The whole-cell I-V curve of the TRPC1/TRPC5 current is not changed upon removal of intra- and extracellular Mg^{2+}, indicating that the rectification mechanisms in TRPC1/TRPC5 may be different from TRPC5 homomers. The recombinant TRPC1/TRPC5 channel is activated by Gq-coupled receptors, and its activity is independent of calcium store depletion, which supports the hypothesis that TRPC heteromers form receptor-modulated currents in the mammalian brain. Moreover, biochemical analyses have identified TRPC1, TRPC4, and TRPC5 heteromultimers in rat embryonic brains, which form channels with novel conductances when expressed in vitro [83].

Together, heteromerization of TRPCs can form channels that have unique properties not exhibited by homomeric TRPC channels. Such a mechanism of homo- and heteromultimeric channel formation should create an incredible diversity of channels with an array of distinct biophysical properties and biological functions.

2.3 Prospectives

More efforts are required on structure analysis and in vivo study to have a better understanding of the TRPC channels. The good news is that there is great progress made by cryo-EM. Therefore, determining the structures of TRPCs is more than promising.

Though lots of efforts have been made to figure out the properties of TRPC channels, conflict results obtained from varied systems make the whole field confusing. Many reasons can be listed to explain the contradiction. Firstly, due to various splice variants of TRPC homologs, it is possible that the expressed TRPC homolog may differ from that utilized to make the native channel. Each cell line has its own preference of protein expression, and even the same cell line cultured in various labs or by different people may have its unique expression profile. Secondly, it is not the exogenously expressed TRPC channels but the endogenous regulatory proteins that cause the difference. Thirdly, even when channel activities detected by electrophysiology are not present in the control cells, it does not necessarily imply that the activity belongs to the exogenously expressed protein. Fourthly, characterization of TRPC channels relies on "transient" overexpression or downregulation TRPC proteins. However, the "transient" time is enough to change the expression of numerous endogenous proteins or the activation of endogenous channels in response to the overexpression protein. In addition, for TRPCs, the universal expressed channels which have relatively small current amplitudes, it is hard to tell how much contribution belongs to these endogenous channels. Generating good antibody, especially functional antibody, and screening specific agonists and inhibitors, along with determining the structures, may be helpful for resolving the differences.

References

1. Acharya JK, Jalink K, Hardy RW, Hartenstein V, Zuker CS (1997) InsP3 receptor is essential for growth and differentiation but not for vision in Drosophila. Neuron 18(6):881–887
2. Amiri H, Schultz G, Schaefer M (2003) FRET-based analysis of TRPC subunit stoichiometry. Cell Calcium 33(5–6):463–470
3. Anderson M, Kim EY, Hagmann H, Benzing T, Dryer SE (2013) Opposing effects of podocin on the gating of podocyte TRPC6 channels evoked by membrane stretch or diacylglycerol. Am J Physiol Cell Physiol 305(3):C276–C289. doi:10.1152/ajpcell.00095.2013
4. Beck A, Speicher T, Stoerger C, Sell T, Dettmer V, Jusoh SA, Abdulmughni A, Cavalie A, Philipp SE, Zhu MX, Helms V, Wissenbach U, Flockerzi V (2013) Conserved gating elements in TRPC4 and TRPC5 channels. J Biol Chem 288(27):19471–19483. doi:10.1074/jbc.M113.478305
5. Boulay G, Zhu X, Peyton M, Jiang M, Hurst R, Stefani E, Birnbaumer L (1997) Cloning and expression of a novel mammalian homolog of Drosophila transient receptor potential (Trp) involved in calcium entry secondary to activation of receptors coupled by the Gq class of G protein. J Biol Chem 272(47):29672–29680
6. Chang Q, Gyftogianni E, van de Graaf SF, Hoefs S, Weidema FA, Bindels RJ, Hoenderop JG (2004) Molecular determinants in TRPV5 channel assembly. J Biol Chem 279(52):54304–54311. doi:10.1074/jbc.M406222200
7. Clapham DE, Runnels LW, Strubing C (2001) The TRP ion channel family. Nat Rev Neurosci 2(6):387–396. doi:10.1038/35077544
8. Cordero-Morales JF, Gracheva EO, Julius D (2011) Cytoplasmic ankyrin repeats of transient receptor potential A1 (TRPA1) dictate sensitivity to thermal and chemical stimuli. Proc Natl Acad Sci U S A 108(46):E1184–E1191. doi:10.1073/pnas.1114124108
9. Engelke M, Friedrich O, Budde P, Schafer C, Niemann U, Zitt C, Jungling E, Rocks O, Luckhoff A, Frey J (2002) Structural domains required for channel function of the mouse transient receptor potential protein homologue TRP1beta. FEBS Lett 523(1-3):193–199
10. Erler I, Al-Ansary DM, Wissenbach U, Wagner TF, Flockerzi V, Niemeyer BA (2006) Trafficking and assembly of the cold-sensitive TRPM8 channel. J Biol Chem 281(50):38396–38404. doi:10.1074/jbc.M607756200
11. Erler I, Hirnet D, Wissenbach U, Flockerzi V, Niemeyer BA (2004) Ca2+-selective transient receptor potential V channel architecture and function require a specific ankyrin repeat. J Biol Chem 279(33):34456–34463. doi:10.1074/jbc.M404778200
12. Feske S, Gwack Y, Prakriya M, Srikanth S, Puppel SH, Tanasa B, Hogan PG, Lewis RS, Daly M, Rao A (2006) A mutation in orai1 causes immune deficiency by abrogating CRAC channel function. Nature 441(7090):179–185. doi:10.1038/nature04702
13. Fujiwara Y, Minor DL Jr (2008) X-ray crystal structure of a TRPM assembly domain reveals an antiparallel four-stranded coiled-coil. J Mol Biol 383(4):854–870. doi:10.1016/j.jmb.2008.08.059
14. Garcia-Sanz N, Fernandez-Carvajal A, Morenilla-Palao C, Planells-Cases R, Fajardo-Sanchez E, Fernandez-Ballester G, Ferrer-Montiel A (2004) Identification of a tetramerization domain in the C terminus of the vanilloid receptor. J Neurosci: Off J Soc Neurosci 24(23):5307–5314. doi:10.1523/JNEUROSCI.0202-04.2004

15. Gaudet R (2007) Structural insights into the function of TRP channels. In: Liedtke WB, Heller S (eds) TRP ion channel function in sensory transduction and cellular signaling cascades, Frontiers in neuroscience. CRC Press, Boca Raton
16. Gaudet R (2008) TRP channels entering the structural era. J Physiol 586(15):3565–3575. doi:10.1113/jphysiol.2008.155812
17. Gillo B, Chorna I, Cohen H, Cook B, Manistersky I, Chorev M, Arnon A, Pollock JA, Selinger Z, Minke B (1996) Coexpression of Drosophila TRP and TRP-like proteins in Xenopus oocytes reconstitutes capacitative Ca2+ entry. Proc Natl Acad Sci U S A 93(24):14146–14151
18. Grisshammer R (2006) Understanding recombinant expression of membrane proteins. Curr Opin Biotechnol 17(4):337–340. doi:10.1016/j.copbio.2006.06.001
19. Harteneck C (2003) Proteins modulating TRP channel function. Cell Calcium 33(5–6):303–310
20. Harteneck C, Plant TD, Schultz G (2000) From worm to man: three subfamilies of TRP channels. Trends Neurosci 23(4):159–166
21. Hellmich UA, Gaudet R (2014) Structural biology of TRP channels. Handb Exp Pharmacol 223:963–990. doi:10.1007/978-3-319-05161-1_10
22. Hellwig N, Albrecht N, Harteneck C, Schultz G, Schaefer M (2005) Homo- and heteromeric assembly of TRPV channel subunits. J Cell Sci 118(Pt 5):917–928. doi:10.1242/jcs.01675
23. Hilgemann DW, Feng S, Nasuhoglu C (2001) The complex and intriguing lives of PIP2 with ion channels and transporters. Sci STKE: Sig Transduct Knowl Environ 2001(111):re19. doi:10.1126/stke.2001.111.re19
24. Hirschler-Laszkiewicz I, Tong Q, Waybill K, Conrad K, Keefer K, Zhang W, Chen SJ, Cheung JY, Miller BA (2011) The transient receptor potential (TRP) channel TRPC3 TRP domain and AMP-activated protein kinase binding site are required for TRPC3 activation by erythropoietin. J Biol Chem 286(35):30636–30646. doi:10.1074/jbc.M111.238360
25. Hoenderop JG, Voets T, Hoefs S, Weidema F, Prenen J, Nilius B, Bindels RJ (2003) Homo- and heterotetrameric architecture of the epithelial Ca2+ channels TRPV5 and TRPV6. EMBO J 22(4):776–785. doi:10.1093/emboj/cdg080
26. Hofmann T, Obukhov AG, Schaefer M, Harteneck C, Gudermann T, Schultz G (1999) Direct activation of human TRPC6 and TRPC3 channels by diacylglycerol. Nature 397(6716):259–263. doi:10.1038/16711
27. Hofmann T, Schaefer M, Schultz G, Gudermann T (2002) Subunit composition of mammalian transient receptor potential channels in living cells. Proc Natl Acad Sci U S A 99(11):7461–7466. doi:10.1073/pnas.102596199
28. Huang GN, Zeng W, Kim JY, Yuan JP, Han L, Muallem S, Worley PF (2006) STIM1 carboxyl-terminus activates native SOC, I(crac) and TRPC1 channels. Nat Cell Biol 8(9):1003–1010. doi:10.1038/ncb1454
29. Jahnel R, Dreger M, Gillen C, Bender O, Kurreck J, Hucho F (2001) Biochemical characterization of the vanilloid receptor 1 expressed in a dorsal root ganglia derived cell line. Eur J Biochem/FEBS 268(21):5489–5496
30. Jeon JP, Hong C, Park EJ, Jeon JH, Cho NH, Kim IG, Choe H, Muallem S, Kim HJ, So I (2012) Selective Galphai subunits as novel direct activators of transient receptor potential canonical (TRPC)4 and TRPC5 channels. J Biol Chem 287(21):17029–17039. doi:10.1074/jbc.M111.326553
31. Jin X, Touhey J, Gaudet R (2006) Structure of the N-terminal ankyrin repeat domain of the TRPV2 ion channel. J Biol Chem 281(35):25006–25010. doi:10.1074/jbc.C600153200
32. Jung S, Muhle A, Schaefer M, Strotmann R, Schultz G, Plant TD (2003) Lanthanides potentiate TRPC5 currents by an action at extracellular sites close to the pore mouth. J Biol Chem 278(6):3562–3571. doi:10.1074/jbc.M211484200
33. Jungnickel MK, Marrero H, Birnbaumer L, Lemos JR, Florman HM (2001) Trp2 regulates entry of Ca2+ into mouse sperm triggered by egg ZP3. Nat Cell Biol 3(5):499–502. doi:10.1038/35074570
34. Kamouchi M, Philipp S, Flockerzi V, Wissenbach U, Mamin A, Raeymaekers L, Eggermont J, Droogmans G, Nilius B (1999) Properties of heterologously expressed hTRP3 channels in bovine pulmonary artery endothelial cells. J Physiol 518(Pt 2):345–358
35. Kedei N, Szabo T, Lile JD, Treanor JJ, Olah Z, Iadarola MJ, Blumberg PM (2001) Analysis of the native quaternary structure of vanilloid receptor 1. J Biol Chem 276(30):28613–28619. doi:10.1074/jbc.M103272200
36. Kerstein PC, Jacques-Fricke BT, Rengifo J, Mogen BJ, Williams JC, Gottlieb PA, Sachs F, Gomez TM (2013) Mechanosensitive TRPC1 channels promote calpain proteolysis of talin to regulate spinal axon outgrowth. J Neurosci: Off J Soc Neurosci 33(1):273–285. doi:10.1523/JNEUROSCI.2142-12.2013
37. Kim H, Kim J, Jeon JP, Myeong J, Wie J, Hong C, Kim HJ, Jeon JH, So I (2012) The roles of G proteins in the activation of TRPC4 and TRPC5 transient receptor potential channels. Channels 6(5):333–343. doi:10.4161/chan.21198
38. Kiselyov K, Xu X, Mozhayeva G, Kuo T, Pessah I, Mignery G, Zhu X, Birnbaumer L, Muallem S (1998) Functional interaction between InsP3 receptors and store-operated Htrp3 channels. Nature 396(6710):478–482. doi:10.1038/24890
39. Kiselyov KI, Shin DM, Wang Y, Pessah IN, Allen PD, Muallem S (2000) Gating of store-operated

channels by conformational coupling to ryanodine receptors. Mol Cell 6(2):421–431
40. Latorre R, Zaelzer C, Brauchi S (2009) Structure-functional intimacies of transient receptor potential channels. Q Rev Biophys 42(3):201–246. doi:10.1017/S0033583509990072
41. Lee KP, Choi S, Hong JH, Ahuja M, Graham S, Ma R, So I, Shin DM, Muallem S, Yuan JP (2014) Molecular determinants mediating gating of Transient Receptor Potential Canonical (TRPC) channels by stromal interaction molecule 1 (STIM1). J Biol Chem 289(10):6372–6382. doi:10.1074/jbc.M113.546556
42. Lepage PK, Boulay G (2007) Molecular determinants of TRP channel assembly. Biochem Soc Trans 35(Pt 1):81–83. doi:10.1042/BST0350081
43. Lepage PK, Lussier MP, Barajas-Martinez H, Bousquet SM, Blanchard AP, Francoeur N, Dumaine R, Boulay G (2006) Identification of two domains involved in the assembly of transient receptor potential canonical channels. J Biol Chem 281(41):30356–30364. doi:10.1074/jbc.M603930200
44. Lepage PK, Lussier MP, McDuff FO, Lavigne P, Boulay G (2009) The self-association of two N-terminal interaction domains plays an important role in the tetramerization of TRPC4. Cell Calcium 45(3):251–259. doi:10.1016/j.ceca.2008.11.002
45. Li M, Yu Y, Yang J (2011) Structural biology of TRP channels. Adv Exp Med Biol 704:1–23. doi:10.1007/978-94-007-0265-3_1
46. Liao M, Cao E, Julius D, Cheng Y (2013) Structure of the TRPV1 ion channel determined by electron cryo-microscopy. Nature 504(7478):107–112. doi:10.1038/nature12822
47. Liao Y, Erxleben C, Abramowitz J, Flockerzi V, Zhu MX, Armstrong DL, Birnbaumer L (2008) Functional interactions among Orai1, TRPCs, and STIM1 suggest a STIM-regulated heteromeric Orai/TRPC model for SOCE/Icrac channels. Proc Natl Acad Sci U S A 105(8):2895–2900. doi:10.1073/pnas.0712288105
48. Liao Y, Erxleben C, Yildirim E, Abramowitz J, Armstrong DL, Birnbaumer L (2007) Orai proteins interact with TRPC channels and confer responsiveness to store depletion. Proc Natl Acad Sci U S A 104(11):4682–4687. doi:10.1073/pnas.0611692104
49. Liao Y, Plummer NW, George MD, Abramowitz J, Zhu MX, Birnbaumer L (2009) A role for orai in TRPC-mediated Ca2+ entry suggests that a TRPC:orai complex may mediate store and receptor operated Ca2+ entry. Proc Natl Acad Sci U S A 106(9):3202–3206. doi:10.1073/pnas.0813346106
50. Liman ER, Corey DP, Dulac C (1999) TRP2: a candidate transduction channel for mammalian pheromone sensory signaling. Proc Natl Acad Sci U S A 96(10):5791–5796
51. Lintschinger B, Balzer-Geldsetzer M, Baskaran T, Graier WF, Romanin C, Zhu MX, Groschner K (2000) Coassembly of Trp1 and Trp3 proteins generates diacylglycerol- and Ca2+-sensitive cation channels. J Biol Chem 275(36):27799–27805. doi:10.1074/jbc.M002705200
52. Lishko PV, Procko E, Jin X, Phelps CB, Gaudet R (2007) The ankyrin repeats of TRPV1 bind multiple ligands and modulate channel sensitivity. Neuron 54(6):905–918. doi:10.1016/j.neuron.2007.05.027
53. Liu X, Singh BB, Ambudkar IS (2003) TRPC1 is required for functional store-operated Ca2+ channels. Role of acidic amino acid residues in the S5-S6 region. J Biol Chem 278(13):11337–11343. doi:10.1074/jbc.M213271200
54. Lucas P, Ukhanov K, Leinders-Zufall T, Zufall F (2003) A diacylglycerol-gated cation channel in vomeronasal neuron dendrites is impaired in TRPC2 mutant mice: mechanism of pheromone transduction. Neuron 40(3):551–561
55. Lussier MP, Cayouette S, Lepage PK, Bernier CL, Francoeur N, St-Hilaire M, Pinard M, Boulay G (2005) MxA, a member of the dynamin superfamily, interacts with the ankyrin-like repeat domain of TRPC. J Biol Chem 280(19):19393–19400. doi:10.1074/jbc.M500391200
56. Lussier MP, Lepage PK, Bousquet SM, Boulay G (2008) RNF24, a new TRPC interacting protein, causes the intracellular retention of TRPC. Cell Calcium 43(5):432–443. doi:10.1016/j.ceca.2007.07.009
57. Maroto R, Raso A, Wood TG, Kurosky A, Martinac B, Hamill OP (2005) TRPC1 forms the stretch-activated cation channel in vertebrate cells. Nat Cell Biol 7(2):179–185. doi:10.1038/ncb1218
58. Mei ZZ, Mao HJ, Jiang LH (2006) Conserved cysteine residues in the pore region are obligatory for human TRPM2 channel function. Am J Physiol Cell Physiol 291(5):C1022–C1028. doi:10.1152/ajpcell.00606.2005
59. Mery L, Strauss B, Dufour JF, Krause KH, Hoth M (2002) The PDZ-interacting domain of TRPC4 controls its localization and surface expression in HEK293 cells. J Cell Sci 115(Pt 17):3497–3508
60. Minke B, Parnas M (2006) Insights on TRP channels from in vivo studies in Drosophila. Annu Rev Physiol 68:649–684. doi:10.1146/annurev.physiol.68.040204.100939
61. Mio K, Ogura T, Kiyonaka S, Hiroaki Y, Tanimura Y, Fujiyoshi Y, Mori Y, Sato C (2007) The TRPC3 channel has a large internal chamber surrounded by signal sensing antennas. J Mol Biol 367(2):373–383. doi:10.1016/j.jmb.2006.12.043
62. Montell C (2001) Physiology, phylogeny, and functions of the TRP superfamily of cation channels. Sci STKE: Sig Transduct Knowl Environ 2001(90):re1. doi:10.1126/stke.2001.90.re1
63. Mosavi LK, Cammett TJ, Desrosiers DC, Peng ZY (2004) The ankyrin repeat as molecular architecture for protein recognition. Protein Sci: Publ Protein Soc 13(6):1435–1448. doi:10.1110/ps.03554604

64. Nooren IM, Kaptein R, Sauer RT, Boelens R (1999) The tetramerization domain of the Mnt repressor consists of two right-handed coiled coils. Nat Struct Biol 6(8):755–759. doi:10.1038/11531
65. Okada T, Inoue R, Yamazaki K, Maeda A, Kurosaki T, Yamakuni T, Tanaka I, Shimizu S, Ikenaka K, Imoto K, Mori Y (1999) Molecular and functional characterization of a novel mouse transient receptor potential protein homologue TRP7. Ca(2+)-permeable cation channel that is constitutively activated and enhanced by stimulation of G protein-coupled receptor. J Biol Chem 274(39):27359–27370
66. Okada T, Shimizu S, Wakamori M, Maeda A, Kurosaki T, Takada N, Imoto K, Mori Y (1998) Molecular cloning and functional characterization of a novel receptor-activated TRP Ca2+ channel from mouse brain. J Biol Chem 273(17):10279–10287
67. Owsianik G, Talavera K, Voets T, Nilius B (2006) Permeation and selectivity of TRP channels. Annu Rev Physiol 68:685–717. doi:10.1146/annurev.physiol.68.040204.101406
68. Paulsen CE, Armache JP, Gao Y, Cheng Y, Julius D (2015) Structure of the TRPA1 ion channel suggests regulatory mechanisms. Nature 525(7570):552. doi:10.1038/nature14871
69. Philipp S, Cavalie A, Freichel M, Wissenbach U, Zimmer S, Trost C, Marquart A, Murakami M, Flockerzi V (1996) A mammalian capacitative calcium entry channel homologous to Drosophila TRP and TRPL. EMBO J 15(22):6166–6171
70. Putney JW Jr (2007) Multiple mechanisms of TRPC activation. In: Liedtke WB, Heller S (eds) TRP ion channel function in sensory transduction and cellular signaling cascades, Frontiers in neuroscience. CRC Press/Taylor & Francis, Boca Raton
71. Quick K, Zhao J, Eijkelkamp N, Linley JE, Rugiero F, Cox JJ, Raouf R, Gringhuis M, Sexton JE, Abramowitz J, Taylor R, Forge A, Ashmore J, Kirkwood N, Kros CJ, Richardson GP, Freichel M, Flockerzi V, Birnbaumer L, Wood JN (2012) TRPC3 and TRPC6 are essential for normal mechanotransduction in subsets of sensory neurons and cochlear hair cells. Open Biol 2(5):120068. doi:10.1098/rsob.120068
72. Raghu P, Colley NJ, Webel R, James T, Hasan G, Danin M, Selinger Z, Hardie RC (2000) Normal phototransduction in Drosophila photoreceptors lacking an InsP(3) receptor gene. Mol Cell Neurosci 15(5):429–445. doi:10.1006/mcne.2000.0846
73. Ramsey IS, Delling M, Clapham DE (2006) An introduction to TRP channels. Annu Rev Physiol 68:619–647. doi:10.1146/annurev.physiol.68.040204.100431
74. Ramsey IS, Moran MM, Chong JA, Clapham DE (2006) A voltage-gated proton-selective channel lacking the pore domain. Nature 440(7088):1213–1216. doi:10.1038/nature04700
75. Rohacs T, Lopes CM, Michailidis I, Logothetis DE (2005) PI(4,5)P2 regulates the activation and desensitization of TRPM8 channels through the TRP domain. Nat Neurosci 8(5):626–634. doi:10.1038/nn1451
76. Saimi Y, Kung C (2002) Calmodulin as an ion channel subunit. Annu Rev Physiol 64:289–311. doi:10.1146/annurev.physiol.64.100301.111649
77. Schaefer M (2005) Homo- and heteromeric assembly of TRP channel subunits. Pflugers Arch - Eur J Physiol 451(1):35–42. doi:10.1007/s00424-005-1467-6
78. Schaefer M, Plant TD, Obukhov AG, Hofmann T, Gudermann T, Schultz G (2000) Receptor-mediated regulation of the nonselective cation channels TRPC4 and TRPC5. J Biol Chem 275(23):17517–17526
79. Schindl R, Frischauf I, Kahr H, Fritsch R, Krenn M, Derndl A, Vales E, Muik M, Derler I, Groschner K, Romanin C (2008) The first ankyrin-like repeat is the minimum indispensable key structure for functional assembly of homo- and heteromeric TRPC4/TRPC5 channels. Cell Calcium 43(3):260–269. doi:10.1016/j.ceca.2007.05.015
80. Schindl R, Fritsch R, Jardin I, Frischauf I, Kahr H, Muik M, Riedl MC, Groschner K, Romanin C (2012) Canonical transient receptor potential (TRPC) 1 acts as a negative regulator for vanilloid TRPV6-mediated Ca2+ influx. J Biol Chem 287(42):35612–35620. doi:10.1074/jbc.M112.400952
81. Schindl R, Romanin C (2007) Assembly domains in TRP channels. Biochem Soc Trans 35(Pt 1):84–85. doi:10.1042/BST0350084
82. Song X, Zhao Y, Narcisse L, Duffy H, Kress Y, Lee S, Brosnan CF (2005) Canonical transient receptor potential channel 4 (TRPC4) co-localizes with the scaffolding protein ZO-1 in human fetal astrocytes in culture. Glia 49(3):418–429. doi:10.1002/glia.20128
83. Strubing C, Krapivinsky G, Krapivinsky L, Clapham DE (2003) Formation of novel TRPC channels by complex subunit interactions in embryonic brain. J Biol Chem 278(40):39014–39019. doi:10.1074/jbc.M306705200
84. Tai Y, Feng S, Ge R, Du W, Zhang X, He Z, Wang Y (2008) TRPC6 channels promote dendritic growth via the CaMKIV-CREB pathway. J Cell Sci 121(Pt 14):2301–2307. doi:10.1242/jcs.026906
85. Tang J, Lin Y, Zhang Z, Tikunova S, Birnbaumer L, Zhu MX (2001) Identification of common binding sites for calmodulin and inositol 1,4,5-trisphosphate receptors on the carboxyl termini of trp channels. J Biol Chem 276(24):21303–21310. doi:10.1074/jbc.M102316200
86. Tang Y, Tang J, Chen Z, Trost C, Flockerzi V, Li M, Ramesh V, Zhu MX (2000) Association of mammalian trp4 and phospholipase C isozymes with a PDZ domain-containing protein, NHERF. J

Biol Chem 275(48):37559–37564. doi:10.1074/jbc.M006635200
87. Tsuruda PR, Julius D, Minor DL Jr (2006) Coiled coils direct assembly of a cold-activated TRP channel. Neuron 51(2):201–212. doi:10.1016/j.neuron.2006.06.023
88. Vannier B, Peyton M, Boulay G, Brown D, Qin N, Jiang M, Zhu X, Birnbaumer L (1999) Mouse trp2, the homologue of the human trpc2 pseudogene, encodes mTrp2, a store depletion-activated capacitative Ca2+ entry channel. Proc Natl Acad Sci U S A 96(5):2060–2064
89. Venkatachalam K, Ma HT, Ford DL, Gill DL (2001) Expression of functional receptor-coupled TRPC3 channels in DT40 triple receptor InsP3 knockout cells. J Biol Chem 276(36):33980–33985. doi:10.1074/jbc.C100321200
90. Venkatachalam K, Montell C (2007) TRP channels. Annu Rev Biochem 76:387–417. doi:10.1146/annurev.biochem.75.103004.142819
91. Venkatachalam K, Zheng F, Gill DL (2003) Regulation of canonical transient receptor potential (TRPC) channel function by diacylglycerol and protein kinase C. J Biol Chem 278(31):29031–29040. doi:10.1074/jbc.M302751200
92. Vig M, Peinelt C, Beck A, Koomoa DL, Rabah D, Koblan-Huberson M, Kraft S, Turner H, Fleig A, Penner R, Kinet JP (2006) CRACM1 is a plasma membrane protein essential for store-operated Ca2+ entry. Science 312(5777):1220–1223. doi:10.1126/science.1127883
93. Villereal ML (2006) Mechanism and functional significance of TRPC channel multimerization. Semin Cell Dev Biol 17(6):618–629. doi:10.1016/j.semcdb.2006.10.010
94. Voets T, Nilius B (2003) The pore of TRP channels: trivial or neglected? Cell Calcium 33(5-6):299–302
95. Wang J, Lu R, Yang J, Li H, He Z, Jing N, Wang X, Wang Y (2015) TRPC6 specifically interacts with APP to inhibit its cleavage by gamma-secretase and reduce Abeta production. Nat Commun 6:8876. doi:10.1038/ncomms9876
96. Woo JS, Lee KJ, Huang M, Cho CH, Lee EH (2014) Heteromeric TRPC3 with TRPC1 formed via its ankyrin repeats regulates the resting cytosolic Ca2+ levels in skeletal muscle. Biochem Biophys Res Commun 446(2):454–459. doi:10.1016/j.bbrc.2014.02.127
97. Xu XZ, Li HS, Guggino WB, Montell C (1997) Coassembly of TRP and TRPL produces a distinct store-operated conductance. Cell 89(7):1155–1164
98. Yuan JP, Zeng W, Huang GN, Worley PF, Muallem S (2007) STIM1 heteromultimerizes TRPC channels to determine their function as store-operated channels. Nat Cell Biol 9(6):636–645. doi:10.1038/ncb1590
99. Yuasa K, Matsuda T, Tsuji A (2011) Functional regulation of transient receptor potential canonical 7 by cGMP-dependent protein kinase Ialpha. Cell Signal 23(7):1179–1187. doi:10.1016/j.cellsig.2011.03.005
100. Zeng W, Yuan JP, Kim MS, Choi YJ, Huang GN, Worley PF, Muallem S (2008) STIM1 gates TRPC channels, but not orai1, by electrostatic interaction. Mol Cell 32(3):439–448. doi:10.1016/j.molcel.2008.09.020
101. Zhang SL, Yu Y, Roos J, Kozak JA, Deerinck TJ, Ellisman MH, Stauderman KA, Cahalan MD (2005) STIM1 is a Ca2+ sensor that activates CRAC channels and migrates from the Ca2+ store to the plasma membrane. Nature 437(7060):902–905. doi:10.1038/nature04147
102. Zhang W, Cheng LE, Kittelmann M, Li J, Petkovic M, Cheng T, Jin P, Guo Z, Gopfert MC, Jan LY, Jan YN (2015) Ankyrin repeats convey force to gate the NOMPC mechanotransduction channel. Cell 162(6):1391–1403. doi:10.1016/j.cell.2015.08.024
103. Zhu X, Jiang M, Peyton M, Boulay G, Hurst R, Stefani E, Birnbaumer L (1996) trp, a novel mammalian gene family essential for agonist-activated capacitative Ca2+ entry. Cell 85(5):661–671
104. Zitt C, Obukhov AG, Strubing C, Zobel A, Kalkbrenner F, Luckhoff A, Schultz G (1997) Expression of TRPC3 in Chinese hamster ovary cells results in calcium-activated cation currents not related to store depletion. J Cell Biol 138(6):1333–1341
105. Zitt C, Zobel A, Obukhov AG, Harteneck C, Kalkbrenner F, Luckhoff A, Schultz G (1996) Cloning and functional expression of a human Ca2+-permeable cation channel activated by calcium store depletion. Neuron 16(6):1189–1196
106. Zubcevic L, Herzik MA Jr, Chung BC, Liu Z, Lander GC, Lee SY (2016) Cryo-electron microscopy structure of the TRPV2 ion channel. Nat Struct Mol Biol 23(2):180–186. doi:10.1038/nsmb.3159

TRPC Channel Downstream Signaling Cascades

3

Zhuohao He

Abstract

The family of TRP channel is comprised of a large group of cation-permeable channels, displaying as signaling integrators for sensing extracellular stimulus and initiating intracellular signaling cascades. This chapter offers a brief review of the signaling molecules related to TRPC channels, the first identified mammalian TRP family. Besides the signaling molecules involved in TRPC activation, I will focus on their upstream and downstream signaling cascades and the molecules involved in their intracellular trafficking.

Keywords

TRPC • Signaling cascades • Intracellular trafficking

TRP, a large family of membrane channels with the common structure of six transmembrane segments and two cytoplasmic terminals, was first identified and cloned in *Drosophila* [37] and later found to be evolutionarily conserved with diverse functions in a wide range of metazoan organisms [22, 32, 57–60]. They have attracted more attentions for their important roles in neuronal development; sensory functions such as sensing temperature, touch, pain, osmolarity, pheromones, taste, and other stimulus; cardiovascular functions; and even in tumor cell proliferation. Their diverse functions are probably due to their central roles in the cellular signaling transduction, associating both extracellular and intracellular signaling cascades. I will take TRPC channels as the examples to review their integrating roles for both extracellular signal sensors and intracellular signaling initiators. The following parts will first give a brief review about the two common signaling cascades directly involved in TRPC channel activation and then offer a review of the related signaling cascades for each TRPC subfamily member, respectively.

Z. He (✉)
Department of Pathology and Laboratory Medicine, Institute on Aging and Center for Neurodegenerative Disease Research, University of Pennsylvania School of Medicine, Philadelphia, PA 19104, USA
e-mail: hezhuohao@gmail.com

Y. Wang (ed.), *Transient Receptor Potential Canonical Channels and Brain Diseases*, Advances in Experimental Medicine and Biology 976, DOI 10.1007/978-94-024-1088-4_3

3.1 Common Signaling Cascade Involved in TRPC Activation

Generally, there are two major mechanisms involved in TRPC activation. They are receptor-operated calcium entry (ROCE) and store-operated calcium entry (SOCE). In ROCE, G protein-coupled receptors (GPCR) and receptor tyrosine kinase (RTK) are responsible for TRPC activation. In the presence of environmental stimuli, i.e., ligands/agonists, these receptors will be activated and subsequently recruit phospholipase C (PLCβ/γ) to hydrolyze phosphatidylinositol 4,5-bisphosphate (PIP_2) to produce diacylglycerol (DAG) and inositol 1,4,5-trisphosphate (IP3). DAG can directly interact with TRPC channels, leading to its activation. The produced IP_3 would bind to IP_3 receptor (IP_3R) located in endoplasmic reticulum (ER), releasing intracellular calcium and subsequently triggering extracellular calcium influx through TRPC channels, the process of which is called SOCE [11].

3.2 Diverse Signaling Cascades of TRPC Channels

TRPC channel family has several subgroups: TRPC1, TRPC4 and TRPC5, TRPC3 and TRPC6, and TRPC7. The different channel-related signaling cascades in the order of different subgroups are summarized in Table 3.1 and will be discussed in details in the following sessions.

3.2.1 TRPC1

Growth factors are important for the activation of TRPC1 channels. It has been reported that TRPC1 mediates bFGF-induced Ca^{2+} entry and the consequent proliferation in embryonic rat neural stem cells (NSCs). The expression of both TRPC1 and FGFR-1 is found in the embryonic rat telencephalon. Knocking down TRPC1 decreases bFGF-induced Ca^{2+} entry, the proliferation of NSC progeny, and an inward cation current evoked by bFGF in proliferating NSCs without affecting the release of Ca^{2+} from intracellular stores or 1-oleoyl-2-acetyl-sn-glycerol-induced Ca^{2+} entry [43]. Netrin-1, brain-derived neurotrophic factor (BDNF), and myelin-associated glycoprotein have been also reported to activate TRPC1 in the process of neuronal growth cone turning. XTRPC1, the *Xenopus* homolog of mammalian TRPC1, are required for proper growth cone turning for *Xenopus* spinal neurons in responses to microscopic gradients of netrin-1, brain-derived neurotrophic factor, and myelin-associated glycoprotein, but not to semaphorin 3A [47]. The TRPC1 channel opening associated with cis/trans isomerization is regulated by FK506-binding proteins (FKBP)12 and 52. But these two signaling molecules, respectively, mediate distinct functional signaling pathways. KFBP52 mediates TRPC1 gating through isomerization, the signaling of which is required for axonal growth cone chemotropic turning in response to netrin-1, myelin-associated glycoprotein, and netrin-1 and its receptor DCC-dependent midline axon guidance of commissural interneurons in the developing spinal cord. On the other hand, FKBP12 mediates spontaneous opening of TRPC1 through isomerization and is not required for growth cone in responses to netrin-1 [48].

TRPC1 is also expressed in perisynaptic regions of the cerebellar parallel fiber–Purkinje cells and physically associated with group I metabotropic glutamate receptors (mGluR1). The mGluR1-evoked slow excitatory postsynaptic current (EPSC) is mediated by TRPC1. Interfering TRPC1 blocks the mGluR1-evoked slow EPSC in Purkinje cells (Kim, [23]). In non-nerve systems, TRPC1 acts synergistically with vascular endothelial growth factor A (VEGF-a) in controlling intersegmental vessel (ISV) growth in zebrafish larvae and appears as the downstream to VEGF-a in controlling angiogenesis. The Ca^{2+} influx through TRPC1 in response to VEGF-a is important for the phosphorylation of extracellular signal-regulated kinase (Erk) [65].

TRPC1 also interacts with a wide range of scaffolding and trafficking proteins. The interaction between TRPC1 and IP_3R is mediated by RhoA in endothelial cells [33] and by Homer1 in HEK293 cells [25]. Additionally, RhoA [33], Cav-1 [1, 6, 29], and β-tubulin [5] affect the

Table 3.1 Summary of molecular signaling associated with TRPC channels

Name	Upstream signaling	Downstream signaling	Modulators
TRPC1	bFGF [43]	Erk [65]	FKBP12, FKBP52 [47]
	Netrin-1, BDNF, and myelin-associated glycoprotein [47]		Rho A [33, 42]
			Homer1 [25]
	mGluR1 [23]		Cav-1 [1, 6, 29, 42]
	VEGF-α [65]		β-tubulin [5]
TRPC4/5	NT3-TrkC [17]	CaMKIIα [17]	Protein 4.1 [10]
	Leptin [44, 45]	Rac1 [54]	SESTD1 [35]
	NO [64]		NHERF, ZO1 [39, 53]
	Thioredoxin [62]		Fyn [40]
			Caveolae [56]
			Stathmin-2, SNARE [16]
			Dynamin, clathrin, and MxA [15]
			PI_3K, PIP_5K [4]
			NCS-1 [19]
TRPC3/6	BDNF [20]	CREB, Erk [20]	VAMP2 [49]
	NT4 [17]	CaMKIV, CREB [17, 51, 68]	RACK1, IP_3R, Orai1 [2, 59]
	PDGF [14]		Rab9, Rab11 [8]
			PI_3K, PTEN [24, 36]
			MxA [30]
			Syntaxin [3]
	UTP [46]	CDK1, Cdc25c [14, 46]	Clathrin, dynamin [15]
	Hypoxia [26]		
	TGF-β and Ang II [12]	HIF-1α, α-ketoglutarate, PHD, GLUT1 [26]	
		Calcineurin-NFAT [12, 41]	
		Rho A [54]	
TRPC7			PRKG1α [67]
			CaM, IP_3R, PIP_2 [21, 34, 49, 66]
TRPC2			RTP1 [31]
			STIM1 [18]
			CaM [52, 63]
			Anoctamin 1 [13] Enkurin [50]

surface expression of TRPC1. Cav-1 and RhoA mediate TRPC1 localization in lipid raft domains where TRPC1 channels are assembled and activated in response to store depletion [42].

3.2.2 TRPC4 and TRPC5

The amino acid sequences of TRPC4 and TRPC5 are highly similar, with 65% identity and only diverge in the last 220 amino acids. These two subunits usually coexist in the same heterotetramers and function similarly. TRPC5 channels could sense extracellular neurotrophic factors such as NT-3 by physically interacting with NT-3 receptor tyrosine receptor kinase C (TrkC). When NT-3 binds to TrkC, whereby PLCγ is recruited to activate TRPC5 in ROCE pathway, the resulted calcium influx through TRPC5 channel specifically activates CaMKIIα, which inhibits dendritic development in primary rat hippocampal neurons [17].

Other molecules, for example, leptin, can activate TRPC4 and TRPC5, in hypothalamic proopiomelanocortin (POMC) neurons and kisspeptin neurons [44, 45]. TRPC4 and TRPC5 channels also sense nitric oxide (NO) directly through the nitrosylation at Cys553 and nearby Cys558 sites on the N-terminal side of pore-forming region. In endothelial cells, nitrosylation of native TRPC5 upon G protein-coupled ATP receptor stimulation elicits entry of Ca^{2+} [64]. TRPC5 homomultimeric and TRPC5–TRPC1 heteromultimeric channels could be activated by extracellular reduced thioredoxin, which acts by breaking a disulfide bridge in the predicted extracellular loop adjacent to the ion-selectivity filter of TRPC5 [62].

The calcium influx through TRPC5 could regulate actin remodeling and cell motility in fibroblasts and kidney podocytes. It has been shown that TRPC5 is in a molecular complex with Rac1 and TRPC5-mediated Ca^{2+} influx induces Rac1 activation, thereby promoting cell migration [54]. It is interesting to notice that Rac1 could also promote dendritic development in neurons [38, 55], whereas the Ca^{2+} signaling through TRPC5 also inhibited dendritic development in cultured rat hippocampal neurons [17]. The different consequential results through the similar signaling pathways in different cell types indicate that TRPC channels would assemble with different signaling molecules into distinct micro-signaling modules, leading to different effects within different cellular contents.

TRPC4/5 channel activity and trafficking can be regulated by different molecules. Protein 4.1 functionally links TRPC4 to the actin cytoskeleton and spectrin in endothelial cells [10]. SESTD1 associates with both TRPC4 and TRPC5 via the CIRB domain and functions to couple TRPC channel activity to lipid signaling [35]. Signaling proteins involved in interactions with TRPC4/5 also include the PDZ-domain proteins NHERF and ZO1 via the "VTTRL" sequence in the C-terminus of TRPC4/5 and PLC [39, 53] and Fyn [40]. The dynamic interplay between tyrosine kinases, TRPC4/5 and NHERF, regulates cell surface expression and activation of these channels. TRPC4 also associates with the caveolae in the region where growth factor receptor signaling proteins and NHERF-binding proteins, such as ezrin, are localized [56]. It has been proposed that the interaction with NHERF and Z01 provides a scaffold to position the channel in the apical or lateral regions of polarized cells, such as endothelial cells. The trafficking of TRPC5 to specific sites in the hippocampal neurons is regulated by its interaction with the stathmin-2, SNARE proteins, and other trafficking proteins, such as dynamin, clathrin, and MxA [15, 16]. Insertion of TRPC5 into the plasma membrane requires phosphatidylinositide 3-kinase (PI_3K), Rac1, and phosphatidylinositol 4-phosphate 5-kinase (PIP_5K) [4]. Moreover, the neuronal calcium sensor-1 (NCS-1) binds to the C-terminus of TRPC5 [19] and is involved in retardation of neurite outgrowth by TRPC5 homomeric channels [4].

3.2.3 TRPC3 and TRPC6

TRPC3 and TRPC6 are the most studied TRPC family members, and different signaling cascades have been identified in their activation, downstream signaling, and trafficking. In the central nerve system, BDNF can activate TRPC3 and 6. In cerebellar granule neuron (CGN) cultures, blocking or downregulating TRPC3/6 channel proteins decreases BDNF-triggered intracellular Ca^{2+} elevation. The Ca^{2+} influx through TRPC3/6 is important for the activation of Erk and cAMP response element-binding protein (CREB), promoting CGN survival [20]. Additionally, NT-4 can also activate TRPC6 channels [17]. The Ca^{2+} influx through TRPC6 specifically activates CaMKIV-CREB signaling pathway, which is critical for TRPC6 effects on dendritic development [17, 51] and synaptogenesis [68].

TRPC3/6 channels also mediate important signaling cascades in non-nerve systems. In glioma cells, inhibition of TRPC6 activity attenuates the increase in intracellular Ca^{2+} induced by platelet-derived growth factor (PDGF). The Ca^{2+} influx through TRPC6 is important for glioma cell growth and clonogenic ability. Through regulating cyclin-dependent kinase 1 and cell division cycle 25 homolog C expression, such signaling pathway helps glioma cells quickly bypass the cell cycle checkpoint in G2/M phase and reduces their sensitivity to ionizing radiation [14]. This signaling cascade is not specific for glioma cells, as TRPC6 also initiates the similar signaling pathway in esophageal cancer cells [46]. Moreover, TRPC6 could respond to hypoxia by the IGF-1R-PLCγ-IP_3R signaling pathway and mediate intracellular Ca^{2+} elevation during hypoxia in glioma cells. TRPC6 promotes HIF-1α hydroxylation to inhibit HIF-1α accumulation. Inhibition of TRPC6 increases α-ketoglutarate levels and promotes PHD activities, leading to HIF-1α degradation. Furthermore, TRPC6 regulates GLUT1 expression, so as to affect glucose uptake during hypoxia, indicating the involvement of TRPC6-related signaling cascades in controlling glucose metabolism [26].

In the cultured cardiomyocytes cells, the activation of nuclear factor of activated T cells (NFAT) and cardiomyocyte hypertrophy requires the Ca^{2+} influx through TRPC3 and TRPC6. The activation of angiotensin (Ang) II receptors could activate PLC, producing DAG, which directly activates TRPC3 and TRPC6, the resulting cation (Na^+ and Ca^{2+}) influx changes the membrane potential to positive, leading to the further activation of voltage-gated Ca^{2+} channels (VGCC). The Ca^{2+} influx through VGCC activates calcineurin-NFAT pathway, to trigger the hypertrophic responses in cardiomyocytes [41].

TRPC6 also senses transforming growth factor β (TGF-β) and angiotensin II in the process of myofibroblast transdifferentiation. Knocking out TRPC6 in mice impairs dermal and cardiac wound healing after injury. The profibrotic ligands TGF-β and angiotensin II induce TRPC6 expression through p38 mitogen-activated protein kinase (MAPK)/serum response factor (SRF) signaling. Once induced, TRPC6 activates the Ca^{2+}-responsive protein phosphatase calcineurin, which itself induces myofibroblast transdifferentiation. Moreover, inhibition of calcineurin prevents TRPC6-dependent transdifferentiation and dermal wound healing [12].

Unlike TRPC5, the Ca^{2+} influx through TRPC6 shows opposite regulations on actin remodeling and cell motility in fibroblasts and kidney podocytes. TRPC6 is in a molecular complex with RhoA, and the Ca^{2+} influx through TRPC6 increases RhoA activity and inhibits fibroblast cell migration [54]. Thus the Ca^{2+} influx through TRPC6 channels in different cell types functions differently, probably due to different signaling partners.

Diverse signaling pathways are involved in regulating the surface localization of TRPC3 and TRPC6. The cell surface expression of TRPC3 is regulated by VAMP2-mediated fusion of TRPC3-containing intracellular vesicles with the plasma membrane. Expression of TRPC3 in the plasma membrane is increased following stimulation with carbachol (CCh), and this increase is abolished by treatment with tetanus toxin, which inhibits VAMP2 activity [49]. The receptor for activated C-kinase-1 (RACK1) interacts with the TRPC3 channel and regulates its plasma membrane localization and activation. The interactions of TRPC3 with IP_3R, Orai1, and RACK1 are important for CCh-induced intracellular calcium elevation and the increase of surface TRPC3 expression [2, 59]. It is regarded that actin cytoskeleton affects TRPC3 localization and function. Enhancing or stabilizing cortical actin layer, via the treatment with jasplakinolide or calyculin A, promotes the internalization of TRPC3 signaling complex and inhibits TRPC3 function [28].

Surface expression of TRPC6 is enhanced following cell stimulation by muscarinic receptor agonists or passive depletion of the ER-Ca^{2+} stores by thapsigargin [9]. The GTPases, Rab9 and Rab11, regulate the intracellular trafficking of TRPC6 in HeLa cells [8]. In cells cotransfected with Rab9, TRPC6 is partially co-localized in the Rab9-containing vesicles. However, when Rab11 is overexpressed, TRPC6 is predominantly presented at the cell periphery. The surface expression and the channel activity of TRPC6 are increased following the expression of a dominant-negative mutant of Rab9 (S21 N) and Rab11, whereas the channel activity is decreased when dominant-negative mutant of Rab11 (S25N) was expressed. Together, the intracellular trafficking of TRPC6 is through early endosomes and late endosomes, where the channel protein interacts with Rab9-containing vesicles and is translocated to the plasma membrane via Rab11-containing vesicles [8]. PI_3K and PTEN are also reported to regulate the trafficking and activation of TRPC6 channels. PTEN-dependent inhibition of PI_3K reduces the translocation of TRPC6 to the plasma membrane and consequently decreases TRPC6-mediated Ca^{2+} influx in T6.11 cells [24, 36]. Other proteins also interact with TRPC6 and regulate its trafficking, such as MxA [30], syntaxin [3], clathrin, and dynamin [15].

3.2.4 TRPC2

The human Trpc2 is a pseudogene, but in other mammals, TRPC2 protein forms functional channels in the vomeronasal organ, testis, spleen, and liver. TRPC2 forms a signaling complex with the receptor-transporting protein 1 (RTP1), Homer1, and IP_3R in the vomeronasal organ. In cells cotransfected with RTP1, the surface expression of TRPC2, as well as the channel activity, is increased compared to cells expressing TRPC2 alone [31]. Other signaling proteins that interact with TRPC2 include STIM1 [18] and CaM [52, 63]. TRPC2 co-localizes with anoctamin 1 in the vomeronasal epithelium [13] and interacts with enkurin in sperm [50].

3.2.5 TRPC7

TRPC7 is probably the least studied TRPC channel. The mechanisms of activation regulation, trafficking, and localization are largely unknown. The function of TRPC7 has been reported to be modulated by cGMP-dependent protein kinase 1α (PRKG1α) [67], CaM, IP_3R, and PIP_2 [21, 34, 49, 66].

3.3 Perspectives

There are different signaling molecules involved in TRPC channel activation, trafficking, subcellular localization, and downstream cascades, enabling their diverse functions in different systems and cells. To better understand their biology and identify potential targets for TRP-related diseases, it is important to identify the specific signaling cascades related to specific TRP channels in different systems. However, the lack of good research tools, such as validated good antibodies and specific agonists/antagonists for each specific TRP channels, impedes the fast development of this field. Some exciting progresses have been made, for example, the detailed structure information of TRP channels has been revealed [7, 27], which will greatly benefit the design and production of more useful antibodies as well as more specific pharmacological agonists/antagonists. These in turn will accelerate the study of TRP channels and the discoveries of more related signaling cascades.

References

1. Ambudkar IS, Bandyopadhyay BC, Liu X, Lockwich TP, Paria B, Ong HL (2006) Functional organization of TRPC-Ca2+ channels and regulation of calcium microdomains. Cell Calcium 40(5–6)495–504. doi:10.1016/j.ceca.2006.08.011
2. Bandyopadhyay BC, Ong HL, Lockwich TP, Liu X, Paria BC, Singh BB, Ambudkar IS (2008) TRPC3 controls agonist-stimulated intracellular Ca2+ release by mediating the interaction between inositol 1,4,5-trisphosphate receptor and RACK1. J Biol Chem 283(47)32821–32830. doi:10.1074/jbc.M805382200
3. Bandyopadhyay BC, Swaim WD, Liu X, Redman RS, Patterson RL, Ambudkar IS (2005) Apical localization of a functional TRPC3/TRPC6-Ca2+-signaling complex in polarized epithelial cells. Role in apical Ca2+ influx. J Biol Chem 280(13)12908–12916. doi:10.1074/jbc.M410013200
4. Bezzerides VJ, Ramsey IS, Kotecha S, Greka A, Clapham DE (2004) Rapid vesicular translocation and insertion of TRP channels. Nat Cell Biol 6(8)709–720. doi:10.1038/ncb1150
5. Bollimuntha S, Cornatzer E, Singh BB (2005) Plasma membrane localization and function of TRPC1 is dependent on its interaction with beta-tubulin in retinal epithelium cells. Vis Neurosci 22(2)163–170. doi:10.1017/S0952523805222058
6. Brazer SC, Singh BB, Liu X, Swaim W, Ambudkar IS (2003) Caveolin-1 contributes to assembly of store-operated Ca2+ influx channels by regulating plasma membrane localization of TRPC1. J Biol Chem 278(29)27208–27215. doi:10.1074/jbc.M301118200
7. Cao E, Liao M, Cheng Y, Julius D (2013) TRPV1 structures in distinct conformations reveal activation

mechanisms. Nature 504(7478)113–118. doi:10.1038/nature12823
8. Cayouette S, Bousquet SM, Francoeur N, Dupre E, Monet M, Gagnon H, Guedri YB, Lavoie C, Boulay G (2010) Involvement of Rab9 and Rab11 in the intracellular trafficking of TRPC6. Biochim Biophys Acta 1803(7)805–812. doi:10.1016/j.bbamcr.2010.03.010
9. Cayouette S, Lussier MP, Mathieu EL, Bousquet SM, Boulay G (2004) Exocytotic insertion of TRPC6 channel into the plasma membrane upon Gq protein-coupled receptor activation. J Biol Chem 279(8)7241–7246. doi:10.1074/jbc.M312042200
10. Cioffi DL, Wu S, Alexeyev M, Goodman SR, Zhu MX, Stevens T (2005) Activation of the endothelial store-operated ISOC Ca2+ channel requires interaction of protein 4.1 with TRPC4. Circ Res 97(11)1164–1172. doi:10.1161/01.RES.0000193597.65217.00
11. Clapham DE, Runnels LW, Strubing C (2001) The trp ion channel family. Nat Rev Neurosci 2(6)387–396
12. Davis J, Burr Adam R, Davis Gregory F, Birnbaumer L, Molkentin Jeffery D (2012) A TRPC6-dependent pathway for myofibroblast transdifferentiation and wound healing in vivo. Dev Cell 23 (4):705–715. doi:http://dx.doi.org/10.1016/j.devcel.2012.08.017
13. Dibattista M, Amjad A, Maurya DK, Sagheddu C, Montani G, Tirindelli R, Menini A (2012) Calcium-activated chloride channels in the apical region of mouse vomeronasal sensory neurons. J Gen Physiol 140(1)3–15. doi:10.1085/jgp.201210780
14. Ding X, He Z, Zhou K, Cheng J, Yao H, Lu D, Cai R, Jin Y, Dong B, Xu Y, Wang Y (2010) Essential role of TRPC6 channels in G2/M phase transition and development of human glioma. J Natl Cancer Inst. doi:10.1093/jnci/djq217
15. Goel M, Sinkins W, Keightley A, Kinter M, Schilling WP (2005) Proteomic analysis of TRPC5- and TRPC6-binding partners reveals interaction with the plasmalemmal Na(+)/K(+)-ATPase. Pflugers Arch - Eur J Physiol 451(1)87–98. doi:10.1007/s00424-005-1454-y
16. Greka A, Navarro B, Oancea E, Duggan A, Clapham DE (2003) TRPC5 is a regulator of hippocampal neurite length and growth cone morphology. Nat Neurosci 6(8)837–845. doi:10.1038/nn1092
17. He Z, Jia C, Feng S, Zhou K, Tai Y, Bai X, Wang Y (2012) TRPC5 channel is the mediator of neurotrophin-3 in regulating dendritic growth via CaMKIIα in rat hippocampal neurons. J Neurosci 32(27)9383–9395. doi:10.1523/jneurosci.6363-11.2012
18. Huang GN, Zeng W, Kim JY, Yuan JP, Han L, Muallem S, Worley PF (2006) STIM1 carboxyl-terminus activates native SOC, Icrac and TRPC1 channels. Nat Cell Biol 8 (9):1003–1010. doi:http://www.nature.com/ncb/journal/v8/n9/suppinfo/ncb1454_S1.html
19. Hui H, McHugh D, Hannan M, Zeng F, Xu SZ, Khan SU, Levenson R, Beech DJ, Weiss JL (2006) Calcium-sensing mechanism in TRPC5 channels contributing to retardation of neurite outgrowth. J Physiol 572(Pt 1)165–172. doi:10.1113/jphysiol.2005.102889
20. Jia Y, Zhou J, Tai Y, Wang Y (2007) TRPC channels promote cerebellar granule neuron survival. Nat Neurosci 10 (5):559–567. doi:http://www.nature.com/neuro/journal/v10/n5/suppinfo/nn1870_S1.html
21. Ju M, Shi J, Saleh SN, Albert AP, Large WA (2010) Ins(1,4,5)P3 interacts with PIP2 to regulate activation of TRPC6/C7 channels by diacylglycerol in native vascular myocytes. J Physiol 588(9):1419–1433. doi:10.1113/jphysiol.2009.185256
22. Kahn-Kirby AH, Bargmann CI (2006) TRP channels in *C. elegans*. Ann Rev Physiol 68(1)719–736. doi:10.1146/annurev.physiol.68.040204.100715
23. Kim SJ, Kim YS, Yuan JP, Petralia RS, Worley PF, Linden DJ (2003) Activation of the TRPC1 cation channel by metabotropic glutamate receptor mGluR1. Nature 426 (6964):285–291. doi:http://www.nature.com/nature/journal/v426/n6964/suppinfo/nature02162_S1.html
24. Kini V, Chavez A, Mehta D (2010) A new role for PTEN in regulating transient receptor potential canonical channel 6-mediated Ca2+ entry, endothelial permeability, and angiogenesis. J Biol Chem 285(43)33082–33091. doi:10.1074/jbc.M110.142034
25. Kiselyov K, Shin DM, Kim JY, Yuan JP, Muallem S (2007) TRPC channels: interacting proteins. Handb Exp Pharmacol 179:559–574. doi:10.1007/978-3-540-34891-7_33
26. Li S, Wang J, Wei Y, Liu Y, Ding X, Dong B, Xu Y, Wang Y (2015) Crucial role of TRPC6 in maintaining the stability of HIF-1α in glioma cells under hypoxia. J Cell Sci 128(17)3317–3329. doi:10.1242/jcs.173161
27. Liao M, Cao E, Julius D, Cheng Y (2013) Structure of the TRPV1 ion channel determined by electron cryo-microscopy. Nature 504(7478)107–112. doi:10.1038/nature12822
28. Lockwich T, Singh BB, Liu X, Ambudkar IS (2001) Stabilization of cortical actin induces internalization of transient receptor potential 3 (Trp3)-associated caveolar Ca2+ signaling complex and loss of Ca2+ influx without disruption of Trp3-inositol trisphosphate receptor association. J Biol Chem 276(45)42401–42408. doi:10.1074/jbc.M106956200
29. Lockwich TP, Liu X, Singh BB, Jadlowiec J, Weiland S, Ambudkar IS (2000) Assembly of Trp1 in a signaling complex associated with caveolin-scaffolding lipid raft domains. J Biol Chem 275(16)11934–11942. doi:10.1074/jbc.275.16.11934
30. Lussier MP, Cayouette S, Lepage PK, Bernier CL, Francoeur N, St-Hilaire M, Pinard M, Boulay G (2005) MxA, a member of the dynamin superfamily, interacts with the ankyrin-like repeat domain of TRPC. J Biol Chem 280(19)19393–19400. doi:10.1074/jbc.M500391200
31. Mast TG, Brann JH, Fadool DA (2010) The TRPC2 channel forms protein-protein interactions with Homer and RTP in the rat vomeronasal organ. BMC Neurosci 11(1)1–16. doi:10.1186/1471-2202-11-61
32. Matsuura H, Sokabe T, Kohno K, Tominaga M, Kadowaki T (2009) Evolutionary conservation and

changes in insect TRP channels. BMC Evol Biol 9:228. doi:10.1186/1471-2148-9-228
33. Mehta D, Ahmmed GU, Paria BC, Holinstat M, Voyno-Yasenetskaya T, Tiruppathi C, Minshall RD, Malik AB (2003) RhoA interaction with inositol 1,4,5-trisphosphate receptor and transient receptor potential channel-1 regulates Ca2+ entry. Role in signaling increased endothelial permeability. J Biol Chem 278(35)33492–33500. doi:10.1074/jbc.M302401200
34. Mery L, Magnino F, Schmidt K, Krause K-H, Dufour J-F (2001) Alternative splice variants of hTrp4 differentially interact with the C-terminal portion of the inositol 1,4,5-trisphosphate receptors. FEBS Lett 487 (3):377–383. doi:10.1016/s0014-5793(00)02362-0
35. Miehe S, Bieberstein A, Arnould I, Ihdene O, Rutten H, Strubing C (2010) The phospholipid-binding protein SESTD1 is a novel regulator of the transient receptor potential channels TRPC4 and TRPC5. J Biol Chem 285(16)12426–12434. doi:10.1074/jbc.M109.068304
36. Monet M, Francoeur N, Boulay G (2012) Involvement of phosphoinositide 3-Kinase and PTEN protein in mechanism of activation of TRPC6 protein in vascular smooth muscle cells. J Biol Chem 287(21)17672–17681. doi:10.1074/jbc.M112.341354
37. Montell C, Rubin GM (1989) Molecular characterization of the drosophila trp locus: a putative integral membrane protein required for phototransduction. Neuron 2 (4):1313–1323. doi:http://dx.doi.org/10.1016/0896-6273(89)90069-X
38. Nakayama AY, Harms MB, Luo L (2000) Small GTPases rac and rho in the maintenance of dendritic spines and branches in hippocampal pyramidal neurons. J Neurosci 20(14)5329–5338
39. Obukhov AG, Nowycky MC (2004) TRPC5 activation kinetics are modulated by the scaffolding protein ezrin/radixin/moesin-binding phosphoprotein-50 (EBP50) J Cell Physiol 201(2)227–235. doi:10.1002/jcp.20057
40. Odell AF, Scott JL, Van Helden DF (2005) Epidermal growth factor induces tyrosine phosphorylation, membrane insertion, and activation of transient receptor potential channel 4. J Biol Chem 280(45)37974–37987. doi:10.1074/jbc.M503646200
41. Onohara N, Nishida M, Inoue R, Kobayashi H, Sumimoto H, Sato Y, Mori Y, Nagao T, Kurose H (2006) TRPC3 and TRPC6 are essential for angiotensin II-induced cardiac hypertrophy. EMBO J 25(22)5305–5316. doi:10.1038/sj.emboj.7601417
42. Pani B, Ong HL, Liu X, Rauser K, Ambudkar IS, Singh BB (2008) Lipid rafts determine clustering of STIM1 in endoplasmic reticulum-plasma membrane junctions and regulation of store-operated Ca2+ entry (SOCE) J Biol Chem 283(25)17333–17340. doi:10.1074/jbc.M800107200
43. Pla AF, Maric D, Brazer S-C, Giacobini P, Liu X, Chang YH, Ambudkar IS, Barker JL (2005) Canonical transient receptor potential 1 plays a role in basic fibroblast growth factor (bFGF)/FGF receptor-1-induced Ca2+ entry and embryonic rat neural stem cell proliferation. J Neurosci 25(10)2687–2701. doi:10.1523/jneurosci.0951-04.2005
44. Qiu J, Fang Y, Bosch MA, Ronnekleiv OK, Kelly MJ (2011) Guinea pig kisspeptin neurons are depolarized by leptin via activation of TRPC channels. Endocrinology 152(4)1503–1514. doi:10.1210/en.2010-1285
45. Qiu J, Fang Y, Rønnekleiv OK, Kelly MJ (2010) Leptin excites proopiomelanocortin neurons via activation of TRPC channels. J Neurosci 30(4)1560–1565. doi:10.1523/jneurosci.4816-09.2010
46. Shi Y, Ding X, He Z-H, Zhou K-C, Wang Q, Wang Y-Z (2009) Critical role of TRPC6 channels in G2 phase transition and the development of human oesophageal cancer. Gut 58(11)1443–1450. doi:10.1136/gut.2009.181735
47. Shim S, Goh EL, Ge S, Sailor K, Yuan JP, Roderick HL, Bootman MD, Worley PF, Song H, Ming GL (2005) XTRPC1-dependent chemotropic guidance of neuronal growth cones. Nat Neurosci 8 (6):730–735. doi:http://www.nature.com/neuro/journal/v8/n6/suppinfo/nn1459_S1.html
48. Shim S, Yuan JP, Kim JY, Zeng W, Huang G-1, Milshteyn A, Kern D, Muallem S, Ming G, Worley PF (2009) Peptidyl-prolyl isomerase FKBP52 controls chemotropic guidance of neuronal growth cones via regulation of TRPC1 channel opening. Neuron 64 (4):471–483. doi:http://dx.doi.org/10.1016/j.neuron.2009.09.025
49. Singh BB, Lockwich TP, Bandyopadhyay BC, Liu X, Bollimuntha S, Brazer S-c, Combs C, Das S, Leenders AGM, Sheng Z-H, Knepper MA, Ambudkar SV, Ambudkar IS (2004) VAMP2-dependent exocytosis regulates plasma membrane insertion of TRPC3 channels and contributes to agonist-stimulated Ca influx. Mol Cell 15(4):635–646. doi:10.1016/j.molcel.2004.07.010
50. Sutton KA, Jungnickel MK, Wang Y, Cullen K, Lambert S, Florman HM (2004) Enkurin is a novel calmodulin and TRPC channel binding protein in sperm. Dev Biol 274 (2):426–435. doi:http://dx.doi.org/10.1016/j.ydbio.2004.07.031
51. Tai Y, Feng S, Ge R, Du W, Zhang X, He Z, Wang Y (2008) TRPC6 channels promote dendritic growth via the CaMKIV-CREB pathway. J Cell Sci 121(14)2301–2307. doi:10.1242/jcs.026906
52. Tang J, Lin Y, Zhang Z, Tikunova S, Birnbaumer L, Zhu MX (2001) Identification of common binding sites for calmodulin and inositol 1,4,5-trisphosphate receptors on the carboxyl termini of trp channels. J Biol Chem 276(24)21303–21310. doi:10.1074/jbc.M102316200
53. Tang Y, Tang J, Chen Z, Trost C, Flockerzi V, Li M, Ramesh V, Zhu MX (2000) Association of mammalian trp4 and phospholipase C isozymes with a PDZ domain-containing protein, NHERF. J Biol Chem 275(48)37559–37564. doi:10.1074/jbc.M006635200

54. Tian D, Jacobo SM, Billing D, Rozkalne A, Gage SD, Anagnostou T, Pavenstadt H, Hsu HH, Schlondorff J, Ramos A, Greka A (2010) Antagonistic regulation of actin dynamics and cell motility by TRPC5 and TRPC6 channels. Sci Signal 3(145)ra77. doi:10.1126/scisignal.2001200
55. Tolias KF, Bikoff JB, Burette A, Paradis S, Harrar D, Tavazoie S, Weinberg RJ, Greenberg ME (2005) The Rac1-GEF Tiam1 couples the NMDA receptor to the activity-dependent development of dendritic arbors and spines. Neuron 45 (4):525–538. doi:http://dx.doi.org/10.1016/j.neuron.2005.01.024
56. Torihashi S, Fujimoto T, Trost C, Nakayama S (2002) Calcium oscillation linked to pacemaking of interstitial cells of Cajal: requirement of calcium influx and localization of TRP4 in caveolae. J Biol Chem 277(21)19191–19197. doi:10.1074/jbc.M201728200
57. Venkatachalam K, Montell C (2007) TRP channels. Annu Rev Biochem 76:387–417. doi:10.1146/annurev.biochem.75.103004.142819
58. Wolstenholme AJ, Williamson SM, Reaves BJ (2011) TRP channels in parasites. Adv Exp Med Biol 704:359–371. doi:10.1007/978-94-007-0265-3_20
59. Woodard GE, López JJ, Jardín I, Salido GM, Rosado JA (2010) TRPC3 regulates agonist-stimulated Ca2+ mobilization by mediating the interaction between type I inositol 1,4,5-trisphosphate receptor, RACK1, and Orai1. J Biol Chem 285(11)8045–8053. doi:10.1074/jbc.M109.033605
60. Wu LJ, Sweet TB, Clapham DE (2010) International union of basic and clinical pharmacology. LXXVI Curr Prog Mamm TRP ion Channel Fam Pharmacol Rev 62(3)381–404. doi:10.1124/pr.110.002725
61. Xiao R, Xu XZ (2009) Function and regulation of TRP family channels in *C. elegans*. Pflugers Arch - Eur J Physiol 458(5)851–860. doi:10.1007/s00424-009-0678-7
62. Xu S-Z, Sukumar P, Zeng F, Li J, Jairaman A, English A, Naylor J, Ciurtin C, Majeed Y, Milligan CJ, Bahnasi YM, Al-Shawaf E, Porter KE, Jiang L-H, Emery P, Sivaprasadarao A, Beech DJ (2008) TRPC channel activation by extracellular thioredoxin. Nature 451 (7174):69–72. doi:http://www.nature.com/nature/journal/v451/n7174/suppinfo/nature06414_S1.html
63. Yildirim E, Dietrich A, Birnbaumer L (2003) The mouse C-type transient receptor potential 2 (TRPC2) channel: alternative splicing and calmodulin binding to its N terminus. Proc Natl Acad Sci 100(5)2220–2225. doi:10.1073/pnas.0438036100
64. Yoshida T, Inoue R, Morii T, Takahashi N, Yamamoto S, Hara Y, Tominaga M, Shimizu S, Sato Y, Mori Y (2006) Nitric oxide activates TRP channels by cysteine S-nitrosylation. Nat Chem Biol 2 (11):596–607. doi:http://www.nature.com/nchembio/journal/v2/n11/suppinfo/nchembio821_S1.html
65. Yu P-c, S-y G, J-w B, J-l D (2010) TRPC1 is essential for in vivo angiogenesis in zebrafish. Circ Res 106(7)1221–1232. doi:10.1161/circresaha.109.207670
66. Yuan JP, Kiselyov K, Shin DM, Chen J, Shcheynikov N, Kang SH, Dehoff MH, Schwarz MK, Seeburg PH, Muallem S, Worley PF (2003) Homer binds TRPC family channels and is required for gating of TRPC1 by IP3 receptors. Cell 114 (6):777–789. doi:10.1016/s0092-8674(03)00716-5
67. Yuasa K, Matsuda T, Tsuji A (2011) Functional regulation of transient receptor potential canonical 7 by cGMPdependent protein kinase Iα. Cell Signal 23(7):1179–1187
68. Zhou J, Du W, Zhou K, Tai Y, Yao H, Jia Y, Ding Y, Wang Y (2008) Critical role of TRPC6 channels in the formation of excitatory synapses. Nat Neurosci 11 (7):741–743. doi:http://www.nature.com/neuro/journal/v11/n7/suppinfo/nn.2127_S1.html

TRPC Channels in Health and Disease

4

Yilin Tai, Shenglian Yang, Yong Liu, and Wei Shao

Abstract

This chapter offers a brief introduction of the functions of TRPC channels in non-neuronal systems. We focus on three major organs of which the research on TRPC channels have been most focused on: kidney, heart, and lung. The chapter highlights on cellular functions and signaling pathways mediated by TRPC channels. It also summarizes several inherited diseases in humans that are related to or caused by TRPC channel mutations and malfunction. A better understanding of TRPC channels functions and the importance of TRPC channels in health and disease should lead to new insights and discovery of new therapeutic approaches for intractable disease.

Keywords

TRPCs • Kidney • Cardiovascular • Lung

With the fact that TRPC channels are universally expressed in most of the major organs, it is not surprising that they contribute to normal development, and their malfunction leads to diseases of these organs. We will discuss in depth the physiological and pathological functions of TRPC channels in nervous system in the following chapters. In this chapter, we will give a general introduction of the roles of TRPC channels in the kidney, cardiovascular system, and lung, the three major organs that the functions of TRPC channels have been most extensively studied in the past decades.

Y. Tai (✉)
Cold Spring Harbor Laboratory,
1 Bungtown Road, Cold Spring Harbor, NY 11724, USA
e-mail: ytai@cshl.edu

S. Yang (✉) • Y. Liu • W. Shao
Center of Cognition and Brain Science,
Institute of Basic Medical Science, Beijing 100039, China
e-mail: slyang0824@hotmail.com

Y. Wang (ed.), *Transient Receptor Potential Canonical Channels and Brain Diseases*,
Advances in Experimental Medicine and Biology 976, DOI 10.1007/978-94-024-1088-4_4

4.1 TRPCs in Kidney Health and Disease

The key function of the kidney is to filter the plasma to dispose metabolic end products, excess electrolytes, and water. It is accomplished by a structure called glomerulus or renal corpuscle. The glomerulus is the functional blood filtration unit and is the first component involved in regulating the composition of urine. Disruption of the glomerular filtration barrier is a common outcome of many kidney diseases, including focal segmental glomerulosclerosis (FSGS), diabetic nephropathy, and lupus nephritis [1]. Proteinuria is a hallmark of dysfunction of glomerular filtration barrier [2]. Persistent dysfunction leads to progressive renal failure and needs for dialysis or kidney transplantation.

The basic unit of the glomerulus tuft is a single capillary with the glomerular basement membrane (GBM) as primary structure scaffold. Endothelial and mesangial cells providing capillary support are located inside GBM, whereas podocytes are attached to the outside the GBM. There are thus four major cell types in the glomerulus: endothelial cells, mesangial cells, parietal epithelial cells of Bowman's capsule, and podocytes. The expression of TRPC channels has been found mostly in mesangial cells and podocytes in the glomerulus. Several lines of evidence show that TRPC1, TRPC3, TRPC4, TRPC5, and TRPC6 were all expressed in the kidney [3–8]. TRPC1 is exclusively expressed in mesangial cells, whereas TRPC3 and TRPC6 have broader expressions. TRPC3 and TRPC6 are confined to podocytes and mesangial cells. They are also expressed in the collecting duct which connects the nephrons to the ureter. In the following part of this chapter, we will provide an overview of the current knowledge of TRPC channels on mesangial cells and podocytes and functions of TRPC channels in the collecting duct which plays an important role in reabsorption and excretion.

4.1.1 TRPCs in Mesangial Cells

Mesangial cells are specialized cells around blood vessels in glomerulus. Major functions of mesangial cells are to remove trapped residues and aggregated protein from the GBM, thus keeping the filter free of debris. They are contractile cells that regulate filtration rate by altering surface area of the capillaries. Ca^{2+} influx across the plasma membrane is critical for mesangial cell contraction in response to vasoactive peptides. Altered responses of mesangial cells to vasoactive peptides is one of the major causes that leads to various renal diseases, such as diabetic nephropathy [9]. Several types of Ca^{2+} channels are involved in this physiological process. These channels include voltage-gated Ca^{2+} channels, receptor-operated Ca^{2+} channels, and store-operated Ca^{2+} channels [10–14]. Both TRPC1 and TRPC4 are key components of store-operated Ca^{2+} channels in mesangial cells [5]. TRPC1 contributes to contractile function of mesangial cells by mediating vasoconstrictor-stimulated Ca^{2+} responses, whereas TRPC4 is activated by store depletion. In high glucose-treated cultured mesangial cells, an in vitro model for diabetes, TRPC6 expression is reduced. TRPC6 knockdown in high glucose-treated mesangial cells shows reduced Ca^{2+} entry in response to angiotensin II, suggesting that deficiency of TRPC6 might contribute to the impaired Ca^{2+} signaling of mesangial cells seen in diabetes [7, 15, 16].

Mesangial cell proliferation and apoptosis are involved in the maintenance of glomerular integrity. Perturbation of glomerular integrity provides pathophysiological mechanisms that underlie kidney disease. Mesangial cell excessive proliferation and extracellular deposition is a pathological condition commonly found in chronic kidney diseases. Mesangial cell apoptosis contributes to the resolution of glomerulosclerosis [17]; however, it is associated with proteinuria and hypertension in diabetic nephropathy [18]. TRPC6 activation has been shown to be involved in inhibiting proliferation and triggering apoptotic cell death in primary neonatal pig mesangial cells. It is achieved by induction of calcineurin/NFAT, FasL/Fas, and caspase signaling cascade [19]. Interestingly, angiotensin II, which can stimulate mesangial cell proliferation, affects TRPC6 protein level and distribution [20]. Nevertheless, whether angiotensin II-stimulated mesangial cell proliferation is mediated by Ca^{2+} influx through TRPC6 needs to be further validated.

4.1.2 TRPCs in Podocyte

Podocytes are pericyte-like cells with a complex cellular organization consisting of a cell body, major processes, and foot processes. Their foot processes elaborate into a characteristic pattern with foot processes of neighboring podocytes, forming in between the filtration slits. Podocyte foot processes play a major role in establishing the selective permeability of the glomerular filtration barrier [21]. Therefore, podocyte injury is associated with marked albuminuria [22].

Disruption of Ca^{2+} signaling and homeostasis were postulated as early events in podocyte injury. Since TRPC6 mutations are found in patients with FSGS, the molecular mechanisms involving TRPC channels have been studied extensively in podocyte biology [13]. Within podocytes, TRPC6 appears to localize in both major processes and foot processes, and at least some TRPC6 colocalizes to the slit diaphragm (SD), suggesting that it is the abnormal function of TRPC6 within the podocyte that ultimately leads to disease in families with FSGS-associated TRPC6 mutations [12]. Mounting evidences have been shown that proteinuria and podocyte foot processes effacement are mediated by rearrangement of the actin cytoskeleton [23]. Recently, angiotensin receptor-activated TRPC5 and TRPC6 channels have been shown as antagonistic regulators of actin dynamics and cell motility through the regulation of Rac1 and RhoA, respectively [24, 25]. The later study shows that inhibition of TRPC6 results in the loss of stress fibers, Rac1 activation, and increased mobility. On the contrary, inhibition of TRPC5 leads to enhanced stress fiber formation, RhoA activation, and decreased motility. Thus, there are two distinct signaling microdomains emerged in podocytes, one with the TRPC5 which specifically interacts with and activates Rac1 and the other with TRPC6 specifically interacts with and activates RhoA. Consistent with previous studies, CsA restores synaptopodin expression in TRPC6-depleted cells, whereas synaptopodin expression is preserved in TRPC5-depleted podocytes [6, 26].

Transgenic mice overexpressing wild-type TRPC6 and TRPC6 gain-of-function mutants develop albuminuria and FSGS-type lesions [27]. In keeping with this, TRPC6 knockout mice are protected from the proteinuria effects of angiotensin II [28]. In the light of the antagonistic effects of TRPC5 and TRPC6 on podocyte actin dynamics, one would assume that they might have opposite effect in the biology of proteinuria development. Surprisingly, a recent study has shown that depletion of TRPC5 or pharmacological inhibition of TRPC5 protects mice from proteinuria [6]. One possible explanation is that the motility of the foot processes needs to be increased fast enough in response to environmental changes but also to be stable enough in the stationary state. Breaking the balance in either direction will lead to leakage of the filter.

4.1.3 TRPCs in Collecting Duct

The collecting duct of the kidney connects the nephron to the ureter. It plays a role in electrolyte and fluid balance through reabsorption and secretion. Both TRPC3 and TRPC6 are expressed in the principle cells of the collecting duct [8, 10]. TRPC3 is primarily localized to the apical membrane, whereas TRPC6 is found in both apical and basolateral domains. Diffuse TRPC3 and TRPC6 are also found in cytoplasm, presumably localized to intracellular vesicles. Arginine-vasopressin (AVP), which is an antidiuretic hormone that controls water homeostasis and urine concentration by controlling water reabsorption in the collecting duct, can selectively translocate

TRPC3, but not TRPC6, to the apical membrane [29]. Furthermore, AVP-induced increase of intracellular Ca^{2+} is attenuated by expressing a dominant-negative TRPC3. These results suggest that TRPC3 targeting to the apical membrane in collecting duct principle cells can contribute to the AVP-induced Ca^{2+} reabsorption in this region of nephron [29].

4.2 TRPCs in Heart and Vasculature

Like in other tissues, Ca^{2+} plays an important role in maintaining the physiological functions of cardiovascular system, such as cardiac contractility, hemodynamic stretch, dilatation, and repair. TRPC channels, which are ubiquitously expressed in almost all cell types in heart and vasculature, work with other membrane receptors and ion channels to regulate intracellular calcium concentration spatiotemporally. Dysfunctions of TRPC channels are involved in many types of cardiovascular diseases; therefore, TRPC channels have been proposed as therapeutic targets for drug development [30, 31].

4.2.1 TRPCs in Heart

TRPC channels are localized to the peripheral plasma membrane in cardiomyocytes. It is reported that the expression and activation of TRPC channels are both increased during cardiac hypertrophy and heart failure. In cultured cardiomyocytes and in vivo models, the hypertrophic factors, such as endothelin-1 (ET-1), angiotensin II (Ang II), or pressure overload, increase the expression of TRPC1 [32, 33] and TRPC3 [34]. In animal models, upregulation of TRPC1 and TRPC7 is observed in myocardium of Dahl salt-sensitive hypertensive rats [33, 35]. In human patients, the expression of TRPC6 is increased in cardiac hypertrophy and heart failure [36], and TRPC5 is found to be increased in human failing heart samples [34]. Cardiac hypertrophy is a thickening of myocardium which results from several pathological conditions, such as hypertension, excess neurohormones, valvular abnormalities, and myocardial infarction remodeling. Dysregulation of Ca^{2+} is one of the mechanisms proposed to be involved in formation of cardiac hypertrophy. The substantial and low increased of $[Ca^{2+}]_i$ elicited by SOCE or ROCE activates calcineurin, a calcium and calmodulin-dependent serine/threonine protein phosphatase, which dephosphorylates nuclear factor of activated T cell (NFAT). Subsequently, activated NFAT translocates into nucleus and induces the transcription of several hypertrophic genes [37]. Recent studies suggest that TRPC channels are responsible for the substantial and low increased of $[Ca^{2+}]_i$ elicited by SOCE or ROCE in cardiomyocyte and contribute to cardiac hypertrophy through Calcineurin-NFAT pathway [38].

In hypertrophied myocytes, the expression of TRPC1 and $[Ca^{2+}]_i$ induced by SOCE are both significantly increased compared to normal myocytes [33]. Overexpression of TRPC1 in cultured cardiomyocytes elevates $[Ca^{2+}]_i$ elicited by SOCE and activates calcineurin/NFAT pathway [33]. *Trpc1* gene silencing inhibits NFAT activation and 5-HT2A receptor-mediated hypertrophic response induced by ET-1 and Ang II [39]. Moreover, TRPC1$^{-/-}$ mice was protected from cardiac hypertrophy and maintained preserved cardiac function after hemodynamic stress and excess neurohormone insults [32]. In contrast, transgenic mice with cardiomyocyte-specific expression of TRPC3 or TRPC6 show enhanced calcineurin/NFAT signaling and are more sensitive to pressure overload or agonist-induced cardiac hypertrophy [36, 40]. Additionally, the hypertrophic phenotype in TRPC3 transgenic mice was abolished by deletion of the *calcineurin A* gene, which further supports the idea that the hypertrophic effect of TRPC channels is associated with enhanced calcineurin/NFAT signaling [41]. Interestingly, it is found that NFAT also increases the expression of TRPC1, TRPC3, and TRPC6 to form a positive feedback loop, which is proposed to be involved in the development of cardiac hypertrophy [33, 34, 36]. Transgenic mice with a dominant-negative form of TRPC3 or TRPC6 show attenuated hypertrophic response after pressure overload or neurohormone stimulations [38]. Consistently, a new report shows that phenylephrine (PE) that caused pathologic cardiac hypertrophy in wild-type mice was prevented by deletion of TPRC3 gene [42]. In addition, deletion of *trpc6*gene prevents stress-induced exaggerated cardiac remodeling in Klotho-deficient mice [43]. Moreover, TRPC3/TRPC6 antagonists (GSK2332255B and GSK2833503A) block cell hypertrophy in neonatal and adult cardiac myocytes following ET-1 or Ang II stimulation in a dose-dependent manner [44], and TRPC3-selective inhibitor Ryr3 attenuates cardiac hypertrophy in mice subjected to pressure overload [45]. The N-terminal fragment of TRPC4, which disturbs the functions of TRPC4 homomeric and TRPC4/TRPC5 heteromeric channels, protects the mice from hypertrophic stimulations [38, 46]. All these findings raise the possibility that TRPC channels might serve as therapeutic targets to prevent cardiac hypertrophy.

Over time, hypertrophic heart eventually ends up with heart failure. Though the transition from cardiac hypertrophy to heart failure is not clear, myocardial apoptosis is proposed to be an important step in between. Intracellular Ca^{2+} overload induces apoptosis in many cell types. It is reported that overexpression of TRPC3 increases apoptosis in adult mouse cardiomyocytes subjected to ischemia-reperfusion [47], which suggests that TRPC3 may be involved in heart failure. Besides TRPC3, TRPC7 acts as a G protein-activated Ca^{2+} channel mediating Ang II-induced myocardial apoptosis [35]. The expression level of TRPC7 and cell apoptosis increased simultaneously in the failing myocardium of Dahl salt-sensitive hypertension rats, and temocapril, an angiotensin-converting enzyme inhibitor, suppressed both [35]. Inconsistent with previous reports, TRPC7, unlike its close homologues TRPC3 and TRPC6, undergoes remarkable downregulation during the establishment of cardiac hypertrophy [48]. Furthermore, TRPC6 activation might suppress heart failure via inhibition of myofibroblast differentiation [49]. Thus, how TRPC channels involved in the transition from cardiac hypertrophy to heart failure still need to be further investigated.

4.2.2 TRPCs in Vasculature

The extracellular Ca^{2+} entrance in vascular smooth muscle cells (VSMC) and endothelial cells regulates various functions in pulmonary and systematic circulation, such as artery remodeling, vasoconstriction, and vasodilatation. All subunits of TRPCs are expressed in VSMC and vascular endothelial cells to form functional channels that are permeable to Ca^{2+}, which suggests that TRPC channels may also play important roles in vascular system [50–52].

Abnormal VSMC proliferation in vascular remodeling is associated with development of hypertension and atherosclerosis [53]. It is shown that the elevation of $[Ca^{2+}]_i$ is critical for VSMC growth. Chelating extracellular or intracellular Ca^{2+} both inhibit the cell proliferation [54]. Upregulation of TRPC1and TRPC4 has been reported in VSMC and contributes to cell growth subjected to various stimulation, such as Ang II, ATP or pressure load insults, by phosphorylation of cyclic AMP response element-binding protein (CREB) through elevation of $[Ca^{2+}]_i$ [55, 56]. Excessive proliferation of pulmonary artery smooth muscle cell (PASMC) has been observed in patients with idiopathic pulmonary arterial hypertension (IPAH). The expression of TRPC3 and TRPC6 is increased in PASMC in the pulmonary artery tissue from IPAH patients. Downregulating the expression of TRPC6 by siRNA attenuates cultured PASMC proliferation from IPAH patients [57]. However, deletion of TRPC6 does not protect mice from chronic pulmonary hypertension and vascular remodeling [58]. Besides, TRPC1 is thought to be critical for cell proliferation in human PASMC from non-pulmonary hypertension [55, 59].

VSMC contraction caused by Ca^{2+} entry through Ca^{2+} permeable channels is important for regulation of blood pressure. Attenuating the function of TRPC1 by anti-TRPC1 antibody inhibits the SOCE-induced cell contraction. Consistently, overexpression of TRPC1 in rat pulmonary artery increases $[Ca^{2+}]_i$ elicited by SOCE and promotes contraction [60]. It's also reported that TRPC6 is the essential component of vascular α_1-adrenoceptor-activated Ca^{2+}-permeable cation channel in rabbit portal vein smooth muscle cell (SMC). SMC contraction induced by α_1-adrenergic agonists can be blocked by suppressing TRPC6 expression [61]. In addition, activation of TRPC6 has been found in vasopressin, a vasoconstrictor, stimulated A7r5 aortic SMC [62]. Unexpectedly, TRPC6$^{-/-}$ mice show elevated blood pressure, hyperactivity of airway smooth muscle cells, and increased contractility in isolated tracheal and aortic rings [63, 64]. Furthermore, SMC from TRPC6$^{-/-}$ aorta or cerebral arteries are more depolarized with enhanced spontaneous and agonist-induced Ca^{2+} entry [63]. These phenomena can be explained by compensatory expression of constitutive active TRPC3 channels in TRPC6$^{-/-}$ mice. It is reported that UTP-induced depolarization of rat cerebral arteries and subsequent contraction of SMC can be blocked by suppressing the expression of TRPC3, not TRPC6, in these cells [65]. In spontaneous hypertension rats, the expression of TRPC3 is abnormally high compare with normotensive Wistar-Kyoto rats [66]. ET-1, which also works as potent vasoconstrictor in controlling blood pressure, activates aCa^{2+}-permeable cation channel with TRPC7 and TRPC3 in rabbit coronary artery myocytes [67]. Regional alveolar hypoxia induces constriction of pulmonary arteries and redirects blood flow to alveoli with higher oxygen content to ensure maximal oxygenation of the venous blood [68]. The phenomenon is called hypoxic pulmonary vasoconstriction, and $[Ca^{2+}]_i$ elevation is suggested to play a key role in this process [69]. TRPC6$^{-/-}$ mice completely lost acute hypoxic pulmonary vasoconstriction, and hypoxia-induced $[Ca^{2+}]_i$ elevation is absent in PASMC of TRPC6$^{-/-}$ mice [58]. Upregulation of TRPC1 and TRPC6 has been reported in hypoxic pulmonary arteries accompanied with increased $[Ca^{2+}]_i$ elevation induced by SOCE or ROCE as well as the basal level of $[Ca^{2+}]_I$ [70]. The increased expression level of TRPC1 and TRPC6 in PASMC is mediated by the activation of oxygen-sensitive transcription factor hypoxia-inducible factor 1 (HIF1) [71].

Endothelial cells are involved in many aspects of vascular biology such as barrier function, angiogenesis, vasoconstriction, and vasodilatation. The endothelium acts as a semi-selective barrier between the vessel lumen and surrounding tissue. Chronic inflammation in vessels changes the shapes of endothelial cells and increases the permeability of endothelium which may lead to tissue edema or swelling [72]. It's suggested RhoA activation and Ca^{2+} entry through TRPC1, TRPC4, and TRPC6 channels both contribute to the thrombin-induced increase in endothelial cell contraction, to the cell shape change, and consequently to the mechanism of increased endothelial permeability [73–75]. TRPC1, TRPC3, and TRPC6, together with vascular epithelial growth factor (VEGF) receptor 2, mediate VEGF-induced Ca^{2+} entry and permeability of human microvascular endothelial cells [76, 77]. TRPC6 channels mediate VEGF-induced angiogenesis in human umbilical cord vascular endothelial cells (HUVEC) [78, 79], and TRPC1 and TRPC4 are required for tubular formation in primary HUVEC in another report [80]. Additionally, hypoxia sensed by endothelial cells leads to growth factor production and vascular remodeling. TRPC3/TRPC4 heteromeric channels in endothelial cells and HEK293 cells are responsible for hypoxia-induced Ca^{2+} entry [81]. Ca^{2+} entry through TRPC channels plays an important role in agonist-induced vasoactivation. Endothelial cells in TRPC3$^{-/-}$ mesenteric arteries showed attenuated PE-stimulated vasoconstriction, impaired acetylcholine-induced nitro oxygen (NO) production, and increased vasodilatation [82, 83]. Similarly, in aortic endothelial cells of TRPC4$^{-/-}$ mice, acetylcholine-induced Ca^{2+} entry and vasodilatation are both reduced [74].

4.3 TRPCs in Lung Health and Disease

The lung is composed of multiple structural cell types including epithelial cells, airway smooth muscle, pulmonary vascular smooth muscle, and endothelial cells. Inflammatory lung diseases, such as asthma and chronic obstructive pulmonary disease (COPD), feature alterations in the morphology and function of structural cells. For example, there is epithelial hyperplasia and development of an epithelial hypersecretory phenotype in asthma and chronic bronchitis and airway smooth muscle hypertrophy and hyperreactivity in asthmatics [84, 85]. Studies of expressions of TRPC channels suggest several TRPCs are highly expressed in different cell types and that their expression pattern levels are distinct [86, 87, 88], suggesting their unique functions in different cell types.

4.3.1 TRPCs in Lung Epithelial Cells

Little is known about TRPC expression in primary airway epithelial cells. TRPC1, TRPC4, and TRPC6 mRNA and TRPC6 protein are expressed in human bronchial epithelium and submucosal gland epithelium [86]. On the other hand, much more is known about the expression of TRPC channels in primary lung endothelial cells. There are several recent studies showing that TRPC1, TRPC3, TRPC4, TRPC6, and TRPC7 are expressed in either human or mouse pulmonary artery endothelial cells [57, 89].

Calcium ion influx through plasmalemmal calcium channels can impact the integrity of lung endothelial barrier and thus its permeability of fluid and protein. Store-operated Ca^{2+} entry increases lung endothelial permeability, both in vivo and in vitro [90]. It has been reported that store-operated calcium channels in culture pulmonary endothelium and caveolar fractions harvested from intact lung epithelium consist of TRPC1 and TRPC4 [91]. The interaction of TRPC4 and Orai1 is responsible for channel's calcium selectivity. Furthermore, thrombin-induced store-operated Ca^{2+} entry is reduced in lung endothelial cells isolated from TRPC4$^{-/-}$ mice [74]. However, in another study, activation of Ca^{2+} entry by OAG or thrombin in human pulmonary artery endothelium is reduced by treating the cells with siRNA against TRPC6 [75]. In concert with attenuated Ca^{2+} entry, RhoA activity, myosin light chain phosphorylation, actin stress fiber formation, and monolayer permeability are all decreased [75]. Ischemia-induced intracellular Ca^{2+} overload and subsequent increase of monolayer permeability are attenuated in endothelial cells isolated from TRPC6$^{-/-}$ mice. Thus, TRPC6$^{-/-}$ mice are protected from ischemia-induced increases in lung permeability and edema [92].

4.3.2 TRPCs in Airway Smooth Muscle Cells

Airway smooth muscles control the passage of air in airways. The dysfunction of airway smooth muscles is implicated in asthma. Excessive contraction of airway smooth muscle will cause airway narrowing, which is the primary mechanism of morbidity and mortality in asthma [93, 94]. Extracellular Ca^{2+} influx has been shown to play a critical role in smooth muscle contraction [95]. Multiple TRPC channels are expressed in smooth muscle cells, of which TRPC1, TRPC3, and TRPC6 have been shown to be expressed consistently across species [87, 96, 97].

TRPC3 is the major component of the native constitutively active nonselective cation channels in airway smooth muscle cells, of which the activity is increased in response to agonists [96]. They play an important role in various cellular responses including contraction, proliferation, migration, and gene expression in airway smooth muscle cells. TRPC3-encoded nonselective cation channels are also important for controlling the resting membrane potential and intracellular Ca^{2+} concentration in airway smooth muscle cells. Knocking down of TRPC3 results in a pronounced hyperpolarization by ~14 mV [96]. Moreover, *trpc3* gene silencing inhibits methacholine-, acetylcholine-, and tumor necrosis factor α (TNFα)-evoked $[Ca^{2+}]_i$, suggesting TRPC3 mediates agonist-induced $[Ca^{2+}]_i$ elevation in smooth muscle cells [98]. In reminiscent of these results, TRPC3 mRNA and protein level are significantly increased in airway smooth muscles following treatment with TNFα, an important asthma mediator [98]. These lines of evidence suggest that TRPC3 plays a fundamental role in smooth muscle physiology, and it is a prominent candidate for treatment of asthma.

4.4 Perspectives

In the past few years, TRPC channels have emerged as central players in various physiological processes. Mutations in these proteins are frequently associated with human diseases. As more information from the in vivo role of TRPC channels in animal models and clinical data from patients carrying mutations become available, our knowledge of the role of TRPC channels in disease pathogenesis will expand considerably. TRPC channels are expressed universally among most cell types. Studies from one system can be referenced to another. Further progress in mechanistic understanding of TRPC channels may help in identification of novel therapeutic targets.

References

1. Suh JH, Miner JH (2013) The glomerular basement membrane as a barrier to albumin. Nat Rev Nephrol 9:470–477
2. Brinkkoetter PT, Ising C, Benzing T (2013) The role of the podocyte in albumin filtration. Nat Rev Nephrol 9:328–336
3. Niehof M, Borlak J (2008) HNF4 alpha and the Ca-channel TRPC1 are novel disease candidate genes in diabetic nephropathy. Diabetes 57:1069–1077
4. Goel M, Zuo CD, Schilling WP (2010) Role of cAMP/PKA signaling cascade in vasopressin-induced trafficking of TRPC3 channels in principal cells of the collecting duct. Am J Physiol Ren Physiol 298:F988–F996
5. Sours-Brothers S, Ding M, Graham S, Ma R (2009) Interaction between TRPC1/TRPC4 assembly and STIM1 contributes to store-operated Ca2+ entry in mesangial cells. Exp Biol Med 234:673–682
6. Schaldecker T et al (2013) Inhibition of the TRPC5 ion channel protects the kidney filter. J Clin Invest 123:5298–5309
7. Sonneveld R et al (2014) Glucose specifically regulates TRPC6 expression in the podocyte in an AngII-dependent manner. Am J Pathol 184:1715–1726
8. Goel M, Sinkins WG, Zuo CD, Estacion M, Schilling WP (2006) Identification and localization of TRPC channels in the rat kidney. Am J Physiol Ren Physiol 290:F1241–F1252
9. Migliorini A, Ebid R, Scherbaum CR, Anders HJ (2013) The danger control concept in kidney disease: mesangial cells. J Nephrol 26:437–449
10. Hsu YJ, Hoenderop JG, Bindels RJ (2007) TRP channels in kidney disease. Biochim Biophys Acta 1772:928–936
11. Kimberling WJ et al (1988) Linkage heterogeneity of autosomal dominant polycystic kidney disease. N Engl J Med 319:913–918
12. Reiser J et al (2005) TRPC6 is a glomerular slit diaphragm-associated channel required for normal renal function. Nat Genet 37:739–744
13. Winn MP et al (2005) A mutation in the TRPC6 cation channel causes familial focal segmental glomerulosclerosis. Science 308:1801–1804
14. Hoenderop JG et al (2002) Modulation of renal Ca2+ transport protein genes by dietary Ca2+ and 1,25-dihydroxyvitamin D3 in 25-hydroxyvitamin D3-1alpha-hydroxylase knockout mice. FASEBJ 16:1398–1406
15. Kim EY, Anderson M, Dryer SE (2012) Insulin increases surface expression of TRPC6 channels in podocytes: role of NADPH oxidases and reactive oxygen species. Am J Physiol Ren Physiol 302: F298–F307
16. Graham S et al (2007) Downregulation of TRPC6 protein expression by high glucose, a possible mechanism for the impaired Ca2+ signaling in glomerular mesangial cells in diabetes. Am J Physiol Ren Physiol 293:F1381–F1390
17. Qian Y, Feldman E, Pennathur S, Kretzler M, Brosius FC 3rd (2008) From fibrosis to sclerosis: mechanisms of glomerulosclerosis in diabetic nephropathy. Diabetes 57:1439–1445
18. Griveas I, Stavianoudakis G, Karanikas E, Gogos K, Nakopoulou L (2009) The role of pure diffuse mesangial hypercellularity in patients with proteinuria. Ren Fail 31:192–195
19. Soni H, Adebiyi A (2016) TRPC6 channel activation promotes neonatal glomerular mesangial cell apoptosis via calcineurin/NFAT and FasL/Fas signaling pathways. Sci Rep 6:29041
20. Qiu G, Ji Z (2014) AngII-induced glomerular mesangial cell proliferation inhibited by losartan via changes in intracellular calcium ion concentration. Clin Exp Med 14:169–176
21. Greka A, Mundel P (2012) Cell biology and pathology of podocytes. Annu Rev Physiol 74:299–323
22. Ilatovskaya DV et al (2015) Podocyte injury in diabetic nephropathy: implications of angiotensin II-dependent activation of TRPC channels. Sci Rep 5:17637
23. Yu H et al (2013) Rac1 activation in podocytes induces rapid foot process effacement and proteinuria. Mol Cell Biol 33:4755–4764
24. Greka A, Mundel P (2011) Balancing calcium signals through TRPC5 and TRPC6 in podocytes. J Am Soc Nephrol: JASN 22:1969–1980
25. Tian D et al (2010) Antagonistic regulation of actin dynamics and cell motility by TRPC5 and TRPC6 channels. Sci Signal 3:ra77
26. Yu H et al (2016) Synaptopodin limits TRPC6 podocyte surface expression and attenuates proteinuria. J Am Soc Nephrol 27(11):3308–3319

27. Moller CC et al (2007) Induction of TRPC6 channel in acquired forms of proteinuric kidney disease. J Am Soc Nephrol 18:29–36
28. Eckel J et al (2011) TRPC6 enhances angiotensin II-induced albuminuria. J Am Soc Nephrol: JASN 22:526–535
29. Goel M, Sinkins WG, Zuo CD, Hopfer U, Schilling WP (2007) Vasopressin-induced membrane trafficking of TRPC3 and AQP2 channels in cells of the rat renal collecting duct. Am J Physiol Ren Physiol 293:F1476–F1488
30. Inoue R et al (2006) Transient receptor potential channels in cardiovascular function and disease. Circ Res 99:119
31. Dietrich A, Gudermann T (2011) TRP channels in the cardiopulmonary vasculature. In: Islam MS (ed) Transient receptor potential channels. Springer Netherlands, Dordrecht, pp 781–810
32. Seth M et al (2009) TRPC1 channels are critical for hypertrophic signaling in the heart. Circ Res 105:1023–1030
33. Ohba T et al (2007) Upregulation of TRPC1 in the development of cardiac hypertrophy. J Mol Cell Cardiol 42:498–507
34. Bush EW et al (2006) Canonical transient receptor potential channels promote cardiomyocyte hypertrophy through activation of calcineurin signaling. J Biol Chem 281:33487–33496
35. Satoh S et al (2007) Transient receptor potential (TRP) protein 7 acts as a G protein-activated Ca2+ channel mediating angiotensin II-induced myocardial apoptosis. Mol Cell Biochem 294:205–215
36. Kuwahara K et al (2006) TRPC6 fulfills a calcineurin signaling circuit during pathologic cardiac remodeling. J Clin Investig 116:3114–3126
37. Molkentin JD et al (1998) A calcineurin-dependent transcriptional pathway for cardiac hypertrophy. Cell 93:215–228
38. Wu X, Eder P, Chang B, Molkentin JD (2010) TRPC channels are necessary mediators of pathologic cardiac hypertrophy. Proc Natl Acad Sci U S A 107:7000–7005
39. Vindis C et al (2010) Essential role of TRPC1 channels in cardiomyoblasts hypertrophy mediated by 5-HT2A serotonin receptors. Biochem Biophys Res Commun 391:979–983
40. Onohara N et al (2006) TRPC3 and TRPC6 are essential for angiotensin II-induced cardiac hypertrophy. EMBO J 25:5305–5316
41. Nakayama H, Wilkin BJ, Bodi I, Molkentin JD (2006) Calcineurin-dependent cardiac hypertrophy is activated by TRPC in the adult mouse heart. FASEB J: Off Publ Fed Am Soc Exp Biol 20:1660–1670
42. Han JW et al (2016) Resistance to pathologic cardiac hypertrophy and reduced expression of CaV1.2 in Trpc3-depleted mice. Mol Cell Biochem 421:55–65
43. Xie J et al (2012) Cardioprotection by Klotho through downregulation of TRPC6 channels in the mouse heart. Nat Commun 3:1238–1238
44. Seo K et al (2014) Combined TRPC3 and TRPC6 blockade by selective small-molecule or genetic deletion inhibits pathological cardiac hypertrophy. Proc Natl Acad Sci U S A 111:1551–1556
45. Kiyonaka S et al (2009) Selective and direct inhibition of TRPC3 channels underlies biological activities of a pyrazole compound. Proc Natl Acad Sci U S A 106:5400–5405
46. Schindl R et al (2008) The first ankyrin-like repeat is the minimum indispensable key structure for functional assembly of homo- and heteromeric TRPC4/TRPC5 channels. Cell Calcium 43:260–269
47. Shan D, Marchase RB, Chatham JC (2008) Overexpression of TRPC3 increases apoptosis but not necrosis in response to ischemia-reperfusion in adult mouse cardiomyocytes. Am J Phys Cell Physiol 294:C833–C841
48. Cui L-B et al (2016) Morphological identification of TRPC7 in cardiomyocytes from normal and renovascular hypertensive rats. J Cardiovasc Pharmacol 67:121–128
49. Nishida M et al (2007) Gα12/13-mediated up-regulation of TRPC6 negatively regulates endothelin-1-induced cardiac myofibroblast formation and collagen synthesis through nuclear factor of activated T cells activation. J Biol Chem 282:23117–23128
50. Yao X, Garland CJ (2005) Recent developments in vascular endothelial cell transient receptor potential channels. Circ Res 97:853
51. Yip H et al (2004) Expression of TRPC homologs in endothelial cells and smooth muscle layers of human arteries. Histochem Cell Biol 122:553–561
52. Earley S, Brayden JE (2015) Transient receptor potential channels in the vasculature. Physiol Rev 95:645–690
53. Bochaton-Piallat ML, Gabbiani G (2005) Modulation of smooth muscle cell proliferation and migration: role of smooth muscle cell heterogeneity. In: von Eckardstein A (ed) Atherosclerosis: diet and drugs. Springer, Berlin/Heidelberg, pp 645–663
54. Leung FP, Yung LM, Yao X, Laher I, Huang Y (2008) Store-operated calcium entry in vascular smooth muscle. Br J Pharmacol 153:846–857
55. Sweeney M et al (2002) Inhibition of endogenous TRP1 decreases capacitative Ca2+entry and attenuates pulmonary artery smooth muscle cell proliferation. Am J Physiol Lung Cell Mol Physiol 283:L144
56. Zhang S, Remillard CV, Fantozzi I, Yuan JXJ (2004) ATP-induced mitogenesis is mediated by cyclic AMP response element-binding protein-enhanced TRPC4 expression and activity in human pulmonary artery smooth muscle cells. Am J Physiol Cell Physiol 287:C1192
57. Yu Y et al (2004) Enhanced expression of transient receptor potential channels in idiopathic pulmonary arterial hypertension. Proc Natl Acad Sci U S A 101:13861–13866
58. Weissmann N et al (2006) Classical transient receptor potential channel 6 (TRPC6) is essential for hypoxic

pulmonary vasoconstriction and alveolar gas exchange. Proc Natl Acad Sci U S A 103:19093–19098

59. Golovina VA et al (2001) Upregulated TRP and enhanced capacitative Ca2+ entry in human pulmonary artery myocytes during proliferation. Am J Physiol Heart Circ Physiol 280:H746
60. Kunichika N et al (2004) Overexpression of TRPC1 enhances pulmonary vasoconstriction induced by capacitative Ca2+ entry. Am J Physiol Lung Cell Mol Physiol 287:L962
61. Inoue R et al (2001) The transient receptor potential protein homologue TRP6 is the essential component of vascular α1-adrenoceptor–activated Ca2 + –permeable cation channel. Circ Res 88:325
62. Jung S, Strotmann R, Schultz G, Plant TD (2002) TRPC6 is a candidate channel involved in receptor-stimulated cation currents in A7r5 smooth muscle cells. Am J Physiol Cell Physiol 282:C347
63. Dietrich A et al (2005) Increased vascular smooth muscle contractility in TRPC6(−)(/)(−) Mice. Mol Cell Biol 25:6980–6989
64. Sel S et al (2008) Loss of classical transient receptor potential 6 channel reduces allergic airway response. Clin Exp Allergy 38:1548–1558
65. Reading SA, Earley S, Waldron BJ, Welsh DG, Brayden JE (2005) TRPC3 mediates pyrimidine receptor-induced depolarization of cerebral arteries. Am J Physiol Heart Circ Physiol 288:H2055
66. Liu D et al (2005) Increased transient receptor potential channel TRPC3 expression in spontaneously hypertensive rats. Am J Hypertens 18:1503–1507
67. Peppiatt-Wildman CM, Albert AP, Saleh SN, Large WA (2007) Endothelin-1 activates a Ca(2+)-permeable cation channel with TRPC3 and TRPC7 properties in rabbit coronary artery myocytes. J Physiol 580:755–764
68. Weissmann N et al (2006) Oxygen sensors in hypoxic pulmonary vasoconstriction. Cardiovasc Res 71:620
69. Weigand L, Foxson J, Wang J, Shimoda LA, Sylvester JT (2005) Inhibition of hypoxic pulmonary vasoconstriction by antagonists of store-operated Ca2+ and nonselective cation channels. Am J Physiol Lung Cell Mol Physiol 289:L5
70. Lin M-J et al (2004) Chronic hypoxia–induced upregulation of store-operated and receptor-operated Ca2+channels in pulmonary arterial smooth muscle cells. Circ Res 95:496
71. Wang J et al (2006) Hypoxia inducible factor 1 mediates hypoxia-induced TRPC expression and elevated intracellular Ca2+ in pulmonary arterial smooth muscle cells. Circ Res 98:1528
72. Tiruppathi C, Ahmmed GU, Vogel SM, Malik AB (2006) Ca2+ signaling, TRP channels, and endothelial permeability. Microcirculation 13:693–708
73. Paria BC et al (2003) Tumor necrosis factor-α induces nuclear factor-κB-dependent TRPC1 expression in endothelial cells. J Biol Chem 278:37195–37203
74. Tiruppathi C et al (2002) Impairment of store-operated Ca2+ entry in TRPC4−/− mice interferes with increase in lung microvascular permeability. Circ Res 91:70–76
75. Singh I et al (2007) Galphaq-TRPC6-mediated Ca2+ entry induces RhoA activation and resultant endothelial cell shape change in response to thrombin. J Biol Chem 282:7833–7843
76. Jho D et al (2005) Angiopoietin-1 opposes VEGF-induced increase in endothelial permeability by inhibiting TRPC1-dependent Ca2+ influx. Circ Res 96:1282
77. Cheng HW, James AF, Foster RR, Hancox JC, Bates DO (2006) VEGF activates receptor-operated cation channels in human microvascular endothelial cells. Arterioscler Thromb Vasc Biol 26:1768
78. Ge R et al (2009) Critical role of TRPC6 channels in VEGF-mediated angiogenesis. Cancer Lett 283:43–51
79. Zadeh MH, Glass CA, Magnussen A, Hancox JC, Bates DO (2008) VEGF-mediated elevated intracellular calcium and angiogenesis in human microvascular endothelial cells in vitro are inhibited by dominant negative TRPC6. Microcirc (New York, NY: 1994) 15:605–614
80. Antigny F, Girardin N, Frieden M (2012) Transient receptor potential canonical channels are required for in vitro endothelial tube formation. J Biol Chem 287:5917–5927
81. Poteser M et al (2006) TRPC3 and TRPC4 associate to form a redox-sensitive cation channel: evidence for expression of native TRPC3-TRPC4 heteromeric channels in endothelial cells. J Biol Chem 281:13588–13595
82. Senadheera S et al (2012) Transient receptor potential canonical type 3 channels facilitate endothelium-derived hyperpolarization-mediated resistance artery vasodilator activity. Cardiovasc Res 95:439–447
83. Yeon S-I et al (2014) Transient receptor potential canonical type 3 channels control the vascular contractility of mouse mesenteric arteries. PLoS One 9:e110413
84. Chiappara G et al (2001) Airway remodelling in the pathogenesis of asthma. Curr Opin Allergy Clin Immunol 1:85–93
85. Jeffery PK (1999) Differences and similarities between chronic obstructive pulmonary disease and asthma. Clin Exp Allergy: J Br Soc Allergy Clin Immunol 29(Suppl 2):14–26
86. Corteling RL et al (2004) Expression of transient receptor potential C6 and related transient receptor potential family members in human airway smooth muscle and lung tissue. Am J Respir Cell Mol Biol 30:145–154
87. Ong HL, Brereton HM, Harland ML, Barritt GJ (2003) Evidence for the expression of transient receptor potential proteins in guinea pig airway smooth muscle cells. Respirology 8:23–32

88. Sweeney M et al (2002) Role of capacitative Ca2+ entry in bronchial contraction and remodeling. J Appl Physiol 92:1594–1602
89. Yu Y et al (2003) PDGF stimulates pulmonary vascular smooth muscle cell proliferation by upregulating TRPC6 expression. Am J Phys Cell Physiol 284:C316–C330
90. Chung YH et al (2007) Immunohistochemical study on the distribution of canonical transient receptor potential channels in rat basal ganglia. Neurosci Lett 422:18–23
91. Sundivakkam PC et al (2012) The Ca(2+) sensor stromal interaction molecule 1 (STIM1) is necessary and sufficient for the store-operated Ca(2+) entry function of transient receptor potential canonical (TRPC) 1 and 4 channels in endothelial cells. Mol Pharmacol 81:510–526
92. Kini V, Chavez A, Mehta D (2010) A new role for PTEN in regulating transient receptor potential canonical channel 6-mediated Ca2+ entry, endothelial permeability, and angiogenesis. J Biol Chem 285:33082–33091
93. Kim HR, Appel S, Vetterkind S, Gangopadhyay SS, Morgan KG (2008) Smooth muscle signalling pathways in health and disease. J Cell Mol Med 12:2165–2180
94. Tliba O, Panettieri RA Jr (2009) Noncontractile functions of airway smooth muscle cells in asthma. Annu Rev Physiol 71:509–535
95. Gerthoffer WT (1986) Calcium dependence of myosin phosphorylation and airway smooth muscle contraction and relaxation. Am J physiol 250:C597–C604
96. Xiao JH, Zheng YM, Liao B, Wang YX (2010) Functional role of canonical transient receptor potential 1 and canonical transient receptor potential 3 in normal and asthmatic airway smooth muscle cells. Am J Respir Cell Mol Biol 43:17–25
97. Godin N, Rousseau E (2007) TRPC6 silencing in primary airway smooth muscle cells inhibits protein expression without affecting OAG-induced calcium entry. Mol Cell Biochem 296:193–201
98. White TA et al (2006) Role of transient receptor potential C3 in TNF-alpha-enhanced calcium influx in human airway myocytes. Am J Respir Cell Mol Biol 35:243–251

TRPC Channels and Programmed Cell Death

5

Jian Zhou and Yichang Jia

Abstract

Neurotrophins, including nerve growth factor (NGF) and brain-derived neurotrophic factor (BDNF), bind to their high-affinity receptors to promote neuronal survival during brain development. One of the key downstream pathways is the phospholipase C (PLC) pathway, which not only plays a central role in calcium release from internal store but also in activation of TRPC channels coupled with neurotrophin receptors. TRPC channels are required for the neurotrophin-mediated neuronal protective effects. In addition, activation of TRPC channels is able to protect neurons in the absence of neurotrophin. In some circumstances, TRPC channels coupled with metabotropic glutamate receptor may mediate the excitotoxicity by calcium overload. One of the key questions in the field is the channel gating mechanisms; understanding of which would help design compounds to modulate the channel properties. The development and identification of TRPC channel agonists or blockers are promising and may unveil new therapeutic drugs for the treatment of neurodegenerative diseases and epilepsy.

Keywords

TRPC channels • Neurotrophins • Programmed cell death

J. Zhou
Laboratory of Neural Signal Transduction, Institute of Neuroscience, Shanghai Institutes for Biological Sciences, Chinese Academy of Sciences, 320 Yueyang Road, Shanghai, 200031 China

Y. Jia (✉)
School of Medicine, Tsinghua University, Medical Science Building, Room D204, Beijing 100084, China

Peking-Tsinghua Joint Center for Life Sciences, Beijing 100084, China

IDG/McGovern Institute for Brain Research at Tsinghua, Beijing 100084, China
e-mail: yichangjia@tsinghua.edu.cn

Y. Wang (ed.), *Transient Receptor Potential Canonical Channels and Brain Diseases*, Advances in Experimental Medicine and Biology 976, DOI 10.1007/978-94-024-1088-4_5

5.1 Introduction

5.1.1 Programmed Cell Death in the Brain

The term of programmed cell death is first coined by Richard Lockshin and Carroll Williams, when they described the phenomenon that the intersegmental muscles of silk moths break down and disappear within a short period during their development [42]. However, it has been a long time for scientists to propose that the programmed cell death during brain development helps to shape the brain by removing the extra neurons which could not find their right targets. In 1892, the founder of modern neuroscience, Ramon y Cajal, proposed the "neurotrophic hypothesis" to explain how the tip of neuronal axons, also known as growth cone, finds their targets [69] (Fig. 5.1). It took almost a century for neuroscientists to accept his prophetic predictions, in which Cajal believed that the target cells must secrete the attractive and repulsive substances to guide the axons to find their desired destination. A hundred years later, the repulsive axon guidance cue, named netrin, is first experimentally identified by Marc Tessier-Lavigne's group [66].

In the early 1950s, a group of scientists, including Rita Levi-Montalcini and Stanley Cohen, together with Viktor Hamburger, discovered nerve growth factor (NGF) as one of the neurotrophic factors to promote nerve growth and neuronal survival during early brain development [13]. Among them, Rita and Stanley won the Nobel Prize in Physiology or Medicine in 1986. In the coculture of the chicken embryos with the tumor cells, Levi-Montalcini observed the overgrowth of nerve fibers from chicken embryo, suggesting the secretary NGF from the tumor cells [13]. Using the assay developed by Levi-Montalcini assessing the nerve growth, Stanley aimed to define the protein identity of NGF. However, the obstacle was the biological materials that contain enough NGF for further purification. During his searching, Stanley happened to find that the snake venom contains the similar substances that can promote nerve growth [13, 56]. Though the snake venom is still not a common resource for the type of research, the accidental finding led Stanley to purify the massive amount of NGF from mouse salivary glands, an organ secreting the saliva similar to venom, and easily to obtain in a large scale, which further led to the characterization of sequence and structure of NGF [7].

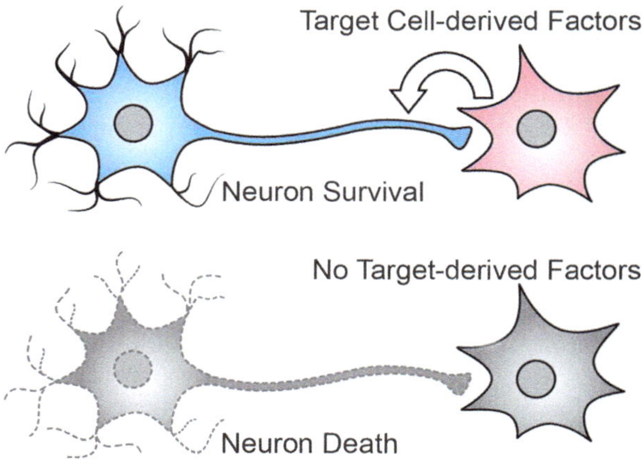

Fig. 5.1 The neurotrophic hypothesis. The hypothesis was originally proposed by Ramon y Cajal [69]. The target tissue secretes trophic factors that may help axon of the innervating neuron find its target. In turn, the target tissue may secrete the trophic factors that promote the neuronal survival, strengthen the synaptic interaction between the neuron and target tissue, and establish the correct neuronal circuits. These trophic factors include NGF (nerve growth factor) and BDNF (brain-derived neurotrophic factor). The picture is modified from the previous literature [67]

In fact, during early brain development, scientists, including Rita Levi-Montalcini and Viktor Hamburger, observed extra neuron death in different brain regions [16, 52]. The processes, scientists believe, must have crucial functions, including but not limited to cleanup of mislocalized or misconnected neurons and establishment of the correct neuronal circuits. The "neurotropic hypothesis" proposed by Ramon y Cajal appears to be rational for explaining the processes—the pro-survival trophic factors released from target tissues and the neuron forming correct connection or finding its correct partners receives the pro-survival factors and survives (Fig. 5.1). Indeed, experimentally, the NGF knockout mice show neuron loss in the sympathetic ganglia [14]. In addition, neuron loss in the sensory ganglia neurons and cerebellar granule cells is also seen in the BDNF (brain-derived neurotrophic factor) knockout mice [20, 21, 35, 63]. Another trophic factor, neurotrophin-3 (NT-3), seems mainly to control neuron survival in the sympathetic superior cervical ganglion, because the approximate 50% of these neurons die during the neurogenesis [20, 21]. Although these results support the "neurotrophic hypothesis," in general, most of trophic factor knockout mice appear that only certain brain regions are affected, suggesting strong functional redundancy of these pro-survival signals.

The specificity of neurotrophins to promote certain neuronal survival has puzzled scientists for a long time, and it could be explained by tight cross talk between neurotrophins and their high-affinity receptors happening at right time and place during brain development [32]. For example, BDNF and its high-affinity receptor TrkB (tyrosine protein kinase receptor-B) are expressed in the similar time window during cerebellar development [34, 39].

5.1.2 Trophic Factors Bound to Their High-Affinity Receptors to Promote Neuron Survival

In the early 1960s, after Stanley Cohen characterized the NGF from mouse salivary glands, he isolated another protein from mouse submaxillary gland and named it epidermal growth factor (EGF) due to its capability to promote epithelial cell proliferation [11, 12]. Using isotope-labeled EGF, Stanley and his colleagues further revealed that the receptor for EGF appears on the cell surface and undergoes phosphorylation after the cell treated with EGF [8, 9]. In addition, Stanley's group showed that the tyrosine phosphorylation on cell surface protein is mediated by EGF application, suggesting the tyrosine phosphorylation after the activation of EGF receptor (EGFR) [76]. After the groundbreaking work of identification of EGFR cDNA sequence, an inspiring study demonstrated that application of insulin is able to activate the tyrosine phosphorylation of a chimeric receptor, which is comprised of the extracellular domain of insulin receptor and the intracellular domain of EGFR [58, 74]. The finding really supports the notion that the shared receptor activation mechanisms underlying the ligand bound to their high-affinity receptors by the receptor tyrosine phosphorylation. Later on these receptors are termed as receptor tyrosine kinase (RTK), including NGF receptor-TrkA (tyrosine protein kinase receptor-A) and BDNF receptor-TrkB. These Trks share high sequence similarity with EGFR and, surprisingly, share the similar downstream signaling pathways with EGF/EGFR as well [4, 25]. The discoveries of EGF and EGFR further lead to the identification and characterization of as many as 58 RTKs in human and mouse genomes [59].

One of the major progresses for RTK and Trk field is the crystal structure of NGF receptor-TrkA in 1999 [79]. Together with the residue mutagenesis studies, the immunoglobulin-like domain proximal to the membrane is identified for NGF binding [75]. The working model is that the two NGF molecules form homodimer to fit into the pocket of the receptors, and the residues at N-terminal of NGF selectively interact with TrkA to determine the ligand/receptor specificity. Due to the high similarity between the Trk family members, including TrkA, TrkB, and TrkC, the structural working model between the extracellular domain of Trk and their ligands appears very similar but with high selectivity [3, 79] (Fig. 5.2).

After the discoveries of EGF and EGFR, the downstream signal cascades after ligand binding

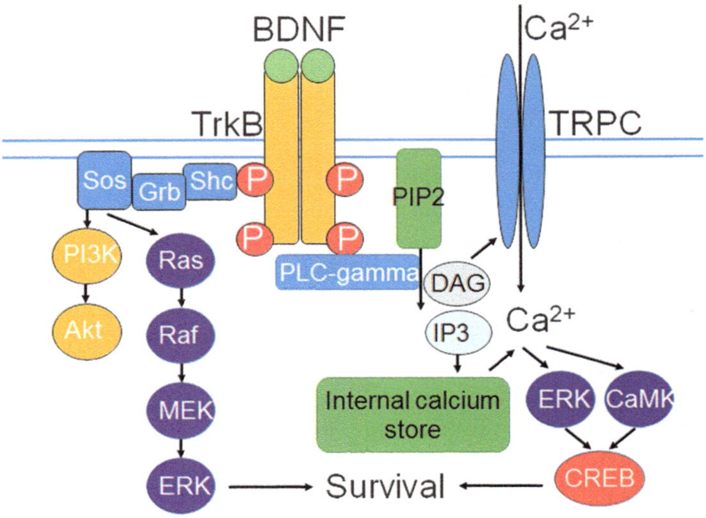

Fig. 5.2 The downstream signaling pathways of BDNF bound to its high-affinity receptor, TrkB. Like NGF, two BDNF molecules form dimmer to fit the pocket of TrkB receptors, which in turn stimulates the receptor dimerization and autophosphorylation. Three major pathways are downstream of BDNF/TrkB activation, the PI3K/Akt, the Ras/Raf/MEK/ERK, and PLC gamma/DAG, IP3 pathways. The Akt and ERK pathways play key roles in supporting the neuronal survival in the presence of BDNF. The PLC gamma pathway resides in a central position in regulating the internal calcium level by IP3 generated from PIP2 hydrolysis. IP3 acts on its ER-localized receptor-IP3R and triggers calcium release from internal calcium store. In addition, PLC gamma also plays a central role in the regulation of TRPC channel property. The PLC activation per se, DAG, and internal calcium release are all proved to be able to open TRPC channels downstream of BDNF bound to TrkB [39, 47]. The calcium influx through TRPC channels may stimulate the ERK and CAMK pathways, which converge to CREB activation, eventually leading to neuronal survival effect [34]. Abbreviations: *BDNF* brain-derived neurotrophic factor, *TrkB* tyrosine protein kinase receptor-B, *NGF* nerve growth factor, *PI3K* phosphatidyl inositol-3 kinase, *Akt* serine threonine kinase, *MEK* MAP/ERK kinase, *ERK* extracellular signal-regulated kinase, *PLC* phospholipase C, *PIP2* phosphatidylinositol 4,5-bisphosphate, *IP3* inositol triphosphate, *DAG* diacylglycerol, *Shc* the Src homology 2 domain containing, *Grb* growth factor receptor-bound protein, *SOS* son of sevenless, *CAMK* Ca^{2+}-calmodulin-dependent protein kinase, *CREB* cAMP response element-binding protein

to its cognate receptor are further characterized, including NGF receptor-TrkA and BDNF receptor-TrkB [4, 25, 33]. As described above, the downstream signals are very similar across all the RTKs. Three major pathways downstream of RTK activation, including the MAPK (microtubule-associated protein kinase), Akt (serine threonine kinase), and PLCγ (phospholipase C gamma) pathways, are proposed. As an example of TrkB activation triggered by BDNF (Fig. 5.2), the ligand/receptor binding induces the dimerization of TrkB, which in turn triggers the autophosphorylation of tyrosine residues in the intracellular domain of TrkB [19, 33]. The 515 tyrosine phosphorylation of TrkB serves as sites to recruit the membrane adaptor protein Shc (the Src homology 2 domain containing), followed by the recruitment of other adaptors like Grb (growth factor receptor-bound protein) and SOS (son of sevenless), leading to the activation of MAPK and Akt pathways. The 816 tyrosine phosphorylation recruits the PLC gamma, which results in the hydrolysis of PIP2 (phosphatidylinositol 4,5-bisphosphate) and generates IP3 (inositol triphosphate) and DAG (diacylglycerol). The PLC gamma pathway plays a central role in the regulation of cytosolic Ca^{2+} level downstream of BDNF/TrkB. The first layer is the yield of IP3 from PIP2 hydrolysis, which acts on its ER (endoplasmic reticulum)-localized receptor-IP3R (IP3 receptor) to induce calcium release from the internal calcium store (ER), resulting in local cytosolic calcium level increase. Activation of PLC gamma pathway, including PLC itself activation, DAG production, and internal calcium release by IP3 acting on IP3R, is essential for the

TRPC3/TRPC6 channel activation and extracellular calcium influx through these channels [39, 47] (Fig. 5.2). All these three pathways are reported to contribute to neuronal survival mediated by BDNF/TrkB activation [34, 68].

5.1.3 Local Calcium Signalings and Their Different Roles in Neuronal Survival

In the resting condition, the cytosolic Ca^{2+} concentration is normally maintained in a low level about 100 nM, which is 10,000 times lower than the extracellular calcium level. Therefore, the calcium influx through the cell surface calcium channels plays key roles in controlling and regulating the cytosolic calcium level. As a second messenger, influxed calcium could stimulate a variety of enzymes, including calcium-calmodulin kinases and calcium-sensitive adenylate cyclases, and fulfill the pivotal functions, including but not limited to activity-dependent neuronal survival and activity-dependent plasticity [23]. However, calcium overload through extracellular calcium influx is proved to be toxic to neurons, especially calcium influx through the ionotropic glutamate receptors, such as NMDAR (N-methyl-D-aspartate receptor) and AMPAR (α-amino-3-hydroxy-5-methyl-4-isoxazolepropionic acid receptor), also termed as excitotoxicity [10]. Therefore, the question is really why calcium influx through different calcium channels or in different conditions results in different or, in some circumstance, completely opposite functions.

In 2002, Dr. Hilmar and his colleagues provide evidence that calcium influx through synaptic or extrasynaptic NMDAR may have completely opposite influence on neuronal survival [28]. In the cultured hippocampal neurons, about 10% inhibitory neurons impose a tonic inhibition on the neuronal network through acting on the GABA (γ-aminobutyric acid)$_A$ receptors, which mediate the extracellular Cl^- influx to lower the membrane potential and neuronal activity. Therefore, blockage of $GABA_A$ receptor in the cultures is able to generate synchronous bursts of action potentials by calcium influx through the synaptic NMDAR [27]. By application of bicuculline, a blocker of $GABA_A$ receptor, in cultured hippocampal neurons, Dr. Hilmar's group showed that synaptic calcium influx through NMDAR, but not the voltage-gated calcium channels, is able to activate CREB (cAMP-responsive element-binding protein), a transcription factor downstream of calcium influx and responsible for BDNF expression and neuronal survival [23]. However, bath application of glutamate in the cultures, which is supposed to activate both synaptic and extrasynaptic NMDAR, did not induce phosphorylated CREB and downstream BDNF expression. Therefore, the data implies that even similar subtype of calcium channels may have opposite effect on neuronal survival due to their different localization in the cellular compartment.

The intracellular local environment underneath the calcium channels seems also to play a key role in conducting the cellular functions of the calcium influx through the corresponding channels. For example, the isoleucine-glutamine (IQ) motif in the carboxyl terminus of the L-type voltage-gated calcium channel, which is responsible for the interaction between the channel and Ca^{2+}-calmodulin, is required for conveying the calcium influx signaling to nucleus and the following CREB phosphorylation [17]. Mutant CaMor IQ motif deletion is not able to convey the signaling to nucleus and the CREB activation, suggesting that the local calcium influx through the channel allows the formation of Ca^{2+}-calmodulin complex, which translocates into nucleus and activates the calcium-dependent CREB phosphorylation and its related functions. The examples we describe here may help to explain the fact that calcium influx through different TRPC channels has various functions (see below).

5.1.4 TRPC Channel Roles in the Brain

Members of the TRPC subfamily have roles in multiple cellular processes inside the brain, including brain development, neuronal survival, and neural stem cell proliferation. For example, TRPC1 is responsible for the mGluR1-mediated slow excitatory postsynaptic conductance (EPSC) in cerebellar Purkinje cells [37]. The activation of

TRPC1 seems to facilitate glutamate-induced cell death in hippocampal slice culture, possibly due to the enhancement of excitotoxicity [50]. The activation of TRPC1 seems also to play a key role in supporting the embryonic neural stem cell proliferation [22]. Although TRPC2 is a pseudo gene in human, it is highly expressed in the mouse vomeronasal system and mediates mouse pheromone transduction and related social behaviors [43]. Gating TRPC3 is coupled with activation of TrkB; therefore, TRPC3 contributes to BDNF/TrkB functions, including BDNF-induced dendritic spine formation [1], cerebellar granule neuron (CGN) survival [34], and motor coordination [29]. TRPC4 is reported to have a role in neurite extension in postmitotic neurons [78]. In hippocampal neurons, TRPC5 is highly expressed and controls neurite extension and growth cone morphology [15, 24]. Therefore, it makes sense that mice without TRPC5 appear long-term potentiation deduction and innate fear deficit [54, 57]. Since TRPC3 and TRPC6 may form functional channels, like TRPC3, TRPC6 also plays essential roles in BDNF/TrkB functions, like BDNF-mediated survival of CGNs during development, BDNF-mediated dendritic growth, as well as BDNF-mediated synapse formation in hippocampal neurons [34, 72, 84]. In addition, the TRPC6 transgenic mouse shows better performance in Morris water maze, underscoring its crucial roles in learning and memory and synaptic plasticity.

5.1.5 TRPC Channels in Neuronal Survival

In the neuronal development, the newborn neurons need to reach appropriate target cells to establish right connection, in order to compete for limited trophic factors and activity-dependent survival signals. As one of the important neurotrophins, BDNF has been recognized for a long time as a potent pro-survival trophic factor secreted by target cells to support a variety of both central and peripheral neuronal survival (Fig. 5.1). It would be interesting to investigate whether TRPC channel can also participate in neuronal survival in the axis of BDNF's function.

As mentioned above, netrin-1 is first identified as axon guidance cue [66]. Strikingly, netrin-1 induces axonal growth cone calcium elevation and growth cone turning depending on the activation of TRPC1 in cultured *Xenopus* neurons [77]. One of the downstreams of netrin-1 bound to its receptor DCC (deleted in colorectal cancer) is PLC gamma; therefore, the TRPC1 activation might be coupled to netrin-1/DCC/PLC signaling pathway [80]. Similarly, in mammal growth cone turning, TRPC channels mediate BDNF-induced growth cone calcium elevation and axon guidance [40].

For TRPC channel function in promoting neuron survival, the first work is done in cultured cerebellar granule neurons (CGNs) [34]. Different from the inside-out cortical neuron maturation pattern in the brain cortex, the CGNs are generated in the external granule layer and migrate into the internal granule layer to become mature inside of the cerebellum [60]. During their postnatal migration, CGNs require excitatory inputs and trophic factors for proper differentiation and maturation; otherwise, they die through apoptosis. This trophic factor-mediated survival can be mimicked in primary CGN cultures in vitro. The cultured neurons undergo spontaneous apoptosis when they are deprived of serum, but are protected by adding BDNF. Knockdown of TRPC3 or TRPC6, but not TRPC1, by specific siRNA, abolished BDNF's neuronal protective effect in cultured CGNs [34]. In addition, like TRPCs responsible for the axon growth cone calcium elevation induced by BDNF, TRPC3 and TRPC6 are required for BDNF-mediated cytosolic calcium elevation in the cultured CGNs. It is well known that ERK/CREB pathway and PI3K/Akt pathway are downstream of BDNF bound to TrkB. However, only the phosphorylation or activation of ERK/CREB pathway triggered by BDNF is dependent on TRPC3 or TRPC6 activations (Fig. 5.2). Furthermore, knockdown of TRPC3 or TRPC6 in CGNs in the cerebellum greatly increased the caspase-3-dependent apoptosis in early cerebellum development, from postnatal days 10–12. The developmental time window is the peak expression of TRPC3, TRPC6, and TrkB inside the cerebellum. These in vitro and in vivo findings indicate that TRPC3 and/or TRPC6 are indispensable for conveying the neutrophin factor message to intracellular calcium level increase and then activate ERK/CREB pathways known to be calcium-dependent, to promote neuronal survival.

The TRPC channels also mediate other trophic factors' protective effect. For example, platelet-derived growth factor-BB (PDGF) can provide trophic support for dopaminergic neurons from neural toxicity induced by HIV Tat (trans-activating regulatory protein)—the viral protein essential for efficient transcription of the viral genome [81, 82]. Knockdown of TRPC5 or TRPC6 expressions by RNAi in primary cultured neurons nearly abolished PDGF-mediated neuronal protective effect against HIV Tat. In addition, the calcium elevation induced by PDGF was also attenuated by TRPC5 or TRPC6 knockdown. Among the downstream signaling pathways triggered by PDGF, only ERK/CREB pathway was dependent on TRPC channels, but not PI3K/Akt pathway, which is similar to the observation seen in cerebellar CGNs [34].

Besides neurotrophic factors, TRPC channel is also involved in chemokine (C-C motif) ligand 2 (CCL2)'s neural protective effect [81, 82]. CCL2 is one of the critical chemokines and functions in leukocyte recruitment. It protects midbrain neurons from HIV-1 Tat-induced neurotoxicity. Like TRPC channels required for PDGF's neural protective effect, blockage of TRPC channels results in suppression of both CCL2-mediated neuroprotection and intracellular Ca^{2+} level elevation. Again, only ERK/CREB pathway activation is downstream of TRPC activation, the signaling pathway responsible for protecting the neurons. In contrast, the CCL2-mediated Akt pathway was independent to TRPC channel activation. Because the CCL2 receptor, C-C chemokine receptor type 2 (CCR2), is a G protein-coupled receptor, the activation of TRPC might be due to the PLC beta-signaling.

Although several studies have excluded the involvement of TRPC1 in the neurotrophin-mediated neuroprotective effect, a couple of studies emphasize the critical role of TRPC1 in neural protection, especially in dopaminergic neurons. In a mouse model for Parkinson's disease (PD), which is induced by MPTP treatment, both sub-acute and sub-chronic treatments of MPTP significantly reduce TRPC1 expression, but not TRPC3, along with increased neuronal death [65]. However, overexpression of TRPC1 is able to protect the neurons from MPTP-induced apoptosis, along with restoration of cytosolic calcium level and mitochondrial membrane potential. The study underscores the potential protective role of TRPC1 channel in the onset and progression of PD. It has been known that the store-operated Ca^{2+} entry (SOCE), which is critical for maintaining Ca^{2+} level inside the endoplasmic reticulum (ER), the internal calcium store, is dependent on TRPC1 activity in certain cell types [6]. Overexpression of TRPC1 can restore the MPTP-induced loss of SOCE and the MPTP-induced ER calcium level decrease [64]. The overexpression of TRPC1 also rescues the MPTP-induced ER stress. It seems that TRPC1 works together with the SOCE modulator stromal interaction molecule 1 (STIM1) and regulates the SOCE and internal Ca^{2+} level, to the survival of dopaminergic neurons in the disease conditions. Indeed, the TRPC1-deficient mice show increased unfolded protein response, along with a decreased number of dopaminergic neurons. Overexpression of TRPC1 in mouse prevented dopaminergic neurons from the MPTP-induced neuron loss. The calcium entry through TRPC1 can activate Akt pathway to fulfill the neural protection, apparently different from ERK/CREB pathway downstream of TRPC3 and TRPC6 for CGN survival. The difference may be due to the different functions of TRPC channels in the regulation of cytosolic calcium level or internal (ER) calcium level.

Among a variety of physiological functions of TRPC channels, promoting neuron survival makes them a promising target for future therapeutic strategies in neurodegenerative diseases. Although trophic factors, including BDNF, can restore many neurological deficits in various disease models or conditions [2, 48], the clinical application for BDNF among other trophic factors is limited by their poor pharmacokinetic profile [55]. Activation of TRPC alone is sufficient to protect neurons from apoptosis [34], suggesting the therapeutic potential in the identification of agonists of TRPC channels. A compound screening for TRPC channel agonists was carried out in human embryonic kidney cells expressing recombinant TRPC channels [61]. Several piperazine-derived compounds emerged as potent activators for TRPC3/TRPC6/TRPC7 channels. In addition, these compounds are able to evoke cation currents and Ca^{2+} influx in rat-cultured neurons. More importantly, the compounds are able to mimic BDNF effects on promoting neurite outgrowth and protecting neuron from apoptosis. All these effects are in a calcium-dependent manner. It could be interesting to see whether the potential TRPC agonists show neural protective effect in vivo and lead to potential drug development for the treatment of various neurodegenerative diseases.

5.1.6 TRPC Channels in Neuronal Death

In contrast to well-established roles of TRPC channels in promoting neuronal survival, lines of evidence support their involvement in neuronal death, mainly due to the calcium overload through the channels.

When TRP channel was first identified in *Drosophila* photoreceptor cells, the mutant TRP resulted in a transient depolarization in response to continuous light [44, 46]. Either loss of TRP function or constitutive TRP activity results in retinal degeneration, which can be suppressed by overexpression or disruption of the Na^{+}/Ca^{2+} exchanger, CalX [45]. These findings indicate that an appropriate TRP activity is important for the maintenance of intracellular calcium homeostasis.

Although TRPC1 has been demonstrated to protect dopaminergic neurons from ER stress, it has also been reported to play roles in the hippocampal neuronal death induced by high concentration of glutamate [50]. The expression of TRPC1 was increased during glutamate-induced toxicity, and knockdown of TRPC1 in hippocampal slice cultures protected neurons from the glutamate-induced cell death, suggesting that TRPC1 may contribute to the massive calcium entry coupling with activation of glutamate receptors. Additional evidence suggests that the coupling could be with the activation of metabotropic glutamate receptor [51]. In a hippocampal cell line, knocking down TRPC1 or blocking of TRPC channels by 2-APB attenuates the mGluR5-mediated cell death. The discrepancy of TRPC1 function in neuronal survival may be explained by the extent of its activation and the level of calcium entry, or the source of calcium—the calcium influx through TRPC1 directly increasing the cytosolic calcium level or replenishing the internal calcium store by TRPC1 and STIM1 interaction. Again, the involvement of TRPC channels in the maintenance of intracellular calcium homeostasis might be the key to keep the balance between neuronal survival and death.

Generally, the epileptic seizure is due to synchronized hyper-neuronal activity inside the brain, and misregulation of many cation channels has been associated with the disease. Dysfunction of TRPC channels could contribute to the abnormal epileptiform burst in theory. Indeed, the large depolarizing plateau potential induced by metabotropic glutamate receptor agonists in lateral septal neurons is completely abolished in the TRPC1/TRPC4 double knockout mice [53]. In addition, neuronal cell death after pilocarpine-induced severe seizures in the lateral septum and hippocampus is also ameliorated in the TRPC1/TRPC4 double knockout mice. These results implicate the notion that neuronal excitotoxicity, an underlying pathogenic process for many neurodegenerative diseases, could be mediated not only by conventional ionotropic glutamate receptors but also the TRPC channels coupled with metabotropic glutamate receptors.

The TRPC5 knockout mice exhibit both reduced seizures induced by pilocarpine and minimal seizure-induced neuronal cell death in the hippocampus [54]. However, TRPC5 may contribute to seizure and excitotoxicity by distinct cellular mechanism compared to TRPC1/TRPC4. The major difference is that the TRPC5 knockout mice exhibited normal epileptiform burst firing induced by metabotropic glutamate receptor agonists, but the similar application on TRPC1 knockout and TRPC1/TRPC4 double knockout mice reduced the epileptiform burst. Therefore, the coupling of metabotropic glutamate receptors with TRPC1/TRPC4 may not appear with TRPC5. In addition, the hippocampal long-term potentiation (LTP) is greatly reduced in TRPC5 knockout mice, but normal in TRPC1 knockout and TRPC1/TRPC4 double knockout mice, suggesting the activation of TRPC5, but not TRPC1/TRPC4, coupled with the LTP induction. It is intriguing to further examine the distinct mechanisms underlying the epileptogenesis and excitotoxicity of TRPC1/TRPC4 and TRPC5, though they have high-protein sequence homology and similar gating properties.

Previous study has shown that the cysteine S-nitrosylation, a type of protein modification, in TRP channels may contribute to the channel activation, implying the oxidative stress in regulation of the channel properties [83]. For example, TRPC5 can be activated by nitric oxide (NO) through a mechanism that requires the S-nitrosylation at extracellular cysteines. The intracellular oxidation rather than extracellular oxidative reaction regulates TRPC5 activation depending on the cytosolic redox state, such as intracellular glutathione disulfide (GSSG) level [31]. Indeed, TRPC5 activation by intracellular GSSG can be reversed by glutathione (GSH) and dithiothreitol, and the Cys176 and Cys178 residues in TRPC5 can be S-glutathionylated by GSSG to activate the channel. The increased levels of endogenous TRPC5 S-glutathionylation are observed in not only Huntington's disease mouse model but also in brain tissues from Huntington's disease patients. Consistently, the TRPC5 blocker can improve striatal cell survival and rearing behavior in Huntington's disease mouse model. Thus, over-activation of TRPC5 through the S-glutathionylated by oxidative stress contributes to neurodegenerative diseases, such as Huntington's disease. However, the gating mechanisms of TRPC channels in oxidative stress may need more examination in cultured neurons and inside the brain in the future.

Another study suggests that TRPC3 and TRPC6 are involved into the epileptogenesis in a similar pilocarpine-induced mouse model [36]. In the hippocampus of the seizure mouse model, TRPC3 expression was elevated in CA1 and CA3 pyramidal cells and dentate granule cells, while TRPC6 expression was reduced in these regions. Application of a TRPC3 inhibitor prevented the upregulation of neuronal TRPC3 expression induced by pilocarpine. However, the TRPC6 activator prevented the downregulation of neuronal TRPC6 expression induced by pilocarpine. In addition, both TRPC3 inhibitor and TRPC6 activator effectively protected neuronal damages from pilocarpine-induced seizure. The underlying mechanism of this functional difference is still unknown. Although heteromeric channel formation is well established for TRPC3/TRPC6/TRPC7 [30, 41, 70], this study suggests that, at least inside of the hippocampus, the TRPC3 and TRPC6 may function separately.

Although TRPC channels protected mammalian dopaminergic neurons from neurotoxin and ER stress-induced cell death, evidence from worms suggests the gain-of-function mutations in the TRP worm ortholog result in their dopaminergic neuronal loss probably through the calcium overload [49]. In a screen for *C. elegans* mutant animals that lack their normal complement of dopaminergic neurons, two strains are identified. In the strains, the dopaminergic neuron loss was progressive from postembryonic to adult. Moreover, the progressive neurodegeneration phenotype of the two strains is semidominant, with heterozygous animals showing degeneration of their dopaminergic neurons, albeit to a lesser degree than the homozygous animals. Furthermore, only the head dopaminergic neurons are affected in these mutant strains, whereas no degeneration is observed in the mid-body and the male tail dopaminergic neurons. Whole-genome sequencing showed that both strains contain gain-of-function mutations in the TRP mechanosensory channel *trp-4*. The mutations affect the same amino acid in the pore-forming sixth transmembrane helix of TRP-4. Substitutions in adjacent amino acids in TRP channel orthologs in yeast are also shown to be gain-of-function mutations resulting in gate destabilization and a higher probability of the channel opening [71, 85]. The neuron loss can be restored by loss-of-function of *trp-4* gene. These two mutants also show defective in basal slowing response, which is a previously characterized dopamine-dependent behavioral response [62]. The neurodegeneration seems dependent on intracellular Ca^{2+} dyshomeostasis, as EGTA suppressed dopaminergic degeneration. These findings further argue that the calcium overload through TRPC channels is neurotoxic, and the mutations per se in the worm TRP-4 may help us gain insights into channel gating.

Similarly, a gain-of-function mutation in other TRPC channel, TRPC3, is found in a mouse mutant strain and results in cerebellar neuron loss [5]. The inherited cerebellar ataxias are a complex group of neurodegenerative disorders characterized by loss of balance and coordination, mainly caused by the degeneration or dysfunction of Purkinje cells, which form the sole output of the cerebellum [18, 38, 73]. A dominant ataxic mutant mouse, moonwalker (*Mwk*), is identified to exhibit impaired motor and coordination control. A slow but progressive loss of Purkinje cells is discovered in the mutant cerebellum. Sequence analysis shows a threonine-to-alanine amino acid change (T635A) in *trpc3* gene. The *Mwk* mutation results in altered TRPC3 gating, which promotes channel opening under conditions of low mGluR1 activation. In addition, overexpression of the mutant TRPC3 but not wild-type TRPC3 induces cell death in NSC-34 cell culture. Inside the cerebellum, the mutation only affects the survival of Purkinje cells but not granule cells, probably due to the distinct receptor/channel coupling property between the Purkinje cells and granule cells. The threonine 635 is a putative phosphorylation site of PKCγ; thus, the mutation may result in a loss of PKCγ-mediated phosphorylation of TRPC3 and the phosphorylation-mediated inhibitory regulation of channel function. However, the follow-up study suggests that threonine 635 is a hydrogen bonding site and the binding capability is crucial for maintaining TRPC3 in a stable and closed state [26]. Nevertheless, these findings uncover previously unknown roles for TRPC3 in both dendritic development and survival of Purkinje cells and provide a unique mechanism underlying cerebellar ataxia. More importantly, the point mutation (T635A) in TRPC3 greatly helps us in understanding TRPC channel gating and the design of compounds to regulate the channel activity.

5.2 Perspectives

Unlike other families of cation channels, the TRPC channels are unique in terms of (1) the receptor-coupling PLC activation and (2) the slow and sustained calcium influx, which may not result in calcium overload in most circumstance. The diverse functions of TRPC channels in neuronal survival or death could be explained by (1) the different receptors coupled with the TRPC channels in different brain regions and cells, (2) the different microenvironment post-channels to convey the calcium signaling, (3) and the amount of calcium loaded into the cells. The gating mechanisms of TRPC channels are still one of the key questions for the future studies, including channel structure and heterogeneity in different brain regions and different brain cells. TRPC channel mutations have been identified in the literature, and to study how these channel mutations affect channel properties could be beneficial to gain new insights into the gating mechanisms. It is promising to identify the channel agonist and blocker, which may have future application on disease intervention, especially neurodegenerative diseases and epilepsy.

Acknowledgment Dr. Yichang Jia acknowledges the funding support from "1000-talents Plan" for young researchers, the Chinese Central Government, from Peking-Tsinghua Joint Center for Life Sciences and IDG/McGovern Institute for Brain Research at Tsinghua, from National Science Foundation of China (31571097, 81371361), from ALS Association (16-IIP-284), and from NIH Pathway to Independence Award (K99/R00, NS079476).

References

1. Amaral MD, Pozzo-Miller L (2007) TRPC3 channels are necessary for brain-derived neurotrophic factor to activate a nonselective cationic current and to induce dendritic spine formation. J Neurosci 27(19):5179–5189. doi:10.1523/JNEUROSCI.5499-06.2007
2. Autry AE, Monteggia LM (2012) Brain-derived neurotrophic factor and neuropsychiatric disorders. Pharmacol Rev 64(2):238–258. doi:10.1124/pr.111.005108
3. Banfield MJ, Naylor RL, Robertson AG, Allen SJ, Dawbarn D, Brady RL (2001) Specificity in Trk receptor: neurotrophin interactions: the crystal structure of TrkB-d5 in complex with neurotrophin-4/5. Structure 9(12):1191–1199
4. Barbacid M (1994) The Trk family of neurotrophin receptors. J Neurobiol 25(11):1386–1403. doi:10.1002/neu.480251107
5. Becker EB, Oliver PL, Glitsch MD, Banks GT, Achilli F, Hardy A, Nolan PM, Fisher EM, Davies KE (2009) A point mutation in TRPC3 causes abnormal Purkinje cell development and cerebellar ataxia in moonwalker mice. Proc Natl Acad Sci U S A 106(16):6706–6711. doi:10.1073/pnas.0810599106
6. Cahalan MD (2009) STIMulating store-operated Ca(2+) entry. Nat Cell Biol 11(6):669–677. doi:10.1038/ncb0609-669
7. Carpenter G, Cohen S (1990) Epidermal growth factor. J Biol Chem 265(14):7709–7712
8. Carpenter G, King L Jr, Cohen S (1978) Epidermal growth factor stimulates phosphorylation in membrane preparations in vitro. Nature 276(5686):409–410
9. Carpenter G, Lembach KJ, Morrison MM, Cohen S (1975) Characterization of the binding of 125-I-labeled epidermal growth factor to human fibroblasts. J Biol Chem 250(11):4297–4304
10. Choi DW (1992) Excitotoxic cell death. J Neurobiol 23(9):1261–1276. doi:10.1002/neu.480230915
11. Cohen S (1962) Isolation of a mouse submaxillary gland protein accelerating incisor eruption and eyelid opening in the new-born animal. J Biol Chem 237:1555–1562
12. Cohen S (1965) The stimulation of epidermal proliferation by a specific protein (EGF). Dev Biol 12(3):394–407
13. Cowan WM (2001) Viktor Hamburger and Rita Levi-Montalcini: the path to the discovery of nerve growth factor. Annu Rev Neurosci 24:551–600. doi:10.1146/annurev.neuro.24.1.551
14. Crowley C, Spencer SD, Nishimura MC, Chen KS, Pitts-Meek S, Armanini MP, Ling LH, McMahon SB, Shelton DL, Levinson AD et al (1994) Mice lacking nerve growth factor display perinatal loss of sensory and sympathetic neurons yet develop basal forebrain cholinergic neurons. Cell 76(6):1001–1011
15. Davare MA, Fortin DA, Saneyoshi T, Nygaard S, Kaech S, Banker G, Soderling TR, Wayman GA (2009) Transient receptor potential canonical 5 channels activate Ca2+/calmodulin kinase Igamma to promote axon formation in hippocampal neurons. J Neurosci 29(31):9794–9808. doi:10.1523/JNEUROSCI.1544-09.2009
16. de la Rosa EJ, de Pablo F (2000) Cell death in early neural development: beyond the neurotrophic theory. Trends Neurosci 23(10):454–458
17. Dolmetsch RE, Pajvani U, Fife K, Spotts JM, Greenberg ME (2001) Signaling to the nucleus by an L-type calcium channel-calmodulin complex through the MAP kinase pathway. Science 294(5541):333–339. doi:10.1126/science.1063395
18. Duenas AM, Goold R, Giunti P (2006) Molecular pathogenesis of spinocerebellar ataxias. Brain 129(Pt 6):1357–1370. doi:10.1093/brain/awl081
19. Duman RS, Voleti B (2012) Signaling pathways underlying the pathophysiology and treatment of depression: novel mechanisms for rapid-acting agents. Trends Neurosci 35(1):47–56. doi:10.1016/j.tins.2011.11.004
20. Ernfors P, Lee KF, Jaenisch R (1994) Mice lacking brain-derived neurotrophic factor develop with sensory deficits. Nature 368(6467):147–150. doi:10.1038/368147a0
21. Ernfors P, Lee KF, Kucera J, Jaenisch R (1994) Lack of neurotrophin-3 leads to deficiencies in the peripheral nervous system and loss of limb proprioceptive afferents. Cell 77(4):503–512
22. Fiorio Pla A, Maric D, Brazer SC, Giacobini P, Liu X, Chang YH, Ambudkar IS, Barker JL (2005) Canonical transient receptor potential 1 plays a role in basic fibroblast growth factor (bFGF)/FGF receptor-1-induced Ca2+ entry and embryonic rat neural stem cell proliferation. J Neurosci 25(10):2687–2701. doi:10.1523/JNEUROSCI.0951-04.2005
23. Ghosh A, Greenberg ME (1995) Calcium signaling in neurons: molecular mechanisms and cellular consequences. Science 268(5208):239–247
24. Greka A, Navarro B, Oancea E, Duggan A, Clapham DE (2003) TRPC5 is a regulator of hippocampal neurite length and growth cone morphology. Nat Neurosci 6(8):837–845. doi:10.1038/nn1092
25. Gschwind A, Fischer OM, Ullrich A (2004) The discovery of receptor tyrosine kinases: targets for cancer therapy. Nat Rev Cancer 4(5):361–370. doi:10.1038/nrc1360
26. Hanson SM, Sansom MS, Becker EB (2015) Modeling suggests TRPC3 hydrogen bonding and not phosphorylation contributes to the ataxia phenotype of the moonwalker mouse. Biochemistry 54(26):4033–4041. doi:10.1021/acs.biochem.5b00235
27. Hardingham GE, Arnold FJ, Bading H (2001) Nuclear calcium signaling controls CREB-mediated gene expression triggered by synaptic activity. Nat Neurosci 4(3):261–267. doi:10.1038/85109
28. Hardingham GE, Fukunaga Y, Bading H (2002) Extrasynaptic NMDARs oppose synaptic NMDARs by triggering CREB shut-off and cell death pathways. Nat Neurosci 5(5):405–414. doi:10.1038/nn835

29. Hartmann J, Dragicevic E, Adelsberger H, Henning HA, Sumser M, Abramowitz J, Blum R, Dietrich A, Freichel M, Flockerzi V, Birnbaumer L, Konnerth A (2008) TRPC3 channels are required for synaptic transmission and motor coordination. Neuron 59(3):392–398. doi:10.1016/j.neuron.2008.06.009
30. Hofmann T, Schaefer M, Schultz G, Gudermann T (2002) Subunit composition of mammalian transient receptor potential channels in living cells. Proc Natl Acad Sci U S A 99(11):7461–7466. doi:10.1073/pnas.102596199
31. Hong C, Seo H, Kwak M, Jeon J, Jang J, Jeong EM, Myeong J, Hwang YJ, Ha K, Kang MJ, Lee KP, Yi EC, Kim IG, Jeon JH, Ryu H, So I (2015) Increased TRPC5 glutathionylation contributes to striatal neuron loss in Huntington's disease. Brain 138(Pt 10):3030–3047. doi:10.1093/brain/awv188
32. Huang EJ, Reichardt LF (2001) Neurotrophins: roles in neuronal development and function. Annu Rev Neurosci 24:677–736. doi:10.1146/annurev.neuro.24.1.677
33. Huang EJ, Reichardt LF (2003) Trk receptors: roles in neuronal signal transduction. Annu Rev Biochem 72:609–642. doi:10.1146/annurev.biochem.72.121801.161629
34. Jia Y, Zhou J, Tai Y, Wang Y (2007) TRPC channels promote cerebellar granule neuron survival. Nat Neurosci 10(5):559–567. doi:10.1038/nn1870
35. Jones KR, Farinas I, Backus C, Reichardt LF (1994) Targeted disruption of the BDNF gene perturbs brain and sensory neuron development but not motor neuron development. Cell 76(6):989–999
36. Kim DS, Ryu HJ, Kim JE, Kang TC (2013) The reverse roles of transient receptor potential canonical channel-3 and -6 in neuronal death following pilocarpine-induced status epilepticus. Cell Mol Neurobiol 33(1):99–109. doi:10.1007/s10571-012-9875-6
37. Kim SJ, Kim YS, Yuan JP, Petralia RS, Worley PF, Linden DJ (2003) Activation of the TRPC1 cation channel by metabotropic glutamate receptor mGluR1. Nature 426(6964):285–291. doi:10.1038/nature02162
38. Klockgether T (2000) Recent advances in degenerative ataxias. Curr Opin Neurol 13(4):451–455
39. Li HS, Xu XZ, Montell C (1999) Activation of a TRPC3-dependent cation current through the neurotrophin BDNF. Neuron 24(1):261–273
40. Li Y, Jia YC, Cui K, Li N, Zheng ZY, Wang YZ, Yuan XB (2005) Essential role of TRPC channels in the guidance of nerve growth cones by brain-derived neurotrophic factor. Nature 434(7035):894–898. doi:10.1038/nature03477
41. Liu X, Bandyopadhyay BC, Singh BB, Groschner K, Ambudkar IS (2005) Molecular analysis of a store-operated and 2-acetyl-sn-glycerol-sensitive non-selective cation channel. Heteromeric assembly of TRPC1-TRPC3. J Biol Chem 280(22):21600–21606. doi:10.1074/jbc.C400492200
42. Lockshin RA, William CM (1965) Programmed cell death. 3. Neural control of the breakdown of the intersegmental muscles of silkmoths. J Insect Physiol 11:601–610
43. Lucas P, Ukhanov K, Leinders-Zufall T, Zufall F (2003) A diacylglycerol-gated cation channel in vomeronasal neuron dendrites is impaired in TRPC2 mutant mice: mechanism of pheromone transduction. Neuron 40(3):551–561
44. Minke B (1977) Drosophila mutant with a transducer defect. Biophys Struct Mech 3(1):59–64
45. Montell C (2005) TRP channels in drosophila photoreceptor cells. J Physiol 567(Pt 1):45–51. doi:10.1113/jphysiol.2005.092551
46. Montell C, Jones K, Hafen E, Rubin G (1985) Rescue of the drosophila phototransduction mutation trp by germline transformation. Science 230(4729):1040–1043
47. Moran MM, Xu H, Clapham DE (2004) TRP ion channels in the nervous system. Curr Opin Neurobiol 14(3):362–369. doi:10.1016/j.conb.2004.05.003
48. Nagahara AH, Tuszynski MH (2011) Potential therapeutic uses of BDNF in neurological and psychiatric disorders. Nat Rev Drug Discov 10(3):209–219. doi:10.1038/nrd3366
49. Nagarajan A, Ning Y, Reisner K, Buraei Z, Larsen JP, Hobert O, Doitsidou M (2014) Progressive degeneration of dopaminergic neurons through TRP channel-induced cell death. J Neurosci 34(17):5738–5746. doi:10.1523/JNEUROSCI.4540-13.2014
50. Narayanan KL, Irmady K, Subramaniam S, Unsicker K, von Bohlen und Halbach O (2008) Evidence that TRPC1 is involved in hippocampal glutamate-induced cell death. Neurosci Lett 446 (2–3):117–122. doi:10.1016/j.neulet.2008.09.034
51. Narayanan KL, Subramaniam S, Bengston CP, Irmady K, Unsicker K, von Bohlen und Halbach O (2014) Role of transient receptor potential channel 1 (TRPC1) in glutamate-induced cell death in the hippocampal cell line HT22. J Mol Neurosci 52 (3):425–433. doi:10.1007/s12031-013-0171-9
52. Oppenheim RW (1991) Cell death during development of the nervous system. Annu Rev Neurosci 14:453–501. doi:10.1146/annurev.ne.14.030191.002321
53. Phelan KD, Mock MM, Kretz O, Shwe UT, Kozhemyakin M, Greenfield LJ, Dietrich A, Birnbaumer L, Freichel M, Flockerzi V, Zheng F (2012) Heteromeric canonical transient receptor potential 1 and 4 channels play a critical role in epileptiform burst firing and seizure-induced neurodegeneration. Mol Pharmacol 81(3):384–392. doi:10.1124/mol.111.075341
54. Phelan KD, Shwe UT, Abramowitz J, Wu H, Rhee SW, Howell MD, Gottschall PE, Freichel M, Flockerzi V, Birnbaumer L, Zheng F (2013) Canonical transient receptor channel 5 (TRPC5) and TRPC1/4 contribute to seizure and excitotoxicity by distinct cellular mech-

anisms. Mol Pharmacol 83(2):429–438. doi:10.1124/mol.112.082271
55. Poduslo JF, Curran GL (1996) Permeability at the blood-brain and blood-nerve barriers of the neurotrophic factors: NGF, CNTF, NT-3, BDNF. Brain Res Mol Brain Res 36(2):280–286
56. Raju TN (2000) The Nobel chronicles. 1986: Stanley Cohen (b 1922); Rita Levi-Montalcini (b 1909). Lancet 355(9202):506
57. Riccio A, Li Y, Moon J, Kim KS, Smith KS, Rudolph U, Gapon S, Yao GL, Tsvetkov E, Rodig SJ, Van't Veer A, Meloni EG, Carlezon WA Jr, Bolshakov VY, Clapham DE (2009) Essential role for TRPC5 in amygdala function and fear-related behavior. Cell 137(4):761–772. doi:10.1016/j.cell.2009.03.039
58. Riedel H, Dull TJ, Schlessinger J, Ullrich A (1986) A chimaeric receptor allows insulin to stimulate tyrosine kinase activity of epidermal growth factor receptor. Nature 324(6092):68–70. doi:10.1038/324068a0
59. Robinson DR, Wu YM, Lin SF (2000) The protein tyrosine kinase family of the human genome. Oncogene 19(49):5548–5557. doi:10.1038/sj.onc.1203957
60. Ryder EF, Cepko CL (1994) Migration patterns of clonally related granule cells and their progenitors in the developing chick cerebellum. Neuron 12(5):1011–1028
61. Sawamura S, Hatano M, Takada Y, Hino K, Kawamura T, Tanikawa J, Nakagawa H, Hase H, Nakao A, Hirano M, Rotrattanadumrong R, Kiyonaka S, Mori MX, Nishida M, Hu Y, Inoue R, Nagata R, Mori Y (2016) Screening of transient receptor potential canonical channel activators identifies novel neurotrophic piperazine compounds. Mol Pharmacol 89(3):348–363. doi:10.1124/mol.115.102863
62. Sawin ER, Ranganathan R, Horvitz HR (2000) *C. elegans* locomotory rate is modulated by the environment through a dopaminergic pathway and by experience through a serotonergic pathway. Neuron 26(3):619–631
63. Schwartz PM, Borghesani PR, Levy RL, Pomeroy SL, Segal RA (1997) Abnormal cerebellar development and foliation in BDNF−/− mice reveals a role for neurotrophins in CNS patterning. Neuron 19(2):269–281
64. Selvaraj S, Sun Y, Watt JA, Wang S, Lei S, Birnbaumer L, Singh BB (2012) Neurotoxin-induced ER stress in mouse dopaminergic neurons involves downregulation of TRPC1 and inhibition of AKT/mTOR signaling. J Clin Invest 122(4):1354–1367. doi:10.1172/JCI61332
65. Selvaraj S, Watt JA, Singh BB (2009) TRPC1 inhibits apoptotic cell degeneration induced by dopaminergic neurotoxin MPTP/MPP(+). Cell Calcium 46(3):209–218. doi:10.1016/j.ceca.2009.07.008
66. Serafini T, Kennedy TE, Galko MJ, Mirzayan C, Jessell TM, Tessier-Lavigne M (1994) The netrins define a family of axon outgrowth-promoting proteins homologous to *C. elegans* UNC-6. Cell 78(3):409–424
67. Skaper SD (2012) The neurotrophin family of neurotrophic factors: an overview. Methods Mol Biol 846:1–12. doi:10.1007/978-1-61779-536-7_1
68. Sossin WS, Barker PA (2007) Something old, something new: BDNF-induced neuron survival requires TRPC channel function. Nat Neurosci 10(5):537–538. doi:10.1038/nn0507-537
69. Sotelo C (2002) The chemotactic hypothesis of Cajal: a century behind. Prog Brain Res 136:11–20
70. Strubing C, Krapivinsky G, Krapivinsky L, Clapham DE (2003) Formation of novel TRPC channels by complex subunit interactions in embryonic brain. J Biol Chem 278(40):39014–39019. doi:10.1074/jbc.M306705200
71. Su Z, Zhou X, Haynes WJ, Loukin SH, Anishkin A, Saimi Y, Kung C (2007) Yeast gain-of-function mutations reveal structure-function relationships conserved among different subfamilies of transient receptor potential channels. Proc Natl Acad Sci U S A 104(49):19607–19612. doi:10.1073/pnas.0708584104
72. Tai Y, Feng S, Ge R, Du W, Zhang X, He Z, Wang Y (2008) TRPC6 channels promote dendritic growth via the CaMKIV-CREB pathway. J Cell Sci 121(Pt 14):2301–2307. doi:10.1242/jcs.026906
73. Taroni F, DiDonato S (2004) Pathways to motor incoordination: the inherited ataxias. Nat Rev Neurosci 5(8):641–655. doi:10.1038/nrn1474
74. Ullrich A, Coussens L, Hayflick JS, Dull TJ, Gray A, Tam AW, Lee J, Yarden Y, Libermann TA, Schlessinger J et al (1984) Human epidermal growth factor receptor cDNA sequence and aberrant expression of the amplified gene in A431 epidermoid carcinoma cells. Nature 309(5967):418–425
75. Urfer R, Tsoulfas P, O'Connell L, Shelton DL, Parada LF, Presta LG (1995) An immunoglobulin-like domain determines the specificity of neurotrophin receptors. EMBO J 14(12):2795–2805
76. Ushiro H, Cohen S (1980) Identification of phosphotyrosine as a product of epidermal growth factor-activated protein kinase in A-431 cell membranes. J Biol Chem 255(18):8363–8365
77. Wang GX, Poo MM (2005) Requirement of TRPC channels in netrin-1-induced chemotropic turning of nerve growth cones. Nature 434(7035):898–904. doi:10.1038/nature03478
78. Weick JP, Austin Johnson M, Zhang SC (2009) Developmental regulation of human embryonic stem cell-derived neurons by calcium entry via transient receptor potential channels. Stem Cells 27(12):2906–2916. doi:10.1002/stem.212
79. Wiesmann C, Ultsch MH, Bass SH, de Vos AM (1999) Crystal structure of nerve growth factor in complex with the ligand-binding domain of the TrkA receptor. Nature 401(6749):184–188. doi:10.1038/43705
80. Xie Y, Hong Y, Ma XY, Ren XR, Ackerman S, Mei L, Xiong WC (2006) DCC-dependent phospholipase C signaling in netrin-1-induced neurite elongation.

J Biol Chem 281(5):2605–2611. doi:10.1074/jbc.M512767200
81. Yao H, Peng F, Dhillon N, Callen S, Bokhari S, Stehno-Bittel L, Ahmad SO, Wang JQ, Buch S (2009) Involvement of TRPC channels in CCL2-mediated neuroprotection against tat toxicity. J Neurosci 29(6):1657–1669. doi:10.1523/JNEUROSCI.2781-08.2009
82. Yao H, Peng F, Fan Y, Zhu X, Hu G, Buch SJ (2009) TRPC channel-mediated neuroprotection by PDGF involves Pyk2/ERK/CREB pathway. Cell Death Differ 16(12):1681–1693. doi:10.1038/cdd.2009.108
83. Yoshida T, Inoue R, Morii T, Takahashi N, Yamamoto S, Hara Y, Tominaga M, Shimizu S, Sato Y, Mori Y (2006) Nitric oxide activates TRP channels by cysteine S-nitrosylation. Nat Chem Biol 2(11):596–607. doi:10.1038/nchembio821
84. Zhou J, Du W, Zhou K, Tai Y, Yao H, Jia Y, Ding Y, Wang Y (2008) Critical role of TRPC6 channels in the formation of excitatory synapses. Nat Neurosci 11(7):741–743. doi:10.1038/nn.2127
85. Zhou X, Su Z, Anishkin A, Haynes WJ, Friske EM, Loukin SH, Kung C, Saimi Y (2007) Yeast screens show aromatic residues at the end of the sixth helix anchor transient receptor potential channel gate. Proc Natl Acad Sci U S A 104(39):15555–15559. doi:10.1073/pnas.0704039104

TRPC Channels and Stroke

6

Junbo Huang

Abstract

TRPC channels play important roles in neuronal death/survival in ischemic stroke, vasospasm in hemorrhagic stroke, thrombin-induced astrocyte pathological changes, and also in the initiation of stroke by affecting blood pressure and atherogenesis. TRPCs' unique channel characters and downstream pathways make them possible new targets for stroke therapy. TRPC proteins have different functions in different cell types. Considering TRPCs' extensive distribution in various tissues and cell types, drugs targeting them could induce more complicated effects. More specific agonists/antagonists and antibodies are required for future study of TRPCs as potential targets for stroke therapy.

Keywords

TRPC channels • Stroke • Ischemia • Hemorrhage • Drug target

6.1 Introduction

Stroke, also known as cerebrovascular accident, is the second leading cause to death in the world, ranking behind coronary artery disease and before cancer, according to the WHO data. It includes ischemic stroke, which is caused by blockage of the main brain artery/arteries, and hemorrhagic stroke, which is caused by bleeding in/around the brain. Stroke affects 15 million individuals annually worldwide, bringing out enormous socioeconomic costs [1]. Researchers and clinicians have devoted great efforts to alleviating damage and improving recovery of stroke patients for decades, and any new stroke intervention could make a substantial impact on our health. So far only limited types of medicine are available for stroke patients, such as recombinant tissue-type plasminogen activator (tPA) for ischemic stroke patients and nimodipine for hemorrhage patients. Failure of the third clinical trial of NX095, which was thought to be a very promising new medicine against ischemic stroke, makes investors more cautious about new stroke drug development. On the other hand, decades of

J. Huang (✉)
Toronto Western Hospital Research Institute, Toronto, ON, Canada
e-mail: jhuang@uhnresearch.ca

Y. Wang (ed.), *Transient Receptor Potential Canonical Channels and Brain Diseases*, Advances in Experimental Medicine and Biology 976, DOI 10.1007/978-94-024-1088-4_6

research uncover many important pathological processes of stroke, like excitotoxicity, oxidative stress, astrogliosis, and apoptosis.

TRPC channels can be activated in different cell types by G-protein-coupled receptors (GPCR) and receptor tyrosine kinases (RTK) through a phospholipase C (PLC)-dependent mechanism [2–4]. These channels have extensive distribution and function in various tissues and cell types, including in neurons, astrocytes, artery endothelial and smooth muscle cells. Neurons are sensitive to extracellular environmental changes, and they are capable of transferring signals among each other rapidly and form a powerful network to deal with information flow. The sensitivity, however, makes them vulnerable to harmful stimuli. Acute injuries like brain trauma and stroke could easily cause irreversible brain damage. A small number of dead cells in the injured area at the beginning could release substances and trigger subsequent reactions in adjacent cells. These pathological reactions would cause secondary damages and expand the infarction to a considerable volume.

6.2 TRPCs in Ischemic Stroke

Ischemic stroke is the main type of stroke, comprising of about 85% stroke cases [5]. When one of the main arteries in the brain gets occluded and the responsive blood supply to the brain tissue stops, ischemic stroke occurs. In ischemic stroke, the brain function loss appears even earlier than dead tissue shows up. If the blood flow could be recovered fast enough, within 4.5 h in average, the damage would be limited, and function would be recovered. Recombinant tissue-type plasminogen activator (tPA) that could dissolve the clots is the only drug that FDA approved to apply on patients who suffered from ischemic stroke. Other clinical approaches, such as stent retrievers, are used to remove the clots physically. The purpose of these approaches and tPA is to recover the cerebral blood flow (CBF). Because of the time limit in using tPA, only a small number of patients can benefit from tPA therapy. For example, in the United States, only fewer than 5% ischemic stroke patients can receive tPA [1].

Penumbra is a conception discussed by researchers and clinicians for more than 30 years. In ischemic stroke, occlusion of a main artery (e.g., middle cerebral artery, MCA) would form an infarct core in the brain region whose blood supply is dependent on the occluded artery. The CBF of infarct core is less than 5 ml/100 g/min. The tissue of infarct core irreversibly dies in minutes after occlusion and would enlarge by continuously recruiting the surrounding tissue to it if CBF did not recover. The infarct core represents the terminal events of the ischemic cascade [5]. Penumbra is the rim of tissue around the infarct core, which still maintains some blood flow (10–18 ml/100 g/min) supplied by collateral circulation [6]. Penumbra tissue is believed to be salvageable because it would die in hours to days, not as fast as the tissues in the infarct core.

The pathological processes in ischemic penumbra have been the focus of researches for decades. A cascade of events starts with energy deprivation since the occlusion, including excitotoxicity together with peri-infarct depolarization, inflammation, necrosis, and apoptosis. These events range from hours, like excitotoxicity, to days, such as inflammation [7]. Excitotoxicity refers to the process that neurons are killed by great amount of extracellular neurotransmitters that is released by dead or damaged cells around. Glutamate is the major trigger of excitotoxicity in the brain ischemia by extensively activating glutamate receptors, which would cause calcium overload to induce a series of detrimental chemical events in neurons directly, such as ion transporter disruption and abnormal activation of enzymes. These direct results would further induce cell membrane and skeleton breakdown, mitochondrial damage and free radical generation, low pH, and DNA breakdown [8].

6.2.1 TRPCs in Neurons Under Ischemic Stroke

Although calcium overload is so destructive to neurons, the diverse routes of Ca^{2+} elevation result in different effects. The Ca^{2+} could play a totally different role depending on the source of the Ca^{2+} entry. Even elevating Ca^{2+} in neurons to the same level, Ca^{2+} entry through NMDA receptors is much more detrimental than that induced by high level of KCl through voltage-gated calcium channels [9]. The role of the Ca^{2+} source is the fundament of neuroprotective research in ischemic stroke. The glutamate receptor inhibitors are extensively tested in cell and animal stroke models, but unfortunately this strategy in clinical trials led to severe side effects [10–12], which is due to the important physiological function of glutamate receptors. These results drive the researchers either to find more detailed downstream pathways of glutamate receptors for specific targeting [13, 14] or to find out new pathways that are independent of glutamate receptors, like TRP channels [15].

TRPC6 is reported to be neuroprotective in both in vitro and in vivo models of ischemic stroke. Primary cultured rat neurons are resistant to the damage caused by oxygen-glucose deprivation (OGD, an in vitro model of ischemia) when they are treated with the TRPC agonist or TRPC6 is overexpressed. TRPC6 transgenic mice also show smaller infarct size 24 h after 2-h MCA occlusion, higher neurological score, and lower mortality in long-term observation [16]. The activation of pERK-pCREB pathway is responsible for TRPC6 protective effects [17]. TRPC6 could also attenuate glutamate-induced Ca^{2+} influx by activating calcineurin, a calcium-dependent phosphatase that could suppress NMDA receptor activation by dephosphorylating NMDA receptors [16]. In ischemic models, TRPC6 in the penumbra at the early stage of ischemia is degraded by calpain, a Ser/Thr protease activated by glutamate-induced Ca^{2+} overload [18–23]. The degradation of TRPC6 seems to start at the N-terminal of the protein. A small peptide with the sequence spanning the calpain cleavage site in TRPC6 inhibits the degradation of TRPC6 and reduces infarct size after ischemia attack [18]. Thus, TRPC6 is a potential intervention target for ischemic stroke therapy. It also supports the idea that different sources of Ca^{2+} entry are differently involved in the pathological process. In another study, TRPC4 expression level is elevated in neurons in the striatum and hippocampus after focal ischemic stroke. The upregulation persists from 12 h to 3 days after rat MCAO model [24]. However, the role of TRPC4 in ischemic stroke is still unknown.

6.2.2 TRPC Channels and Atherosclerosis

Atherosclerosis is a major cause of ischemic stroke. Main arteries become narrow in the persons with atherosclerosis and are easy to get blocked by small clots. The plaque on the vessel wall that grows gradually in atherosclerosis could also break up from the wall to rapidly form the thrombus, leading to blockage of the vessel as well as stroke.

TRPC channels are involved in several steps of atherogenesis—generation of atherosclerosis (including endothelium damage, monocyte adhesion and recruitment to the endothelium), macrophage survival, and efferocytosis. When the endothelium damage happens, monocytes in blood would be recruited to the lesion region and differentiated to macrophages that could initiate inflammatory reactions in intima. In healthy arteries, the macrophages would get into apoptosis and cleaned up by phagocytic cells, the inflammatory reaction would be reduced, and the little damage would get fixed. In arteries with atherosclerosis risk, the macrophage would not enter apoptosis process. The survived macrophage would continue the inflammatory reaction, activate, and recruit more immune cells to the lesion region; thus, the lesion would get enlarged, and the plague starts to be built up. TRPC channel proteins participate in these pathological processes of atherogenesis. TRPC6 mediates oxLDL-induced endothelial cell apoptosis. In human aortic endothelial cells, oxLDL stimulation inhibits the expression of miR-26a, which can suppress TRPC6 expression to protect cells from apoptosis. In contrast, overexpressing TRPC6 abolishes the anti-apoptosis effect of miR-26a [25]. Atherosclerotic lesions prefer to form in arteries with certain geometries, such as bifurcations, which could be due to the special sheer stress the areas receive. Some sheer stress patterns were identified as "atheroprotective" and some others as "atheroprone," which matches the areas that are resistant or susceptible to atherosclerosis formation [26]. High expression level of TRPC6 and TRPV1 in endothelial cells under atheroprone sheer stress is found compared with the cells under atheroprotective sheer stress [27]. Considering the function of TRPC6 in oxLDL-induced endothelial cell apoptosis, it is speculative that TRPC6 high expression contributes to the susceptibility of artery wall under certain sheer stress to oxLDL-induced atherosclerosis. TRPC3 and TRPM7 expression is higher in endothelial cells exposed to shear stress than those grown under static conditions [27]. These results suggest a possible role of TRPC3 in regulation of blood pressure.

Overexpression of TRPC3 in endothelial cells promotes the plague formation in a mouse model of atherosclerosis [28]. In human coronary artery endothelial cells (HCAECs), TRPC3 mediates TNFα-induced monocytes adhesion. TRPC3 is required in ATP- or TNFα-induced upregulation of VCAM-1 expression and monocytes adhesion. Removing extracellular Ca^{2+} or applying TRP channel inhibitors suppresses VCAM-1 expression and monocyte adhesion. Knocking down TRPC3 with siRNA blocks the Ca^{2+} influx induced by ATP [29], as well as the VCAM-1 expression increase and monocytes adhesion induced by ATP or TNFα, while knocking down TRPC6 or TRPC7 has no effect. It is assumed that TRPC3 mediates TNFα-induced VCAM-1 expression by modulating NFκB pathway because TRPC3 is required in TNFα-induced phosphorylation of IκBα [30, 31]. On the other hand, TNFα likely enhances the expression of TRPC3 to have a relatively long-term effect on the intracellular divalent cation level. TRPC6 and TRPC7 are not observed to participate in the process because knocking them down does not affect TNFα-induced VCAM-1 expression or monocyte adhesion. In patients with essential hypertension, high level of TRPC3 was detected in monocytes, and knocking down TRPC3 decreases the ability of monocytes to adhere to the endothelium [32]. These results could be at least partially explained by the fact that in hypertensive rats TRPC3 mediates the store-operated calcium influx in monocytes [33]. In atherogenesis, the replacement of lost endothelial cells is compromised by extra calcium influx into endothelial cells [34]. In response to lysophosphatidylcholine (lysoPC), a component of oxLDL, TRPC6 can translocate to plasma membrane and further initiate externalization of TRPC5. The translocation of TRPC6 is calmodulin (CaM)-dependent [35], and TRPC5 that is translocated to the plasma membrane induces continuous Ca^{2+} influx, which is believed to affect endothelial cell migration and shape change [36]. These reports unmask a process that one member of TRPC channel regulates another member of TRPC channel to enhance or consolidate cellular responses. TRPC3 is necessary for efficient macrophage apoptosis following efferocytosis. Less apoptotic cells are observed in $TRPC3^{-/-}$ groups, and these apoptotic cells are harder to be cleaned by phagocytic cells than $TRPC3^{+/+}$ apoptotic cells [37]. It is not clear yet if any endogenous mechanism is responsible for suppression of TRPC3 on macrophage in the progress of atherogenesis.

6.3 TRPCs and Hemorrhagic Stroke

Hemorrhagic stroke is usually caused by an artery burst. There are two types of hemorrhagic stroke, intracerebral hemorrhage (ICH, bleeding inside the brain) and subarachnoid hemorrhage (SAH, bleeding outside the brain). More than one third of the patients would die in the first month after hemorrhagic stroke [38]. The bleeding could cause damage by different mechanisms. The skull space is so limited that even little amount of blood/hematoma could cause huge pressure change. This increased intracranial pressure could cause brain tissue damage directly. It could also harm the brain by ischemia because the circulation system is also pressed. In the situation of SAH, the vessel around the bleeding spot would spasm, which could also induce brain tissue ischemia. The blood-brain barrier (BBB) is broken down when bleeding occurs. The blood components can kill neurons directly by excitotoxicity or indirectly by inducing neuron to re-enter cell cycle [39]. A series of immune reactions are rapidly initiated. Immune cells from both blood and brain tissues participate in the inflammation process to cause brain tissue breakdown [40]. The brain-derived cells activated by hemorrhage include astrocytes, microglia, and master cells.

6.3.1 TRPC Channels and Activation of Astrocytes

Thrombin, the critical coagulation protease in blood, is involved in various pathological conditions when BBB is damaged, including ischemia, hemorrhage, Alzheimer's disease, Parkinson's disease, and multiple sclerosis [41]. Although low doses of thrombin are neuroprotective (thrombin precondition), high doses of thrombin leaking to the brain, as in cases of hemorrhagic stroke, would cause brutal damage. The activation of astrocytes is one of the most crucial events initiated by thrombin. Thrombin injected to the brain directly could strongly promote astrocyte proliferation, morphological changes, and astrogliosis [42, 43].

By adding thrombin with or without thrombin inhibitors, it is observed that thrombin activation, the oscillation of $[Ca^{2+}]_i$, proliferation, and morphological changes of astrocytes are remarkably related [44, 45]. Thrombin directly binds to and activates the protease-activated receptor (PAR) on the cell membrane to start the cascades in astrocytes. PAR is a G-protein-coupled receptor. When it is activated by thrombin, Gq/11-coupled signaling leads to inositol trisphosphate (IP3)-dependent endoplasmic reticulum Ca^{2+} release and extracellular Ca^{2+} entry to promote the plasma Ca^{2+} elevation [45]. Knocking down TRPC3 by siRNA suppresses astrocyte activation. It is found that TRPC3 is the only TRPC channel protein that is upregulated by thrombin stimulation. This upregulation is correlated with increased Ca^{2+} influx while it is blocked by the TRPC3 inhibitor, pyrazole-3. It looks like that TRPC3 amplifies the thrombin-PAR signal by a positive feedback: elevating the cytosolic calcium level and increasing its own expression in astrocytes. In vivo data show that agonist of PAR could increase TRPC3 expression in mouse brain [46]. The pyrazole-3 improves the neurological outcomes in a mouse intracerebral hemorrhage model, and astrogliosis is also mitigated [47].

6.3.2 TRPC Channels and Hypertension

The cause of hemorrhagic stroke includes hypertension, cerebral amyloid angiopathy, rupture of aneurysm, and vascular malformations. Improper use of medicine/drug could also lead to stroke, such as cocaine and anticoagulants. Hypertension is attributed to more than 25% of ischemic stroke and around 60% of hemorrhagic stroke [48]. High blood pressure weakens the arteries and makes them more likely to tear, while aneurysm, vascular malformations, and amyloid angiopathy all create weak spots on arteries/arterioles. Thus, any kind of combination of these factors with hypertension could be very dangerous because the risk of hemorrhage is largely increased.

TRPC3 and TRPC6 are involved in the pathological process of hypertension. There is a considerable amount of reports about the function of TRPC3 in hypertension. Endogenous TRPC3 is expressed in both endothelial cells and vessel smooth muscle cells (myocytes), while the patterns of the expression are discrepant depending on the type and location of the vessel. It has been reported that in carotid artery TRPC3 is found only in myocyte [49]. In contrast, in mesenteric arteries, TRPC3 is only detected in endothelial cells [50, 51]. Endothelium and smooth muscle are physically and functionally connected to react to the change of the blood flow and adjust the tone of vessel. The expression discrepancy actually provides the chance to generate variant models to uncover the roles of TRPC3 in different cells and in the interaction of endothelium and smooth muscle. Generally, TRPC3 expressed in vascular smooth muscle cells contributes to the contraction of the muscle, while TRPC3 expressed in the endothelial cells contributes to vasodilation/relaxation. Endothelium regulates the tone of smooth muscle by secreting endothelium-derived relaxing factors. When the myocytes contract too much or when its relaxation could not be properly regulated by endothelium, hypertension manifests itself.

Under physiological condition, TRPC3 on myocytes contributes to vessel smooth muscle contraction by activating IP3 receptors on sarcoplasmic reticulum (SR, ER of myocytes). IP3R-induced calcium release from SR to cytoplasm leads to the movement of muscle filament—muscle contraction. In smooth muscle cells, TRPC3 on plasma membrane and IP3 receptors on SR are physically connected by cell scaffolding protein caveolin-1(caveolin-1 connects TRPC3 and IP3R). This connection facilitates TRPC3 coupling to IP3R in smooth muscle cells of resistance-sized cerebral arteries to induce vasocontraction. TRPC3 on myocytes could be activated by endothelin-1 (ET-1) that is released by endothelium. The channel activity could be modified by the level of IP3 in cytoplasm [52].

In endothelial cells, TRPC3 is involved in flow- and bradykinin-induced vasodilation, possibly because of the calcium influx through the channel. Application of TRPC3 antisense oligos attenuates the $[Ca^{2+}]i$ rise in endothelial cells that is stimulated by flow or bradykinin [51]. The calcium elevation in endothelial cells induces the relaxation of myocytes by hyperpolarizing myocytes. This process is termed as endothelium-derived hyperpolarization (EDH). The calcium elevation in endothelial cells opens the intermediate-conductance calcium-activated potassium channels (IK_{Ca}) that locate at specialized myoendothelial microdomain contact sites and cause an efflux of potassium from endothelial cells, which induces the hyperpolarization of adjacent myocytes. The calcium through endothelial TRPC3 is involved in this EDH mechanism [50]. The endothelium could also relax smooth muscle by generating nitric oxide (NO). In rat carotid artery, NO could cause vasorelaxation by inhibiting TRPC3 on myocytes through activating protein kinase G [49].

TRPC3 protein level is increased in animal models and patients with essential hypertension [53–55]. According to the histology results, this upregulation occurs in myocytes [56]. There is no direct evidence about whether the elevation of TRPC expression initiates hypertension or if it is just a pathological result of the disease. In TRPC6 knockout mice ($TRPC6^{-/-}$ mice), expression of TRPC3 is remarkably increased in both aorta and cerebral arteries. $TRPC6^{-/-}$ mice have elevated blood pressure. Compared with aorta rings isolated from wild type of mice, aorta rings of $TRPC6^{-/-}$ mice show increased contraction in response to adrenoceptor agonist stimulation. The vessel myocyte cultures of the knockout mice have higher basal cation entry, increased TRPC-carried cation currents, and more depolarized membrane potentials. The higher basal cation entry could be fully blocked by siRNA against TRPC3 [57]. These results imply the important role of TRPC3 in hypertension. In spontaneously hypertensive rats (SHR), a commonly used animal model of hypertension, TRPC3 protein level is much higher in vascular myocytes and aorta than it is in normotensive Wistar–Kyoto rats. Knocking down TRPC3 with siRNA in SHR vascular myocytes could block angiotensin II-induced calcium increase, while overexpressing TRPC3 enhances it. It is noticeable that angiotensin II stimulation could induce increase of TRPC3 expression in SHR myocytes, which could be blocked by an antagonist of AT1R (angiotensin II receptor) [55]. Angiotensin is a major target for drugs to lower blood pressure, and it is not the only hormone that has interaction with TRPC channels in vascular smooth muscle. TRPC1, TRPC3, and TRPC5 are highly expressed in mesenteric arterioles of SHR and necessary for norepinephrine-induced vasomotion and calcium influx to vascular myocytes [58]. These results suggest that TRPC3 could be a target for hypertension therapy. A specific TRPC3 inhibitor, Pyr3, could block endothelin-1-induced artery constriction more effectively in SHR than in normotensive Wistar–Kyoto rats [56]. Other TRPCs are also associated with hypertension, such as TRPC1 and TRPC6, whose expression is increased in ouabain-induced hypertension model [59] and hypertensive rats [60]. Nevertheless, the elevation of blood pressure in $TRPC6^{-/-}$ mice, which could be mainly caused by increase of TRPC3, suggests that the role of TRPC6 in vessel cells and the way it was modified could be very different from those of TRPC3.

6.3.3 TRPC Channels and Vasospasm After Hemorrhagic Stroke

Subarachnoid hemorrhage (SAH) comprises about 5% of all stroke cases but is associated with very high mortality and morbidity. In the report in 2009, 50% of the SAH patients couldn't survive in the first 30 days after bleeding occurred [61]. In recent reports, the fatality of SAH is still as high as 35% in the first month [62]. Except for the direct damage caused by increased intracranial pressure and BBB breakdown, delayed cerebral ischemia (DCI) could appear in hours or days caused by vasospasm and microemboli followed by hemorrhage, which induces second damage. Around two-thirds of aneurysmal SAH patients would develop vasospasm in 3–14 days after the initial rupture [63]. SAH should receive urgent hemodynamic therapy (blood vessel fixation) and intensive care monitoring. Vasospasm doesn't occur immediately with hemorrhage, which makes it relatively the most treatable component of SAH.

Vasospasm occurs when the balance of constriction and relaxation is broken in vessel wall so that arteries/arterioles are over-constriction. The mechanisms of vasospasm after SAH are complicated and not fully uncovered. Hemoglobin released from subarachnoid blood clot is believed to be the main initiator of vasospasm. It could initiate vasospasm by different mechanisms including nitric oxide (NO) and endothelin level changes, oxidative stress and free radical on smooth muscle cells, and modification of ion channels [64]. NO is a powerful endogenous vasodilator. Low levels of NO after SAH undoubtedly contribute to vasospasm by breaking the balance of vessel constriction vs. relaxation. A mutation on gene coding endothelial NO synthase is found to raise the risk of vasospasm [65]. Endothelin-1 (ET-1) generation increases after SAH in endothelial cells, astrocytes and leucocytes, which causes the elevation of ET-1 in cerebrospinal fluid and plasma. ETs can bind to ET receptors on smooth muscle cells and elevate the $[Ca^{2+}]_i$ to induce myocyte over-constriction and vasospasm [66, 67]. Astrocytes, neurons, and arterioles work as a functional unit to adjust blood supply to activate neurons in normal brain. In SAH, the calcium oscillation patterns in astrocyte endfoot alter and change the $[K^+]$ in the restricted perivascular space by activating calcium-sensitive potassium channel. This $[K^+]$ change could cause abnormal vasoconstriction in SAH [68].

To summarize, blocking the elevation of $[Ca^{2+}]_i$ in smooth muscle cells is an important route to prevent vasospasm after SAH. Nimodipine, an L-type calcium channel blocker, is the only FDA-approved medicine against vasospasm [69]. Except for alleviating vasospasm, nimodipine could also function in SAH brain by increasing fibrinolytic activity, neuroprotection, and inhibition of cortical spreading. Other drugs such as ET receptor antagonists, Rho-kinase inhibitor, and ryanodine receptors are also in clinical trial or in preclinical research [70]. Clazosentan, an ET_A receptor antagonist, is more effective than nimodipine to alleviate vasospasm, while its third-phase clinical trial did not show difference of functional outcome or mortality between drug-treated groups and placebo groups [71]. Side effects of clazosentan include hypotension and pulmonary complications, which could have neutralized the beneficial effects of the drug. The major side effect of current agents against vasospasm, including nimodipine, is hypertension, which could be lethal to patients with SAH by inducing second bleeding [70]. Thus, researching into the details of the $[Ca^{2+}]_i$ elevation in myocytes in vasospasm would be necessary for developing new drugs with higher efficiency and milder side effects.

TRPC channels are involved in ET-1-induced vasospasm in SAH models. In rabbit basilar artery, ET-1 could induce $[Ca^{2+}]_i$ elevation through voltage-dependent calcium channels (VDCC) and nonselective cation channels (NSCC). ET-1-induced $[Ca^{2+}]_i$ elevation in artery myocytes and vasocontraction are sensitive to SKF96365, an inhibitor of the receptor-mediated Ca^{2+} channels (later used as a TRP channel antagonist), and U73122, a PLC inhibitor, but not sensitive to nifedipine [72, 73]. Thus, ET-evoked contractions of the smooth muscle are considered to be mediated by Ca^{2+} influx mostly through nonselective cation channels. The seventh day after SAH in dog, the artery myocytes are more sensitive to ET-1. ET-1 level elevates to twofolds in cerebrospinal fluid compared with before SAH, around 200 pg/ml (0.08 nmol/L) [74]. When even higher dose of ET-1 (10 nmol/L) is applied, in 93% of control animals, basilar artery myocytes would not react, while large inward and outward currents are induced in 64% of SAH myocytes. These currents could be blocked by removing extracellular Ca^{2+} and by antibodies against TRPC1 or TRPC4. The mRNA and protein levels of TRPC1, TRPC4, and STIM1 are increased in SAH models [75, 76]. In a rat SAH model, TRPC1 and STIM1 expression increase is correlated with morphological changes of the arteries spatially and temporally—decrease of lumen area and increase of artery wall thickness [76]. It is not clear when the TRPC expression increase starts after SAH or if it is a reason or result or just paralleled phenomenon of morphological changes of the vessels. Knocking down techniques and more specific TRPC inhibitors could help to figure out function of upregulated TRPCs in vessel change after SAH and also to rule out the involvement of other TRPCs, such as TRPC3, whose expression does not change after SAH [76]. Considering the function of TRPC3 in smooth muscle cells of hypertension individual, more detailed research of TRPC3 in SAH is needed. No matter what the relationship of TRPCs and vessel morphological change is, upregulated TRPC1 and TRPC4 make myocytes more sensitive to ET-1 stimulation and easier to spasm, which makes them the possible targets for therapy of SAH-induced vasospasm.

6.4 Perspectives

Since TRP channel was first time cloned in *Drosophila* decades ago, they have been found to involve in different physiological and pathological processes. Some of them have become promising drug targets of cancer (TRPV6) or pain (TRPV1). For stroke therapy, translational medicine obstacles and big variability of patient situation make high cost of new drug development. TRPC channels are found to be diversely expressed in various cell types of nerve and circulation system and affect stroke progress in unique ways, which make them possible therapy targets. More selective TRPC channel inhibitors and agonists could give exciting opportunity to new drug development. Considering the similarity of members in TRPC family, resolving the crystal structure of TRPC channels could help pharmacological blocker/activator selection. Genetic tools like conditional knockout/knock-in approaches in specific tissues/cells could help to understand more completely about functions of TRPC channels. It is still unclear how exactly channels that are composed of the same/different TRPC proteins modify the component and pattern of ion flow under physiological and pathological states and how TRPC channels interact with cellular parts like cytoskeleton, mitochondrial, and lipid rafts. More comprehensive understanding of the complexity of TRPC channels will provide insight of mechanisms on stroke-induced damage and new insight on its therapy.

References

1. Cook DJ, Tymianski M (2011) Translating promising preclinical neuroprotective therapies to human stroke trials. Expert Rev Cardiovasc Ther 9:433–449
2. Liman ER, Corey DP, Dulac C (1999) TRP2: a candidate transduction channel for mammalian pheromone sensory signaling. Proc Natl Acad Sci U S A A96:5791–5796
3. Nilius B, Owsianik G, Voets T, Peters JA (2007) Transient receptor potential cation channels in disease. Physiol Rev 87:165–217
4. Wu LJ, Sweet TB, Clapham DE (2010) International Union of basic and clinical pharmacology. LXXVI. Current progress in the mammalian TRP ion channel family. Pharmacol Rev 62:381–404
5. Donnan GA, Fisher M, Macleod M, Davis SM (2008) Stroke. Lancet 371:1612–1623
6. Tasker RC, Duncan ED (2015) Focal cerebral ischemia and neurovascular protection: a bench-to-bedside update. Curr Opin Pediatr 27:694–699
7. Fisher M (2004) The ischemic penumbra: identification, evolution and treatment concepts. Cerebrovasc Dis 17(Suppl 1):1–6

8. Szydlowska K, Tymianski M (2010) Calcium, ischemia and excitotoxicity. Cell Calcium 47:122–129
9. Tymianski M, Charlton MP, Carlen PL, Tator CH (1993) Source specificity of early calcium neurotoxicity in cultured embryonic spinal neurons. J Neurosci 13:2085–2104
10. Ginsberg MD (2008) Neuroprotection for ischemic stroke: past, present and future. Neuropharmacology 55:363–389
11. Hoyte L, Barber PA, Buchan AM, Hill MD (2004) The rise and fall of NMDA antagonists for ischemic stroke. Curr Mol Med 4:131–136
12. Prass K, Dirnagl U (1998) Glutamate antagonists in therapy of stroke. Restor Neurol Neurosci 13:3–10
13. Aarts M et al (2002) Treatment of ischemic brain damage by perturbing NMDA receptor- PSD-95 protein interactions. Science 298:846–850
14. Sun HS et al (2009) Suppression of hippocampal TRPM7 protein prevents delayed neuronal death in brain ischemia. Nat Neurosci 12:1300–1307
15. Aarts M et al (2003) A key role for TRPM7 channels in anoxic neuronal death. Cell 115:863–877
16. Li H et al (2012) TRPC6 inhibited NMDA receptor activities and protected neurons from ischemic excitotoxicity. J Neurochem 123:1010–1018
17. Jia Y, Zhou J, Tai Y, Wang Y (2007) TRPC channels promote cerebellar granule neuron survival. Nat Neurosci 10:559–567
18. Du W et al (2010) Inhibition of TRPC6 degradation suppresses ischemic brain damage in rats. J Clin Invest 120:3480–3492
19. Lin Y et al (2013) Neuroprotective effect of resveratrol on ischemia/reperfusion injury in rats through TRPC6/CREB pathways. J Mol Neurosci 50:504–513
20. Lin Y et al (2013) Hyperforin attenuates brain damage induced by transient middle cerebral artery occlusion (MCAO) in rats via inhibition of TRPC6 channels degradation. J Cereb Blood Flow Metab 33:253–262
21. Yao C, Zhang J, Chen F, Lin Y (2013) Neuroprotectin D1 attenuates brain damage induced by transient middle cerebral artery occlusion in rats through TRPC6/CREB pathways. Mol Med Rep 8:543–550
22. Yao C, Zhang J, Liu G, Chen F, Lin Y (2014) Neuroprotection by (-)-epigallocatechin-3-gallate in a rat model of stroke is mediated through inhibition of endoplasmic reticulum stress. Mol Med Rep 9:69–76
23. Zhang J, Mao X, Zhou T, Cheng X, Lin Y (2014) IL-17A contributes to brain ischemia reperfusion injury through calpain-TRPC6 pathway in mice. Neuroscience 274:419–428
24. Gao YQ, Gao H, Zhou ZY, Lu SD, Sun FY (2004) Expression of transient receptor potential channel 4 in striatum and hippocampus of rats is increased after focal cerebral ischemia. Sheng Li Xue Bao 56:153–157
25. Zhang Y et al (2015) MicroRNA-26a prevents endothelial cell apoptosis by directly targeting TRPC6 in the setting of atherosclerosis. Sci Rep 5:9401
26. Dai G et al (2004) Distinct endothelial phenotypes evoked by arterial waveforms derived from atherosclerosis-susceptible and -resistant regions of human vasculature. Proc Natl Acad Sci U S A A101:14871–14876
27. Thilo F et al (2012) Pulsatile atheroprone shear stress affects the expression of transient receptor potential channels in human endothelial cells. Hypertension 59:1232–1240
28. Smedlund KB, Birnbaumer L, Vazquez G (2015) Increased size and cellularity of advanced atherosclerotic lesions in mice with endothelial overexpression of the human TRPC3 channel. Proc Natl Acad Sci U S A A112:E2201–E2206
29. Smedlund K, Vazquez G (2008) Involvement of native TRPC3 proteins in ATP-dependent expression of VCAM-1 and monocyte adherence in coronary artery endothelial cells. Arterioscler Thromb Vasc Biol 28:2049–2055
30. Kim I et al (2001) Vascular endothelial growth factor expression of intercellular adhesion molecule 1 (ICAM-1), vascular cell adhesion molecule 1 (VCAM-1), and E-selectin through nuclear factor-kappa B activation in endothelial cells. J Biol Chem 276:7614–7620
31. Smedlund K, Tano JY, Vazquez G (2010) The constitutive function of native TRPC3 channels modulates vascular cell adhesion molecule-1 expression in coronary endothelial cells through nuclear factor kappaB signaling. Circ Res 106:1479–1488
32. Zhao Z et al (2012) Increased migration of monocytes in essential hypertension is associated with increased transient receptor potential channel canonical type 3 channels. PLoS One 7:e32628
33. Liu DY et al (2007) Monocytes from spontaneously hypertensive rats show increased store-operated and second messenger-operated calcium influx mediated by transient receptor potential canonical Type 3 channels. Am J Hypertens 20:1111–1118
34. Chaudhuri P, Colles SM, Damron DS, Graham LM (2003) Lysophosphatidylcholine inhibits endothelial cell migration by increasing intracellular calcium and activating calpain. Arterioscler Thromb Vasc Biol 23:218–223
35. Chaudhuri P et al (2016) Membrane translocation of TRPC6 channels and endothelial migration are regulated by calmodulin and PI3 kinase activation. Proc Natl Acad Sci U S A A113:2110–2115
36. Chaudhuri P et al (2008) Elucidation of a TRPC6-TRPC5 channel cascade that restricts endothelial cell movement. Mol Biol Cell 19:3203–3211
37. Tano JY et al (2011) Impairment of survival signaling and efferocytosis in TRPC3-deficient macrophages. Biochem Biophys Res Commun 410:643–647
38. Gonzalez-Perez A, Gaist D, Wallander MA, McFeat G, Garcia-Rodriguez LA (2013) Mortality after hemorrhagic stroke: data from general practice (The Health Improvement Network). Neurology 81:559–565

39. Thiex R et al (2007) Addition of intravenous N-methyl-D-aspartate receptor antagonists to local fibrinolytic therapy for the optimal treatment of experimental intracerebral hemorrhages. J Neurosurg 106:314–320
40. Wang J, Tsirka SE (2005) Contribution of extracellular proteolysis and microglia to intracerebral hemorrhage. Neurocrit Care 3:77–85
41. Krenzlin H, Lorenz V, Danckwardt S, Kempski O, Alessandri B (2016) The importance of thrombin in cerebral injury and disease. Int J Mol Sci 17
42. Nicole O et al (2005) Activation of protease-activated receptor-1 triggers astrogliosis after brain injury. J Neurosci 25:4319–4329
43. Nishino A et al (1993) Thrombin may contribute to the pathophysiology of central nervous system injury. J Neurotrauma 10:167–179
44. Cavanaugh KP, Gurwitz D, Cunningham DD, Bradshaw RA (1990) Reciprocal modulation of astrocyte stellation by thrombin and protease nexin-1. J Neurochem 54:1735–1743
45. Ubl JJ, Reiser G (1997) Characteristics of thrombin-induced calcium signals in rat astrocytes. Glia 21:361–369
46. Shirakawa H et al (2010) Transient receptor potential canonical 3 (TRPC3) mediates thrombin-induced astrocyte activation and upregulates its own expression in cortical astrocytes. J Neurosci 30:13116–13129
47. Munakata M et al (2013) Transient receptor potential canonical 3 inhibitor Pyr3 improves outcomes and attenuates astrogliosis after intracerebral hemorrhage in mice. Stroke 44:1981–1987
48. Klungel OH et al (2000) Control of blood pressure and risk of stroke among pharmacologically treated hypertensive patients. Stroke 31:420–424
49. Chen J, Crossland RF, Noorani MM, Marrelli SP (2009) Inhibition of TRPC1/TRPC3 by PKG contributes to NO-mediated vasorelaxation. Am J Physiol Heart Circ Physiol 297:H417–H424
50. Senadheera S et al (2012) Transient receptor potential canonical type 3 channels facilitate endothelium-derived hyperpolarization-mediated resistance artery vasodilator activity. Cardiovasc Res 95:439–447
51. Liu CL, Huang Y, Ngai CY, Leung YK, Yao XQ (2006) TRPC3 is involved in flow- and bradykinin-induced vasodilation in rat small mesenteric arteries. Acta Pharmacol Sin 27:981–990
52. Adebiyi A et al (2010) Isoform-selective physical coupling of TRPC3 channels to IP3 receptors in smooth muscle cells regulates arterial contractility. Circ Res 106:1603–1612
53. Liu DY et al (2007) Increased store-operated and 1-oleoyl-2-acetyl-sn-glycerol-induced calcium influx in monocytes is mediated by transient receptor potential canonical channels in human essential hypertension. J Hypertens 25:799–808
54. Liu D et al (2005) Increased transient receptor potential channel TRPC3 expression in spontaneously hypertensive rats. Am J Hypertens 18:1503–1507
55. Liu D et al (2009) Increased transient receptor potential canonical type 3 channels in vasculature from hypertensive rats. Hypertension 53:70–76
56. Adebiyi A et al (2012) An elevation in physical coupling of type 1 inositol 1,4,5-trisphosphate (IP3) receptors to transient receptor potential 3 (TRPC3) channels constricts mesenteric arteries in genetic hypertension. Hypertension 60:1213–1219
57. Dietrich A et al (2005) Increased vascular smooth muscle contractility in TRPC6-/- mice. Mol Cell Biol 25:6980–6989
58. Chen X et al (2010) Increased rhythmicity in hypertensive arterial smooth muscle is linked to transient receptor potential canonical channels. J Cell Mol Med 14:2483–2494
59. Pulina MV et al (2010) Upregulation of Na+ and Ca2+ transporters in arterial smooth muscle from ouabain-induced hypertensive rats. Am J Physiol Heart Circ Physiol 298:H263–H274
60. Linde CI et al (2012) Increased arterial smooth muscle Ca2+ signaling, vasoconstriction, and myogenic reactivity in Milan hypertensive rats. Am J Physiol Heart Circ Physiol 302:H611–H620
61. Feigin VL, Lawes CM, Bennett DA, Barker-Collo SL, Parag V (2009) Worldwide stroke incidence and early case fatality reported in 56 population-based studies: a systematic review. Lancet Neurol 8:355–369
62. Smith M, Citerio G (2015) What's new in subarachnoid hemorrhage. Intensive Care Med 41:123–126
63. Dorsch N (2011) A clinical review of cerebral vasospasm and delayed ischaemia following aneurysm rupture. Acta Neurochir Suppl 110:5–6
64. Pluta RM et al (2009) Cerebral vasospasm following subarachnoid hemorrhage: time for a new world of thought. Neurol Res 31:151–158
65. Khurana VG et al (2004) Endothelial nitric oxide synthase gene polymorphisms predict susceptibility to aneurysmal subarachnoid hemorrhage and cerebral vasospasm. J Cereb Blood Flow Metab 24:291–297
66. Zimmermann M, Seifert V (1998) Endothelin and subarachnoid hemorrhage: an overview. Neurosurgery 43:863–875
67. Johansson SE, Andersen XE, Hansen RH, Povlsen GK, Edvinsson L (2015) Cerebrovascular endothelin-1 hyper-reactivity is associated with transient receptor potential canonical channels 1 and 6 activation and delayed cerebral hypoperfusion after forebrain ischaemia in rats. Acta Physiol (Oxf) 214:376–389
68. Koide M, Bonev AD, Nelson MT, Wellman GC (2012) Inversion of neurovascular coupling by subarachnoid blood depends on large-conductance Ca2+-activated K+ (BK) channels. Proc Natl Acad Sci U S A A109:E1387–E1395

69. Keyrouz SG, Diringer MN (2007) Clinical review: prevention and therapy of vasospasm in subarachnoid hemorrhage. Crit Care 11:220
70. Macdonald RL (2014) Delayed neurological deterioration after subarachnoid haemorrhage. Nat Rev Neurol 10:44–58
71. Macdonald RL et al (2011) Clazosentan, an endothelin receptor antagonist, in patients with aneurysmal subarachnoid haemorrhage undergoing surgical clipping: a randomised, double-blind, placebo-controlled phase 3 trial (CONSCIOUS-2). Lancet Neurol 10:618–625
72. Kawanabe Y, Hashimoto N, Masaki T (2002) Characterization of Ca2+ channels involved in endothelin-1-induced contraction of rabbit basilar artery. J Cardiovasc Pharmacol 40:438–447
73. Kawanabe Y, Masaki T, Hashimoto N (2006) Involvement of phospholipase C in endothelin 1-induced stimulation of Ca++ channels and basilar artery contraction in rabbits. J Neurosurg 105:288–293
74. Macdonald RL et al (1998) Prevention of vasospasm after subarachnoid hemorrhage in dogs by continuous intravenous infusion of PD156707. Neurol Med Chir (Tokyo) 38(Suppl):138–145
75. Xie A et al (2007) Novel mechanism of endothelin-1-induced vasospasm after subarachnoid hemorrhage. J Cereb Blood Flow Metab 27:1692–1701
76. Song JN et al (2013) Potential contribution of SOCC to cerebral vasospasm after experimental subarachnoid hemorrhage in rats. Brain Res 1517:93–103

7 TRPC Channels and Alzheimer's Disease

Rui Lu, Qian He, and Junfeng Wang

Abstract

Alzheimer's disease (AD) is the most common neurodegenerative disease in the world. The "amyloid hypothesis" is one of the predominant hypotheses for the pathogenesis of AD. Besides, tau protein accumulation, calcium homeostasis disruption, and glial cell activation are also remarkable features in AD. Recently, there are some reports showing that TRPC channels may function in AD development, especially TRPC6. In this chapter, we will discuss the evidence for the involvement of TRPC channels in Alzheimer's disease and the potential of therapeutics for AD based on TRPC channels.

Keywords

Alzheimer's disease • TRPC6 • β-amyloid • Calcium

7.1 Introduction

Alzheimer's disease (AD) is one of the most common neurodegenerative diseases leading to dementia in the aged, affecting 48 million people worldwide in 2015. The prevalence is estimated that 1 in 85 persons would be living with AD by 2050, according to a report by Johns Hopkins University [12]. Clinically, AD is characterized with memory decline, cognitive impairment, emotion swings, language breakdown, and bodily function lost at the final stage [76]. Synaptic dysfunction is suggested to be responsible for the clinical manifestations, and synapse loss is found to be best correlated with the dementia degree [88]. Pathologically, AD is characterized with metabolism decline in the parietal lobe of the cerebral cortex, brain atrophy, senile plaques, and neurofibrillary tangles in autopsied AD brain sections [34]. The presence of senile plaques and neurofibrillary tangles is required for the definitive diagnosis of Alzheimer's disease [23]. Senile plaques are mainly composed of β-amyloid (Aβ)

R. Lu • Q. He
Institute of Neuroscience, Shanghai Institutes for Biological Sciences, Chinese Academy of Sciences, 320 Yueyang Road, Shanghai 200031, China

J. Wang (✉)
Research and Development of Biogen, 225 Binney Street, Cambridge, MA 02142, USA
e-mail: wangjunfeng2023@aliyun.com

Y. Wang (ed.), *Transient Receptor Potential Canonical Channels and Brain Diseases*, Advances in Experimental Medicine and Biology 976, DOI 10.1007/978-94-024-1088-4_7

peptides, and neurofibrillary tangles are mainly composed of hyperphosphorylated tau protein. Due to the toxicity of Aβ peptides or tau protein to AD cell and animal models, Aβ or tau is proposed to be the causative agents for the pathogenesis of the disease, giving rise to "amyloid cascade hypothesis" and "tau hypothesis" [61].

The amyloid hypothesis suggests that due to genetic or environmental factors, the balance between Aβ production and clearance is disrupted, and Aβ is overaccumulated. Aβ would then aggregate into oligomers, which could attack synapses and neurons, leading to the injury of neurites and malfunction of synapses. Aβ would also interrupt the ion homeostasis of neurons, which would generate oxidative stress. Moreover, Aβ would disrupt the balance of kinase and phosphatase, leading to the hyperphosphorylation of tau protein and the formation of neurofibrillary tangles. Finally, Aβ would induce synapse loss and neuronal death, neural circuits' destruction, and dementia [31]. The amyloid hypothesis is accepted in the field and supported by multiple lines of evidence. First, it has been found that in early-onset Alzheimer's patients, who are usually familial inherited, several hundreds of mutations are located within amyloid precursor protein (APP) [5] or presenilin genes [18]. APP encodes the substrates of β-amyloid, and presenilins encode the catalytic subunit of the key enzyme in APP processing. Most of the mutations results in more Aβ production or enhancement of the ratio of Aβ42/Aβ40, leading to more Aβ aggregation [92]. Second, almost half of the late-onset Alzheimer's patients, who are usually sporadic, carry the E4 allele of the ApoE gene [24]. ApoE encodes a protein related with Aβ production and clearance, and ApoE4 has a compromised function in Aβ production and clearance [39]. Third, Aβ is found to be toxic when applying to cultured cells [95] or animal models [35], and inhibition of Aβ generation or enhancement of its clearance is reported to improve the AD-like pathology [46, 65]. Most recently and importantly, treatment of mild to moderate Alzheimer's patients with Aβ antibody led to a delay of the disease progression [20, 79].

Aβ is a peptide composed of 39–43 amino acids, with Aβ40 and Aβ42 as the two major forms. Aβ is generated from sequential cleavage of APP by enzymes named secretase. There are two types of APP cleavage, the "amyloidosis" pathway and "nonamyloidosis" pathway [26]. In the "amyloidosis" pathway, APP is first cleaved by β-secretase (β-site of APP-cleaving enzyme, BACE1), shedding the N terminal of APP (sAPPβ) and leaving the C terminal fragment (β-CTF, C99) on the membrane, where the β-CTF is intramembrane cleaved by γ-secretase (mainly composed of presenilin, Pen2, Aph1, and nicastrin), releasing the 39–43aa Aβ peptides and the APP intracellular domain (AICD). In the "nonamyloidosis" pathway, APP is first cleaved by α-secretases (carried out by ADAM10, 17, and 9) after the 16th aa of Aβ peptide, shedding sAPPα and leaving α-CTF (C83), and then α-CTF is cleaved by γ-secretase, leaving p3 and AICD. As α cleavage precludes Aβ formation and sAPPα is neurotrophic, the nonamyloidosis pathway is proposed to be protective for the disease [26]. The Aβ monomer is mainly secreted into extracellular space and then aggregates to oligomers and fibrils finally under certain conditions [75]. The Aβ fibril is the main component of senile plaques, and the oligomer is the most toxic form of Aβ [17]. Applying Aβ oligomers to primary cultured neurons [44] and mouse models [47, 77] is able to induce AD-like pathologies.

The physiological function of Aβ peptides is basically unclear. A low level of Aβ could promote neuronal survival, enhance synaptic plasticity [69], and stimulate neurotransmitter release in hippocampal neurons [1]. The mechanism underlying the neurotoxicity of Aβ has been widely investigated [81]. It could activate caspase through ER stress [63], death receptors [37], or JNK pathway [89], to induce neuronal apoptosis. Aβ could also bind to α7 nicotinic acetylcholine receptors (α7 nAChRs) [53], regulate synaptic NMDA receptor trafficking [82], induce synaptic protein degradation [41], and lead to the dysfunction of synapses. Moreover, Aβ could bind to the receptors on the astrocytes and microglia, leading to the activation of the glial cells and the release of cytokines and other inflammatory factors, which would induce the death of neurons [58, 94]. There is accumulating evidence suggesting that Aβ can disrupt the cellular Ca^{2+} homeostasis to induce synapse and neuronal loss. Thus, stabilizing cellular Ca^{2+} homeostasis might be the potential preventative and therapeutic strategies. Aβ might form a calcium channel on the membrane [40] or potentiate calcium influx through L-type calcium channels [90], and the cytosolic calcium elevation could activate calcium-dependent protease-calpain, which could cleave p35 to p25 and induce apoptosis [45].

7.2 Genetic Evidence for the Involvement of TRPC Channels in Alzheimer's Disease

Besides APP, presenilin1, presenilin2, and ApoE, several single nucleotide polymorphisms (SNPs) at genes implicated in immune system (CR1, CD33, EPHA1), cholesterol/lipid metabolism (clusterin, ABCA7), and vesicle trafficking (PICAM, BIN1, CD2AP) have been discovered by genome-wide association studies (GWAS) to be associated with late-onset Alzheimer's disease (LOAD) [7, 74]. These findings provide new implications for the understanding and therapy of the disease. The first genetic evidence linking TRPC channel to AD came from a study involving two extended pedigrees, each with 15–16 siblings, and among them 5–6 siblings are affected with LOAD [70, 71]. After genotypic analysis of the microarray data of the AD samples vs. the controls, six SNPs on chromosome 20q11.22 are found to be significant after Bonferroni correction, and all six SNPs are located in the gene of TRPC4AP (TRPC4-associated protein). In addition, a set of ten SNPs, including the above six SNPs, are analyzed, and haplotype analysis reveals that nine out of the ten affected siblings have a specific haplotype, and the genotype is homozygous, while genotypes for the control samples are generally heterozygous or opposite homozygous.

Extending the work to include 199 unrelated patients and 85 unaffected spouses to determine the prevalence of the TRPC4AP haplotypes, it is found that 36% of the patients have the haplotype, while only 26% of the spouse controls have the haplotype [70, 71]. It is also found that those patients with the haplotype might have more behavioral changes as well as psychiatric problems. Thus, the TRPC4AP haplotype is associated with late-onset Alzheimer's disease, although the results are waiting to be replicated.

TRPC4AP is also named TNF-R1 ubiquitous scaffolding and signaling protein (TRUSS) and is originally identified in a yeast two-hybrid screen as a TNF-R1-associated protein. TRPC4AP is previously found to be a component of TNF-R1 signaling complexes involved in NF-κB pathway [83]. As indicated by its name, TRPC4AP also interacts with TRPC4 as well as TRPC1 and 5. In the context of reduced endoplasmic reticulum Ca^{2+} storage induced by enhanced G protein-coupled m1 muscarinic acetylcholine receptor (m1AchR) signaling, TRPC4AP, TRPC4, and TNF-R1 all elevate ER Ca^{2+} loading. Although the physiological or pathological significance of this ER Ca^{2+} loading elevation is unclear, it may indicate the malfunction of TRPC4 or its associated protein could lead to the development of Alzheimer's disease.

In the analysis of the association of SNPs with AD in publicly available GWAS data set consisting of three cohorts using data mining methods [11], 199 SNPs mostly associated with genes in calcium signaling, cell adhesion, endocytosis, immune response, and synaptic function are identified. In the model building with prior biological knowledge, 19 SNPs within six genes are identified, and four SNPs in TRPC1 are relevant in AD. However, there is no direct functional evidence linking TRPC1 and AD. Previous studies have found that TRPC1 might be involved in store-operated calcium entry (SOCE) [66], which might be dysfunctional in AD [84]. However, the functional significance of these SNPs to TRPC1 is still unclear.

7.3 TRPC Channels and Calcium Signal in Alzheimer's Disease

Cleavages of APP by α-/β-/γ-secretases and Aβ generation have been found to be regulated by various cellular signals, including calcium signal. The effects of elevated cytosolic calcium concentration on Aβ production are controversial, depending on the concentrations of drugs and different cell lines [9, 13]. PS1 can modulate capacitative calcium entry (CCE), and FAD-linked PS1 mutation inhibits CCE and promotes Aβ formation [96]. PS1 also interacts with IP3R and SERCA directly, while FAD-linked PS1 mutation enhances IP3R/SERCA activity and Aβ formation [14, 28]. Besides, a new calcium channel CALHM can enhance α-secretase cleavage of APP and suppress Aβ generation [21]. Recently, synaptic NMDA receptor activation elevates α-secretase activity and inhibits Aβ formation [33, 56]. In addition, ionomycin-induced calcium entry activates γ-secretase cleavage of E-cadherin [55]. All these works suggest that calcium signals contribute to APP cleavage and Aβ production.

The first work linking the TRPC channels with APP processing comes from a study on SH-SY5Y neuroblastoma cells, which express abundant M3 muscarinic acetylcholine receptors (mAchR). When stimulated by carbachol or oxoM, mAchR would activate the PLC-IP_3 signal to induce the calcium release from the internal store, followed by CCE [72]. In the study, stimulation of SH-SY5Y cells with M3 mAchR agonist oxoM enhances sAPPα production, which is dependent on extracellular calcium influx, but not dependent on calcium mobilization from intracellular stores. Treatment of the cells with a nonselective inhibitor (Cd^{2+}), an L-type channel inhibitor (nifedipine), an N-type channel inhibitor (conotoxin) (CgTx), an Na^+/K^+-ATPase inhibitor (ouabain), or the Na^+/Ca^{2+} exchanger inhibitor (benzamil) did not affect oxoM-induced calcium entry pattern and sAPPα release. However, treatment with CCE inhibitor Gd^{3+} or SKF96365 dramatically reduces the Ca^{2+} entry or sAPPα release induced by oxoM [42]. The involvement of TRPC channels in the CCE and sAPPα release in SH-SY5Y cells is thus proposed. Moreover, TRPC1 is expressed in SH-SY5Y neuroblastoma cells assayed by RT-PCR analysis [10]. It remains unclear whether specific TRPC channels are involved in the CCE and sAPPα release in SH-SY5Y cells.

7.4 TRPC Channels and Glial Activation in Alzheimer's Disease

Besides senile plaques, neuronal neurofibrillary tangles, and neuronal loss, gliosis is also common in AD brains, suggesting that glial activation contributes to the pathogenesis of AD [60, 62]. Glia are nonneuronal cells of the nervous systems, and their main function is to provide physical support and nutrients to neurons, insulate one neuron from another, and clear pathogens or dead neurons [3, 87]. Glia are more than "glue" in the nervous system, and the interaction between glia and neurons is essential for normal brain function [8, 25]. For instance, astrocytes could clear neurotransmitters within the synaptic cleft and prevent possible excitotoxicity caused by accumulation of neurotransmitters such as glutamate [73, 85, 86]. Astrocytes are also crucial for brain development [16], synaptic plasticity [2, 64], and synaptogenesis [15, 80]. Many diseases such as Alzheimer's disease [57], Parkinson's disease [32], and ALS [67] are accompanied with glial activation.

The Aβ plaques are usually surrounded by activated astrocytes, suggesting the important contribution of astrocytes in AD brains. On one hand, astrocytes could digest Aβ. The cultured adult mouse astrocytes migrate to the plaques in response to monocyte chemoattractant protein-1 (MCP-1), a chemokine present in AD lesions. Then, the astrocytes uptake and degrade Aβ42 [93]. Compared with wild-type astrocytes, astrocytes from the apolipoprotein E (ApoE)-knockout mice are deficient in internalizing and degrading Aβ. These results suggest that ApoE is essential for the astrocyte-mediated degradation of Aβ [43].

On the other hand, Aβ could activate astrocytes and induce their inflammatory responses. In transgenic mice expressing the Swedish double mutation of human amyloid precursor protein 695, interleukin (IL)-1β-positive astrocytes are around Aβ deposits before the age of 13 months. Transforming growth factor (TGF)-β1, TGF-β3, and IL-10- and IL-6-positive astrocytes are detectable in 13-month or older transgenic mouse brain [4]. Moreover, astrocytes release soluble inflammatory factors under Aβ stimulation and exacerbate Aβ-induced caspase3 activation and neuronal death. Once the activated astrocytes are treated with anti-inflammatory agent minocycline, the astrocytic inflammatory responses and the Aβ-induced caspase3 activation, caspase3-cleaved tau, and neuronal death are also reduced [27]. Taken together, these results suggest that inhibition of inflammatory activation of astrocytes might be beneficial for AD treatment.

As the calcium signaling through TRPC channels is necessary for astrocyte activation induced by diverse factors such as lipopolysaccharide (LPS), IL-17, and thrombin, it is possible that TRPC channels might also be involved in the Aβ-induced astrocyte activation. Indeed, Aβ42 treatment could enhance DHPG-induced Ca^{2+} signals and store-operated Ca^{2+} entry (SOCE) in cultured astrocytes. At the same time, Aβ42 treatment upregulates the expression of TRPC1 and TRPC5. Moreover, SOCE is also upregulated in untreated astrocytes from the AD mice compared with astrocytes from wild-type mice. Consistently, in APP KO astrocytes, SOCE activated by ER Ca^{2+} store depletion with CPA is greatly reduced, and the protein levels of TRPC1 and Orai1 are downregulated. Moreover, knockdown of APP in cultured wild-type astrocytes reduces ATP- and CPA-induced ER Ca^{2+} release, extracellular Ca^{2+} influx, and TRPC1 expression level [52].

Several members of TRPC family, including TRPC1, TRPC4, TRPC5, and TRPC6, are expressed in astrocytes [6, 30, 68], and their expression levels are increased with age [54]. Several studies have shown that TPRC channels play important roles in glial activation. When stimulated by exogenous LPS, cultured astrocytes could be activated and proliferate, upregulate glial fibrillary acidic protein, and secrete IL-6 and IL-1β. Simultaneously, LPS induces $[Ca^{2+}]_i$ increase in astrocytes. SKF-96365, the TRPC channel blocker, inhibits the LPS-induced $[Ca^{2+}]_i$ increase and astrocyte activation [51]. These results support the potential involvement of TRPC channels in LPS-induced astrocyte activation. Similarly, TRPC channels also play important roles in IL-17-induced astrocyte activation. IL-17 activates MAPK, PI3K/Akt, and NF-kB, leading to upregulation of MIP-1a in astrocytes, while SKF96365 inhibits IL-17-induced astrocyte activation and upregulation of MIP-1a [97].

Another study shows that TRPC3 is essential for thrombin-induced astrocyte activation. Thrombin, a major blood-derived serine protease, could leak into the brain parenchyma upon blood-brain barrier disruption and induce brain injury and astrogliosis. Thrombin treatment induces morphological changes, upregulation of S100B, and proliferation in cultured cortical astrocytes. Meanwhile, thrombin induces upregulation of TRPC3 at the protein level and increased Ca^{2+} influx after thapsigargin treatment. The TRPC3 upregulation is mediated through protease-activated receptor 1 (PAR-1), extracellular signal-regulated protein kinase, c-Jun NH2-terminal kinase, and NF-kB signaling. Finally, the thrombin-induced astrocyte activation could be inhibited by specific knockdown of TRPC3 using RNA interference and a selective TRPC3 inhibitor, pyrazole-3. These results suggest that calcium signaling through TRPC3 is necessary for thrombin-induced astrocyte activation [78].

7.5 TRPC6 in Alzheimer's Disease

TRPC6 in the neurons could promote neuronal survival [38], synaptogenesis [98], and learning and memory [98]. Under the condition of brain ischemia, neuronal TRPC6 is degraded, and prevention of TRPC6 degradation is beneficial for neuronal survival [22]. Moreover, genetic disruption of TRPC6 in the autism patient leads to abnormal neuronal development, morphology, and function [29]. As neuronal survival, learning, and memory are compromised in AD, TRPC6 might play a role in AD development.

The pharmacological evidence to suggest the potential role of TRPC6 in AD comes from the studies using hyperforin as a TRPC6 channel agonist. Hyperforin, a phytochemical produced by the plant *Hypericum perforatum*, is one of the three main active constituents of St. John's wort and the primary active constituent responsible for the antidepressant properties of its extracts. Some studies have identified TRPC6 as a specific target of hyperforin. The hyperforin-induced cation entry is highly specific and related to TRPC6 and could be inhibited by a dominant negative mutant of TRPC6. Hyperforin elevates the intracellular Ca^{2+} concentration by activating TRPC6 channels without activating the TRPC3 and TRPC4. Furthermore, the stimulative effect of hyperforin on neuronal axonal sprouting is TRPC6 dependent [49]. Further study shows that hyperforin modulates dendritic spine morphology in CA1 and CA3 pyramidal neurons of hippocampal slice cultures through the activation of TRPC6 channels. Hyperforin evokes intracellular Ca^{2+} transients which are sensitive to the TRPC channel blocker La^{3+}, thus mimicking the effects of the BDNF on hippocampal pyramidal neurons [50]. These findings suggest that hyperforin is a selective agonist for TRPC6 channels.

Several studies have showed that hyperforin could reduce Aβ levels and improve behavioral performance in AD models. In rats injected with amyloid fibrils in the hippocampus, hyperforin could reduce amyloid deposit formation and therefore decrease the Aβ-induced neurotoxicity, reactive oxidative species, and behavioral impairments [19]. Moreover, a hyperforin derivative IDN5706 – tetrahydrohyperforin – also prevents the cognitive deficit and synaptic impairment in double transgenic APPswe/PSEN1ΔE9 mice in a dose-dependent manner. Tetrahydrohyperforin decreases the proteolytic processing of APP, total fibrillar and oligomeric forms of Aβ, tau hyperphosphorylation, and astrogliosis [36]. Further studies show that the target of tetrahydrohyperforin appears to be TRPC6 [59]. In this study, mouse hippocampal slices are incubated with tetrahydrohyperforin, the TRPC3/6/7 activator OAG, SKF96365, and Aβ oligomers. Tetrahydrohyperforin and OAG have a similar stimulating effect on fEPSPs, which is inhibited by SKF96365. Aβ oligomers induce fEPSP reduction which could be rescued by tetrahydrohyperforin. In wild-type mice, tetrahydrohyperforin improves the spatial memory, an effect that is neutralized by coadministration of SKF96365. There is a strong pharmacophore similarity of tetrahydrohyperforin and other reported TRPC6 agonists (IDN5522, OAG, and Hyp9). These findings indicate that hyperforin and its derivatives might be effective for AD treatment and highlight the potential protective roles of TRPC6 in AD.

The association of TRPC6 with AD is further implicated by an in vitro study [48]. The effects of two familial Alzheimer's disease-linked presenilin2 mutants (N141I and M239V) and a loss-of-function presenilin2 mutant (D263A) on the TRPC6 channel activity are assessed. The co-expression of presenilin2 or its FAD mutants and TRPC6 in HEK293T cells abolishes agonist-induced TRPC6 activation without affecting agonist-induced Ca^{2+} entry. The inhibitory effect of presenilin2 and its FAD mutants could not be attributed to Aβ increase in the medium because Aβ treatment alone does not affect the TRPC6 channel activity. In contrast, the co-expression of a loss-of-function PS2 mutant and TRPC6 in HEK293T cells enhances agonist-induced Ca^{2+} entry. These results suggest that the wild-type or familial Alzheimer's disease-linked presenilin2 mutants influence TRPC6 channel activity in HEK293T cells and the normal function of TRPC6 might be compromised in AD.

The direct evidence that TRPC6 is protective against AD comes from a work showing that TRPC6 modulates Aβ production [91]. The γ-secretase is a potential therapeutic target for AD, but its potent inhibitors would affect the normal function of γ-secretase which cleaves many substrates and lead to different side effects. The TRPC6 inhibits Aβ production by specifically interacting with APP and C99 to block the cleavage of C99 by the γ-secretase. The inhibitory effects are specific to APP, but not to Notch or other substrates tested. The substrate specificity is conferred by the specific interaction of TRPC6 with APP (C99), but not with other substrates. Once TRPC6 binds to C99, the interaction between C99 and presenilin1 (PS1) is reduced. The TRPC6 domain responsible for the inhibitory effects is identified, and a fusion peptide derived from the second transmembrane domain of TRPC6 could also reduce Aβ levels without effects on Notch cleavage. Moreover, crossing APP/PS1 mice with TRPC6 transgenic mice leads to a marked reduction in both plaque load and Aβ levels and improvement in structural and behavioral impairment. The TRPC6-derived peptide also reduces Aβ levels in APP/PS1 mice.

7.6 Conclusion and Perspectives

Most studies show indirect evidence linking the TRPC channels and Alzheimer's disease. However, direct evidence shows that TRPC6 is protective against AD. Reducing Aβ production by enhancing TRPC6 is specifically valuable under the circumstances that all the γ-secretase inhibitors failed in the clinical trials, largely due to the severe side effects. Inhibiting APP and presenilin interaction by TRPC6 may represent a novel and viable strategy to target Alzheimer's disease. The potential of TRPC6-based therapies such as hyperforin and TRPC6-derived peptide deserves further evaluation.

References

1. Abramov E, Dolev I, Fogel H, Ciccotosto GD, Ruff E, Slutsky I (2009) Amyloid-beta as a positive endogenous regulator of release probability at hippocampal synapses. Nat Neurosci 12(12):1567–1576. doi:10.1038/nn.2433
2. Allen NJ, Barres BA (2005) Signaling between glia and neurons: focus on synaptic plasticity. Curr Opin Neurobiol 15(5):542–548. doi:S0959-4388(05)00125-X [pii]10.1016/j.conb.2005.08.006
3. Allen NJ, Barres BA (2009) Neuroscience: glia – more than just brain glue. Nature 457 (7230):675–677. doi:457675a [pii]10.1038/457675a
4. Apelt J, Schliebs R (2001) Beta-amyloid-induced glial expression of both pro- and anti-inflammatory cytokines in cerebral cortex of aged transgenic Tg2576 mice with Alzheimer plaque pathology. Brain Res 894(1):21–30
5. Benilova I, Karran E, De Strooper B (2012) The toxic Abeta oligomer and Alzheimer's disease: an emperor in need of clothes. Nat Neurosci 15(3):349–357. doi:10.1038/nn.3028
6. Beskina O, Miller A, Mazzocco-Spezzia A, Pulina MV, Golovina VA (2007) Mechanisms of interleukin-1beta-induced Ca2+ signals in mouse cortical astrocytes: roles of store- and receptor-operated Ca2+ entry. Am J Phys Cell Phys 293 (3):C1103–C1111. doi:00249.2007 [pii]10.1152/ajpcell.00249.2007
7. Bettens K, Sleegers K, Van Broeckhoven C (2013) Genetic insights in Alzheimer's disease. Lancet Neurol 12(1):92–104. doi:10.1016/S1474-4422(12)70259-4
8. Bezzi P, Volterra A (2001) A neuron-glia signalling network in the active brain. Curr Opin Neurobiol 11(3):387–394
9. Bojarski L, Herms J, Kuznicki J (2008) Calcium dysregulation in Alzheimer's disease. Neurochem Int 52(4–5):621–633. doi:10.1016/j.neuint.2007.10.002
10. Bollimuntha S, Singh BB, Shavali S, Sharma SK, Ebadi M (2005) TRPC1-mediated inhibition of 1-methyl-4-phenylpyridinium ion neurotoxicity in human SH-SY5Y neuroblastoma cells. J Biol Chem 280(3):2132–2140. doi:10.1074/jbc.M407384200
11. Briones N, Dinu V (2012) Data mining of high density genomic variant data for prediction of Alzheimer's disease risk. BMC Med Genet 13:7. doi:10.1186/1471-2350-13-7
12. Brookmeyer R, Johnson E, Ziegler-Graham K, Arrighi HM (2007) Forecasting the global burden of Alzheimer's disease. Alzheimers Dement J Alzheimers Assoc 3(3):186–191. doi:10.1016/j.jalz.2007.04.381
13. Brzyska M, Elbaum D (2003) Dysregulation of calcium in Alzheimer's disease. Acta Neurobiol Exp 63(3):171–183
14. Cheung KH, Shineman D, Muller M, Cardenas C, Mei L, Yang J, Tomita T, Iwatsubo T, Lee VM, Foskett JK (2008) Mechanism of Ca2+ disruption in Alzheimer's disease by presenilin regulation of InsP3 receptor channel gating. Neuron 58 (6):871–883. doi:S0896-6273(08)00367-X [pii]10.1016/j.neuron.2008.04.015
15. Christopherson KS, Ullian EM, Stokes CC, Mullowney CE, Hell JW, Agah A, Lawler J, Mosher DF, Bornstein P, Barres BA (2005) Thrombospondins are astrocyte-secreted proteins that promote CNS synaptogenesis. Cell 120 (3):421–433. doi:S0092-8674(04)01245-0 [pii]10.1016/j.cell.2004.12.020
16. Clarke LE, Barres BA (2013) Emerging roles of astrocytes in neural circuit development. Nat Rev Neurosci 14 (5):311–321. doi:nrn3484 [pii]10.1038/nrn3484
17. Dahlgren KN, Manelli AM, Stine WB, Jr., Baker LK, Krafft GA, LaDu MJ (2002) Oligomeric and fibrillar species of amyloid-beta peptides differentially affect neuronal viability. J Biol Chem 277 (35):32046–32053. doi:10.1074/jbc.M201750200M201750200 [pii]
18. Dillen K, Annaert W (2006) A two decade contribution of molecular cell biology to the centennial of Alzheimer's disease: are we progressing toward therapy? Int Rev Cytol 254:215–300. doi:10.1016/S0074-7696(06)54005-7
19. Dinamarca MC, Cerpa W, Garrido J, Hancke JL, Inestrosa NC (2006) Hyperforin prevents beta-amyloid neurotoxicity and spatial memory

impairments by disaggregation of Alzheimer's amyloid-beta-deposits. Mol Psychiatry 11 (11):1032–1048. doi:4001866 [pii]10.1038/sj.mp.4001866
20. Doody RS, Thomas RG, Farlow M, Iwatsubo T, Vellas B, Joffe S, Kieburtz K, Raman R, Sun X, Aisen PS, Siemers E, Liu-Seifert H, Mohs R (2014) Phase 3 trials of solanezumab for mild-to-moderate Alzheimer's disease. N Engl J Med 370(4):311–321. doi:10.1056/NEJMoa1312889
21. Dreses-Werringloer U, Lambert JC, Vingtdeux V, Zhao H, Vais H, Siebert A, Jain A, Koppel J, Rovelet-Lecrux A, Hannequin D, Pasquier F, Galimberti D, Scarpini E, Mann D, Lendon C, Campion D, Amouyel P, Davies P, Foskett JK, Campagne F, Marambaud P (2008) A polymorphism in CALHM1 influences Ca2+ homeostasis, Abeta levels, and Alzheimer's disease risk. Cell 133 (7):1149–1161. doi:S0092-8674(08)00751-4 [pii]10.1016/j.cell.2008.05.048
22. Du W, Huang J, Yao H, Zhou K, Duan B, Wang Y (2010) Inhibition of TRPC6 degradation suppresses ischemic brain damage in rats. J Clin Invest 120 (10):3480–3492. doi:43165 [pii]10.1172/JCI43165
23. Dubois B, Feldman HH, Jacova C, Dekosky ST, Barberger-Gateau P, Cummings J, Delacourte A, Galasko D, Gauthier S, Jicha G, Meguro K, O'Brien J, Pasquier F, Robert P, Rossor M, Salloway S, Stern Y, Visser PJ, Scheltens P (2007) Research criteria for the diagnosis of Alzheimer's disease: revising the NINCDS-ADRDA criteria. Lancet Neurol 6(8):734–746. doi:10.1016/S1474-4422(07)70178-3
24. Farrer LA, Cupples LA, Haines JL, Hyman B, Kukull WA, Mayeux R, Myers RH, Pericak-Vance MA, Risch N, van Duijn CM (1997) Effects of age, sex, and ethnicity on the association between apolipoprotein E genotype and Alzheimer disease. A meta-analysis. APOE and Alzheimer Disease Meta Analysis Consortium. JAMA 278(16):1349–1356
25. Fields RD, Stevens-Graham B (2002) New insights into neuron-glia communication. Science 298 (5593):556–562. doi:10.1126/science.298.5593.556298/5593/556 [pii]
26. Gandy S (2005) The role of cerebral amyloid beta accumulation in common forms of Alzheimer disease. J Clin Invest 115(5):1121–1129. doi:10.1172/JCI25100
27. Garwood CJ, Pooler AM, Atherton J, Hanger DP, Noble W (2011) Astrocytes are important mediators of Abeta-induced neurotoxicity and tau phosphorylation in primary culture. Cell Death Dis 2:e167. doi:cddis201150 [pii]10.1038/cddis.2011.50
28. Green KN, Demuro A, Akbari Y, Hitt BD, Smith IF, Parker I, LaFerla FM (2008) SERCA pump activity is physiologically regulated by presenilin and regulates amyloid beta production. J Cell Biol 181 (7):1107–1116. doi:10.1083/jcb.200706171jcb.200706171 [pii]
29. Griesi-Oliveira K, Acab A, Gupta AR, Sunaga DY, Chailangkarn T, Nicol X, Nunez Y, Walker MF, Murdoch JD, Sanders SJ, Fernandez TV, Ji W, Lifton RP, Vadasz E, Dietrich A, Pradhan D, Song H, Ming GL, Gu X, Haddad G, Marchetto MC, Spitzer N, Passos-Bueno MR, State MW, Muotri AR (2015) Modeling non-syndromic autism and the impact of TRPC6 disruption in human neurons. Mol Psychiatry 20 (11):1350–1365. doi:mp2014141 [pii]10.1038/mp.2014.141
30. Grimaldi M, Maratos M, Verma A (2003) Transient receptor potential channel activation causes a novel form of [Ca 2+]I oscillations and is not involved in capacitative Ca 2+ entry in glial cells. J Neurosci 23 (11):4737–4745. doi:23/11/4737 [pii]
31. Hardy J, Selkoe DJ (2002) The amyloid hypothesis of Alzheimer's disease: progress and problems on the road to therapeutics. Science 297(5580):353–356. doi:10.1126/science.1072994
32. Hirsch EC, Breidert T, Rousselet E, Hunot S, Hartmann A, Michel PP (2003) The role of glial reaction and inflammation in Parkinson's disease. Ann N Y Acad Sci 991:214–228
33. Hoey SE, Williams RJ, Perkinton MS (2009) Synaptic NMDA receptor activation stimulates alpha-secretase amyloid precursor protein processing and inhibits amyloid-beta production. J Neurosci 29(14):4442–4460. doi:10.1523/JNEUROSCI.6017-08.2009
34. Holtzman DM, Morris JC, Goate AM (2011) Alzheimer's disease: the challenge of the second century. Sci Transl Med 3(77):77sr71. doi:10.1126/scitranslmed.3002369
35. Hsia AY, Masliah E, McConlogue L, Yu GQ, Tatsuno G, Hu K, Kholodenko D, Malenka RC, Nicoll RA, Mucke L (1999) Plaque-independent disruption of neural circuits in Alzheimer's disease mouse models. Proc Natl Acad Sci U S A 96(6):3228–3233
36. Inestrosa NC, Tapia-Rojas C, Griffith TN, Carvajal FJ, Benito MJ, Rivera-Dictter A, Alvarez AR, Serrano FG, Hancke JL, Burgos PV, Parodi J, Varela-Nallar L (2011) Tetrahydrohyperforin prevents cognitive deficit, Abeta deposition, tau phosphorylation and synaptotoxicity in the APPswe/PSEN1DeltaE9 model of Alzheimer's disease: a possible effect on APP processing. Transl Psychiatry 1:e20. doi:tp201119 [pii]10.1038/tp.2011.19
37. Ivins KJ, Ivins JK, Sharp JP, Cotman CW (1999) Multiple pathways of apoptosis in PC12 cells. CrmA inhibits apoptosis induced by beta-amyloid. J Biol Chem 274(4):2107–2112
38. Jia Y, Zhou J, Tai Y, Wang Y (2007) TRPC channels promote cerebellar granule neuron survival. Nat Neurosci 10 (5):559–567. doi:nn1870 [pii]10.1038/nn1870
39. Jiang Q, Lee CY, Mandrekar S, Wilkinson B, Cramer P, Zelcer N, Mann K, Lamb B, Willson TM, Collins JL, Richardson JC, Smith JD, Comery TA, Riddell D, Holtzman DM, Tontonoz P, Landreth GE (2008) ApoE promotes the proteolytic degradation of Abeta. Neuron 58 (5):681–693. doi:S0896-6273(08)00328-0 [pii]10.1016/j.neuron.2008.04.010
40. Kawahara M, Kuroda Y (2000) Molecular mechanism of neurodegeneration induced by Alzheimer's

beta-amyloid protein: channel formation and disruption of calcium homeostasis. Brain Res Bull 53(4):389–397
41. Kelly BL, Ferreira A (2006) beta-Amyloid-induced dynamin 1 degradation is mediated by N-methyl-D-aspartate receptors in hippocampal neurons. J Biol Chem 281 (38):28079–28089. doi:M605081200 [pii]10.1074/jbc.M605081200
42. Kim JH, Choi S, Jung JE, Roh EJ, Kim HJ (2006) Capacitative Ca2+ entry is involved in regulating soluble amyloid precursor protein (sAPPalpha) release mediated by muscarinic acetylcholine receptor activation in neuroblastoma SH-SY5Y cells. J Neurochem 97(1):245–254. doi:10.1111/j.1471-4159.2006.03734.x
43. Koistinaho M, Lin S, Wu X, Esterman M, Koger D, Hanson J, Higgs R, Liu F, Malkani S, Bales KR, Paul SM (2004) Apolipoprotein E promotes astrocyte colocalization and degradation of deposited amyloid-beta peptides. Nat Med 10 (7):719–726. doi:10.1038/nm105nm1058 [pii]
44. Lacor PN, Buniel MC, Furlow PW, Clemente AS, Velasco PT, Wood M, Viola KL, Klein WL (2007) Abeta oligomer-induced aberrations in synapse composition, shape, and density provide a molecular basis for loss of connectivity in Alzheimer's disease. J Neurosci 27 (4):796–807. doi:27/4/796 [pii]10.1523/JNEUROSCI.3501-06.2007
45. Lee MS, Kwon YT, Li M, Peng J, Friedlander RM, Tsai LH (2000) Neurotoxicity induces cleavage of p35 to p25 by calpain. Nature 405(6784):360–364. doi:10.1038/35012636
46. Leissring MA, Farris W, Chang AY, Walsh DM, Wu X, Sun X, Frosch MP, Selkoe DJ (2003) Enhanced proteolysis of beta-amyloid in APP transgenic mice prevents plaque formation, secondary pathology, and premature death. Neuron 40(6):1087–1093
47. Lesne S, Koh MT, Kotilinek L, Kayed R, Glabe CG, Yang A, Gallagher M, Ashe KH (2006) A specific amyloid-beta protein assembly in the brain impairs memory. Nature 440 (7082):352–357. doi:nature04533 [pii]10.1038/nature04533
48. Lessard CB, Lussier MP, Cayouette S, Bourque G, Boulay G (2005) The overexpression of presenilin2 and Alzheimer's-disease-linked presenilin2 variants influences TRPC6-enhanced Ca2+ entry into HEK293 cells. Cell Signal 17 (4):437–445. doi:S0898-6568(04)00180-9 [pii]10.1016/j.cellsig.2004.09.005
49. Leuner K, Kazanski V, Muller M, Essin K, Henke B, Gollasch M, Harteneck C, Muller WE (2007) Hyperforin – a key constituent of St. John's wort specifically activates TRPC6 channels. FASEB J 21 (14):4101–4111. doi:fj.07-8110com [pii]10.1096/fj.07-8110com
50. Leuner K, Li W, Amaral MD, Rudolph S, Calfa G, Schuwald AM, Harteneck C, Inoue T, Pozzo-Miller L (2013) Hyperforin modulates dendritic spine morphology in hippocampal pyramidal neurons by activating Ca(2+) -permeable TRPC6 channels. Hippocampus 23(1):40–52. doi:10.1002/hipo.22052
51. Li JH, Zhao ST, Wu CY, Cao X, Peng MR, Li SJ, Liu XA, Gao TM (2013) Store-operated Ca2+ channels blockers inhibit lipopolysaccharide induced astrocyte activation. Neurochem Res 38(10):2216–2226. doi:10.1007/s11064-013-1130-0
52. Linde CI, Baryshnikov SG, Mazzocco-Spezzia A, Golovina VA (2011) Dysregulation of Ca2+ signaling in astrocytes from mice lacking amyloid precursor protein. Am J Phys Cell Phys 300(6):C1502–C1512. doi:10.1152/ajpcell.00379.2010
53. Liu Q, Kawai H, Berg DK (2001) beta -Amyloid peptide blocks the response of alpha 7-containing nicotinic receptors on hippocampal neurons. Proc Natl Acad Sci U S A 98(8):4734–4739. doi:10.1073/pnas.081553598
54. Malarkey EB, Ni Y, Parpura V (2008) Ca2+ entry through TRPC1 channels contributes to intracellular Ca2+ dynamics and consequent glutamate release from rat astrocytes. Glia 56(8):821–835. doi:10.1002/glia.20656
55. Marambaud P, Shioi J, Serban G, Georgakopoulos A, Sarner S, Nagy V, Baki L, Wen P, Efthimiopoulos S, Shao Z, Wisniewski T, Robakis NK (2002) A presenilin-1/gamma-secretase cleavage releases the E-cadherin intracellular domain and regulates disassembly of adherens junctions. EMBO J 21(8):1948–1956. doi:10.1093/emboj/21.8.1948
56. Marcello E, Gardoni F, Mauceri D, Romorini S, Jeromin A, Epis R, Borroni B, Cattabeni F, Sala C, Padovani A, Di Luca M (2007) Synapse-associated protein-97 mediates alpha-secretase ADAM10 trafficking and promotes its activity. J Neurosci 27(7):1682–1691. doi:10.1523/JNEUROSCI.3439-06.2007
57. Meda L, Baron P, Scarlato G (2001) Glial activation in Alzheimer's disease: the role of Abeta and its associated proteins. Neurobiol Aging 22 (6):885–893. doi:S0197458001003074 [pii]
58. Meda L, Cassatella MA, Szendrei GI, Otvos L Jr, Baron P, Villalba M, Ferrari D, Rossi F (1995) Activation of microglial cells by beta-amyloid protein and interferon-gamma. Nature 374(6523):647–650. doi:10.1038/374647a0
59. Montecinos-Oliva C, Schuller A, Parodi J, Melo F, Inestrosa NC (2014) Effects of tetrahydrohyperforin in mouse hippocampal slices: neuroprotection, long-term potentiation and TRPC channels. Curr Med Chem 21 (30):3494–3506. doi:CMC-EPUB-61425 [pii]
60. Mrak RE, Griffin WS (1996) Role of activated glia and of glial cytokines in Alzheimer's disease: a review. EOS 16(3–4):80–84
61. Mudher A, Lovestone S (2002) Alzheimer's disease-do tauists and baptists finally shake hands? Trends Neurosci 25(1):22–26
62. Nagele RG, Wegiel J, Venkataraman V, Imaki H, Wang KC (2004) Contribution of glial cells to the development of amyloid plaques in Alzheimer's disease. Neurobiol Aging 25 (5):663–674. doi:10.1016/j.

neurobiolaging.2004.01.007S0197458004001034 [pii]
63. Nakagawa T, Zhu H, Morishima N, Li E, Xu J, Yankner BA, Yuan J (2000) Caspase-12 mediates endoplasmic-reticulum-specific apoptosis and cytotoxicity by amyloid-beta. Nature 403(6765):98–103. doi:10.1038/47513
64. Navarrete M, Perea G, Fernandez de Sevilla D, Gomez-Gonzalo M, Nunez A, Martin ED, Araque A (2012) Astrocytes mediate in vivo cholinergic-induced synaptic plasticity. PLoS Biol 10 (2):e1001259. doi:10.1371/journal.pbio.1001259PBIOLOGY-D-11-02406 [pii]
65. Ohno M, Cole SL, Yasvoina M, Zhao J, Citron M, Berry R, Disterhoft JF, Vassar R (2007) BACE1 gene deletion prevents neuron loss and memory deficits in 5XFAD APP/PS1 transgenic mice. Neurobiol Dis 26(1):134–145. doi:10.1016/j.nbd.2006.12.008
66. Parekh AB, Putney JW Jr (2005) Store-operated calcium channels. Physiol Rev 85(2):757–810. doi:10.1152/physrev.00057.2003
67. Philips T, Robberecht W (2011) Neuroinflammation in amyotrophic lateral sclerosis: role of glial activation in motor neuron disease. Lancet Neurol 10 (3):253–263. doi:S1474-4422(11)70015-1 [pii]10.1016/S1474-4422(11)70015-1
68. Pizzo P, Burgo A, Pozzan T, Fasolato C (2001) Role of capacitative calcium entry on glutamate-induced calcium influx in type-I rat cortical astrocytes. J Neurochem 79(1):98–109
69. Plant LD, Boyle JP, Smith IF, Peers C, Pearson HA (2003) The production of amyloid beta peptide is a critical requirement for the viability of central neurons. J Neurosci 23(13):5531–5535
70. Poduslo SE, Huang R, Huang J (2009) The frequency of the TRPC4AP haplotype in Alzheimer's patients. Neurosci Lett 450(3):344–346. doi:10.1016/j.neulet.2008.11.050
71. Poduslo SE, Huang R, Huang J, Smith S (2009) Genome screen of late-onset Alzheimer's extended pedigrees identifies TRPC4AP by haplotype analysis. Am J Med Genet B Neuropsychiatr Genet Off Publ Int Soc Psychiatr Genet 150B(1):50–55. doi:10.1002/ajmg.b.30767
72. Putney JW Jr (1997) Type 3 inositol 1,4,5-trisphosphate receptor and capacitative calcium entry. Cell Calcium 21(3):257–261
73. Robinson MB (1998) The family of sodium-dependent glutamate transporters: a focus on the GLT-1/EAAT2 subtype. Neurochem Int 33(6):479–491
74. Rosenthal SL, Kamboh MI (2014) Late-onset Alzheimer's disease genes and the potentially implicated pathways. Curr Genet Med Rep 2:85–101. doi:10.1007/s40142-014-0034-x
75. Roychaudhuri R, Yang M, Hoshi MM, Teplow DB (2009) Amyloid beta-protein assembly and Alzheimer disease. J Biol Chem 284 (8):4749–4753. doi:R800036200 [pii]10.1074/jbc.R800036200
76. Selkoe DJ (2004) Alzheimer disease: mechanistic understanding predicts novel therapies. Ann Intern Med 140 (8):627–638. doi:140/8/627 [pii]
77. Shankar GM, Li S, Mehta TH, Garcia-Munoz A, Shepardson NE, Smith I, Brett FM, Farrell MA, Rowan MJ, Lemere CA, Regan CM, Walsh DM, Sabatini BL, Selkoe DJ (2008) Amyloid-beta protein dimers isolated directly from Alzheimer's brains impair synaptic plasticity and memory. Nat Med 14 (8):837–842. doi:nm1782 [pii]10.1038/nm1782
78. Shirakawa H, Sakimoto S, Nakao K, Sugishita A, Konno M, Iida S, Kusano A, Hashimoto E, Nakagawa T, Kaneko S (2010) Transient receptor potential canonical 3 (TRPC3) mediates thrombin-induced astrocyte activation and upregulates its own expression in cortical astrocytes. J Neurosci 30 (39):13116–13129. doi:30/39/13116 [pii]10.1523/JNEUROSCI.1890-10.2010
79. Siemers ER, Sundell KL, Carlson C, Case M, Sethuraman G, Liu-Seifert H, Dowsett SA, Pontecorvo MJ, Dean RA, Demattos R (2016) Phase 3 solanezumab trials: secondary outcomes in mild Alzheimer's disease patients. Alzheimers Dement J Alzheimers Assoc 12(2):110–120. doi:10.1016/j.jalz.2015.06.1893
80. Slezak M, Pfrieger FW (2003) New roles for astrocytes: regulation of CNS synaptogenesis. Trends Neurosci 26 (10):531–535. doi:S0166-2236(03)00258-3 [pii]10.1016/j.tins.2003.08.005
81. Small DH, Mok SS, Bornstein JC (2001) Alzheimer's disease and Abeta toxicity: from top to bottom. Nat Rev Neurosci 2 (8):595–598. doi:10.1038/3508607235086072 [pii]
82. Snyder EM, Nong Y, Almeida CG, Paul S, Moran T, Choi EY, Nairn AC, Salter MW, Lombroso PJ, Gouras GK, Greengard P (2005) Regulation of NMDA receptor trafficking by amyloid-beta. Nat Neurosci 8 (8):1051–1058. doi:nn1503 [pii]10.1038/nn1503
83. Soond SM, Terry JL, Colbert JD, Riches DW (2003) TRUSS, a novel tumor necrosis factor receptor 1 scaffolding protein that mediates activation of the transcription factor NF-kappaB. Mol Cell Biol 23(22):8334–8344
84. Stutzmann GE, Smith I, Caccamo A, Oddo S, Parker I, Laferla F (2007) Enhanced ryanodine-mediated calcium release in mutant PS1-expressing Alzheimer's mouse models. Ann N Y Acad Sci 1097:265–277. doi:10.1196/annals.1379.025
85. Sutherland ML, Delaney TA, Noebels JL (1995) Molecular characterization of a high-affinity mouse glutamate transporter. Gene 162 (2):271–274. doi:037811199500293F [pii]
86. Sutherland ML, Delaney TA, Noebels JL (1996) Glutamate transporter mRNA expression in proliferative zones of the developing and adult murine CNS. J Neurosci 16(7):2191–2207
87. Temburni MK, Jacob MH (2001) New functions for glia in the brain. Proc Natl Acad Sci U S A 98 (7):3631–3632. doi:10.1073/pnas.08107319898/7/3631 [pii]

88. Terry RD, Masliah E, Salmon DP, Butters N, DeTeresa R, Hill R, Hansen LA, Katzman R (1991) Physical basis of cognitive alterations in Alzheimer's disease: synapse loss is the major correlate of cognitive impairment. Ann Neurol 30(4):572–580. doi:10.1002/ana.410300410
89. Troy CM, Rabacchi SA, Xu Z, Maroney AC, Connors TJ, Shelanski ML, Greene LA (2001) beta-Amyloid-induced neuronal apoptosis requires c-Jun N-terminal kinase activation. J Neurochem 77(1):157–164
90. Ueda K, Shinohara S, Yagami T, Asakura K, Kawasaki K (1997) Amyloid beta protein potentiates Ca2+ influx through L-type voltage-sensitive Ca2+ channels: a possible involvement of free radicals. J Neurochem 68(1):265–271
91. Wang J, Lu R, Yang J, Li H, He Z, Jing N, Wang X, Wang Y (2015) TRPC6 specifically interacts with APP to inhibit its cleavage by gamma-secretase and reduce Abeta production. Nat Commun 6:8876. doi:ncomms9876 [pii]10.1038/ncomms9876
92. Wolfe MS (2007) When loss is gain: reduced presenilin proteolytic function leads to increased Abeta42/Abeta40. Talking point on the role of presenilin mutations in Alzheimer disease. EMBO Rep 8(2):136–140. doi:10.1038/sj.embor.7400896
93. Wyss-Coray T, Loike JD, Brionne TC, Lu E, Anankov R, Yan F, Silverstein SC, Husemann J (2003) Adult mouse astrocytes degrade amyloid-beta in vitro and in situ. Nat Med 9 (4):453–457. doi:10.1038/nm838nm838 [pii]
94. Yan SD, Chen X, Fu J, Chen M, Zhu H, Roher A, Slattery T, Zhao L, Nagashima M, Morser J, Migheli A, Nawroth P, Stern D, Schmidt AM (1996) RAGE and amyloid-beta peptide neurotoxicity in Alzheimer's disease. Nature 382(6593):685–691. doi:10.1038/382685a0
95. Yankner BA, Duffy LK, Kirschner DA (1990) Neurotrophic and neurotoxic effects of amyloid beta protein: reversal by tachykinin neuropeptides. Science 250(4978):279–282
96. Yoo AS, Cheng I, Chung S, Grenfell TZ, Lee H, Pack-Chung E, Handler M, Shen J, Xia W, Tesco G, Saunders AJ, Ding K, Frosch MP, Tanzi RE, Kim TW (2000) Presenilin-mediated modulation of capacitative calcium entry. Neuron 27(3):561–572
97. Zhang Y, Huang R, Yi H, Bai Y, Chao J, Yao H (2016) IL-17 induces MIP-1alpha expression in primary mouse astrocytes via TRPC channel. Inflammopharmacology 24 (1):33–42. doi:10.1007/s10787-015-0256-x10.1007/s10787-015-0256-x [pii]
98. Zhou J, Du W, Zhou K, Tai Y, Yao H, Jia Y, Ding Y, Wang Y (2008) Critical role of TRPC6 channels in the formation of excitatory synapses. Nat Neurosci 11 (7):741–743. doi:nn.2127 [pii]10.1038/nn.2127

TRPC Channels and Parkinson's Disease

8

Pramod Sukumaran, Yuyang Sun, Anne Schaar, Senthil Selvaraj, and Brij B. Singh

Abstract

Parkinson's disease (PD) is a common neurodegenerative disorder, which involves degeneration of dopaminergic neurons that are present in the substantia nigra pars compacta (SNpc) region. Many factors have been identified that could lead to Parkinson's disease; however, almost all of them are directly or indirectly dependent on Ca^{2+} signaling. Importantly, though disturbances in Ca^{2+} homeostasis have been implicated in Parkinson's disease and other neuronal diseases, the identity of the calcium channel remains elusive. Members of the transient receptor potential canonical (TRPC) channel family have been identified as a new class of Ca^{2+} channels, and it could be anticipated that these channels could play important roles in neurodegenerative diseases, especially in PD. Thus, in this chapter we have entirely focused on TRPC channels and elucidated its role in PD.

Keywords

Calcium signaling • Dopaminergic neurons • ER stress • Oxidative stress • Parkinson's disease

8.1 Introduction

Calcium (Ca^{2+}) is an important element that functions as a prominent regulator for processes such as gene regulation, neuronal cell growth and differentiation, motility and axonal development, and even neuronal cell death [6, 9, 65]. Thus, it is not surprising that disruption of Ca^{2+} homeostasis in neuronal cells results in decreased neuronal functions leading to neurodegenerative diseases such as Parkinson's, Huntington's, and Alzheimer's [1, 10, 11, 46, 77, 78]. Due to these outcomes, Ca^{2+} homeostasis is strictly maintained in neuronal cells. Cells have evolved a multitude of mechanisms to regulate cellular Ca^{2+} levels and Ca^{2+} channels, and pumps play a key role in this regulation. Cumulative studies

Pramod Sukumaran and Yuyang Sun contributed equally

P. Sukumaran • Y. Sun • A. Schaar • S. Selvaraj
B.B. Singh (✉)
Department of Biomedical Sciences, School of Medicine and Health Sciences, University of North Dakota, Grand Forks, ND 58201, USA
e-mail: brij.singh@med.und.edu

Y. Wang (ed.), *Transient Receptor Potential Canonical Channels and Brain Diseases*,
Advances in Experimental Medicine and Biology 976, DOI 10.1007/978-94-024-1088-4_8

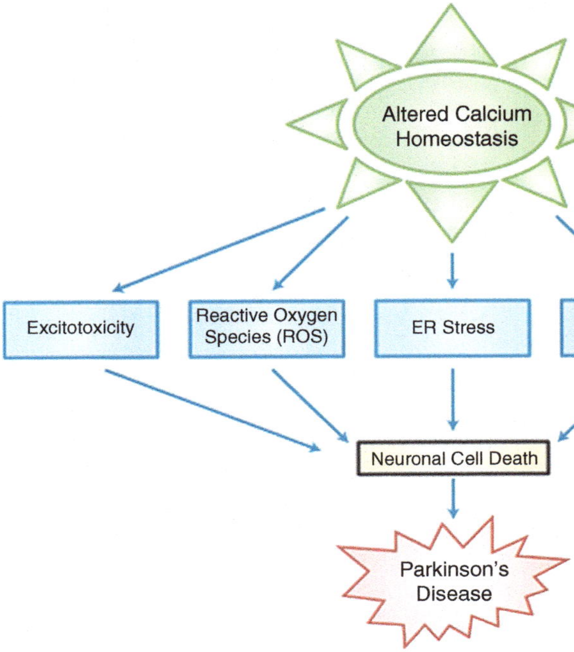

Fig. 8.1 Mechanisms that potentially leads to PD. Schematic model shows altered Ca^{2+} homeostasis in neuronal cells which causes Parkinson's disease via various mechanism(s) such as ER stress, ROS, mitochondrial dysfunction, excitotoxicity, and inflammation

suggest that both excessive elevation and attenuation of intracellular Ca^{2+} will lead to neuronal degeneration through different mechanisms as suggested in this chapter (Fig. 8.1). Increased intracellular calcium $[Ca^{2+}]_i$ concentrations mainly via the AMPA or NMDA channels lead to enhanced activation of Ca^{2+}-dependent processes that are normally inert or functional at low Ca^{2+} levels, thereby causing metabolic imbalances which result in neuronal death [3, 13, 20]. In contrast, decreased $[Ca^{2+}]_i$, upon store depletion, could induce ER stress or inhibit activation of proteins that are essential for cell survival [48]. Thus, different actions of Ca^{2+} in neuronal cells could be dependent not only on its cellular concentration but also on the ion channels that modulate Ca^{2+} entry [44, 48, 74]. Ca^{2+} channels, mainly the transient receptor potential canonical (TRPC) channels, have recently emerged as a key regulator of Ca^{2+} homeostasis in neuronal cell function [9, 57]. Further, TRPC channel function and expression are altered in various neuronal diseases such as Parkinson's disease (PD) [4, 57, 59]. PD is a neurodegenerative disorder, which is characterized with progressive degeneration of dopaminergic neurons and affects the aging population. Thus, in this chapter we will focus on the functional implication of TRPC channels in neuronal cell function and their roles in PD.

8.2 TRPC Functions in Neurons

Neuronal cells are heavily dependent on Ca^{2+} signaling for their function and survival [32]. Initiation of the action potential activates the voltage-gated calcium channels that modulate neurosecretion. Growing evidence suggests that this mode of Ca^{2+} entry is also essential in maintaining the cytosolic, ER, and mitochondrial Ca^{2+} levels. Importantly, as TRPC proteins have been identified as putative calcium channels that are activated by second messenger-mediated store depletion, they could play a critical role in neuronal survival, proliferation, and differentiation. There is evidence that TRPCs are highly expressed in all regions of the central nervous system (CNS) [24], but in some cases the expression of individual TRPC channels is altered during development, especially as observed with TRPC1 and TRPC3, which are more expressed in embryonic CNS than in adult neurons [62]. Furthermore, neuronal growth in the presence of growth factors, such as basic fibroblast growth factor (bFGF), was dependent on Ca^{2+} entry through TRPC channels, especially TRPC1, 4, but not TRPC5, suggesting their important roles in neuronal growth and survival [17]. Importantly, inhibiting the function of TRPC channels or even silencing of TRPC1 alone decreases bFGF-induced intracellular Ca^{2+} increase and proliferation of neuronal stem cells [36, 40, 63]. TRPC3 has also been shown to be associated with BDNF receptors stimulation of neuronal growth [21, 66] and Epo-persuade cell differentiation and proliferation [39]. Consistent with these results, expression of TRPC1, TRPC2, and TRPC4 is observed to be higher, whereas expression of TRPC6 is decreased in neuronal stem cells. This differential expression of various TRPC channels suggests that different TRPC channels play contrasting roles in neuronal cell proliferation and differentiation [63]. In addition, using hippocampal neuronal cells (H19-7), Wu and their colleagues showed that Ca^{2+} influx through TRPC1 and TRPC3 was essential in regulating the shift between proliferation and differentiation [71].

Importantly, Ca^{2+} entry is not always beneficial. Activation by massive glutamate increases $[Ca^{2+}]_i$ mainly through TRPC1 channels, which leads to cell death as observed in hippocampal organotypic slice cultures. TRPC1 expression is enhanced after glutamate treatment, and both inhibition of TRPC channel by 2APB and knockdown of TRPC1 significantly reduce cell death, indicating that TRPC1 is involved in glutamate-induced cell death in the hippocampus. In contrast, studies have also shown that physiological activation of TRPC1 through other G-protein-coupled receptors could protect the neurons from several extracellular stimuli [15, 35, 37, 71]. Similarly, loss of Ca^{2+} entry has also been shown to decrease endoplasmic reticulum (ER) Ca^{2+} levels, induce ER stress, and promote cell death [58]. Together, these studies suggest that TRPC channels have a dual role where normal activation of TRPC channels regulates neuronal development, while excessive Ca^{2+} influx, as observed upon glutamate treatment, can induce neuronal damage (Fig. 8.2). In addition, there appears to be a set point for Ca^{2+} entry, where physiologic concentration of calcium is beneficial, but either excessive or decreased calcium entry is harmful. Consistent with this notion, studies show that equally high Ca^{2+} loads are toxic through the

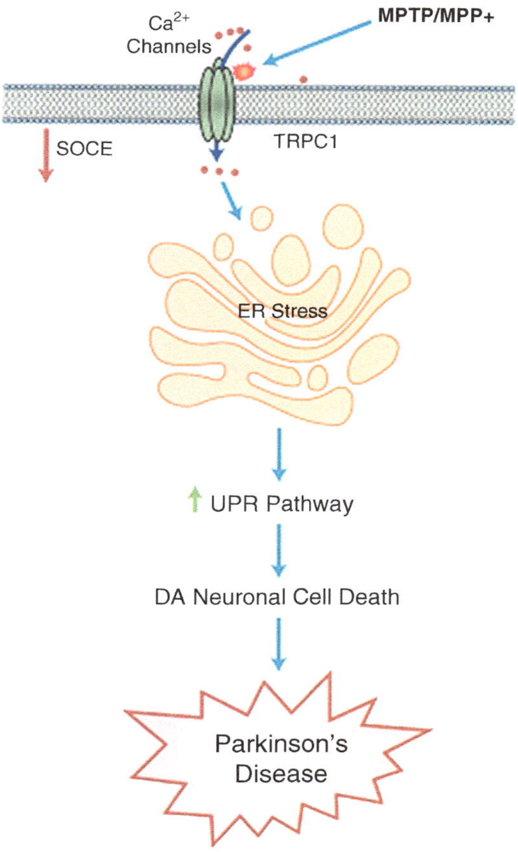

Fig. 8.2 Role of TRPC proteins in PD. Proposed model for MPP^+/MPTP-induced DA loss which could lead to the onset/progression of PD. MPP^+/MPTP attenuates the expression of TRPC1 and SOC-mediated Ca^{2+} influx, which leads to prolonged ER Ca^{2+} depletion and activation of the UPR pathways and subsequent ER stress-mediated neurodegeneration

NMDA channels, compared to through the voltage-dependent Ca^{2+} channels [47, 61] suggesting that diverse Ca^{2+} channels have different roles in deciding the fate of the neuron. Another explanation could be that TRPC multimers that consist of different subunits could have completely different functions. Consistent with this notion, TRPC1, TRPC4, and TRPC5 are observed to be highly expressed in the pyramidal cell layer of the hippocampus, frontal cortex, and dentate gyrus, while TRPC6 is dispersedly expressed only at the molecular layer of the dentate gyrus [14, 61]. Furthermore, both TRPC1 and TRPC3 are observed to protect hippocampal cell lines, and silencing either of these channels inhibits cell development and proliferation [71]. In contrast, the expression of TRPC1 and TRPC6 is found in the substantia nigra region and colocalizes with tyrosine hydroxylase (TH), as well as with mGluR1, suggesting their role in modulating dopaminergic neuron function [9, 16]. Selective degeneration neuron is a common feature in several neurodegenerative diseases including Alzheimer's disease, epilepsy, Huntington's disease, stroke, and Parkinson's disease. Accumulating evidence suggests that glutamate, an excitatory neurotransmitter, is involved in the neurodegeneration [50]. In addition, TRPC channels could also function as a scaffold protein to regulate other proteins. This function is independent of their traditional role of regulating Ca^{2+} influx, and most TRPC have been shown to form large multimers. Thus, exploring the mechanisms underlying expression regulation and Ca^{2+} entry modulation of TRPC channels would be helpful to clarify their roles in neurodegeneration.

8.3 Role of TRPC Channels in Parkinson's Disease

Parkinson's disease (PD) is the second most common neurodegenerative disorder, which is caused by progressive loss of dopaminergic (DA) neurons in the substantia nigra pars compacta (SNpc). Degeneration of DA neurons in the SNpc region is the reason for the observed motor symptoms with this disease [26]. Although the mechanism underlying the selective degeneration of these neurons is largely unknown, a host of pathogenic factors have been suggested to be responsible for the degeneration of DA neurons in the SNpc. These factors include mitochondrial dysfunction, ER and oxidative stress, protein aggregation, excitotoxicity, and inflammation, and each factor could individually or collectively play a role in the pathogenesis of PD (Fig. 8.1). In addition, lysosomal-mediated degradation has recently been suggested to inhibit the protein degradation pathway that could eventually lead to protein aggregation as observed in PD. The role of Ca^{2+} regulation in PD has been of interest as it overlaps with several of the established pathways that lead to neurodegeneration. Importantly, as discussed above, changes in $[Ca^{2+}]_i$ could mediate intracellular events that trigger or inhibit cell death process [7, 53]. Increases in $[Ca^{2+}]_i$ via the Cav1.3 channels has been shown to be necessary to stimulate the release of dopamine (DA) from dopaminergic neurons of the SNpc [12, 49]. Interestingly, Cav1.3 channels are highly expressed in samples obtained from PD patients, indicating the importance of Ca^{2+} in PD. Furthermore, disturbances in Ca^{2+} homeostasis have been implicated in PD [1, 10, 11, 46, 77, 78]. As many factors involved in neuronal functions are dependent on Ca^{2+} signaling, it could be anticipated that loss of these critical functions could contribute to PD [7, 53]. In addition, $[Ca^{2+}]i$ is maintained by the removal of Ca^{2+} through plasma membrane Ca^{2+}-ATPase pump and the Na^+/Ca^{2+} exchanger or sequestration into intracellular organelle stores by the sarco-endoplasmic reticulum ATPase pump (SERCA), and most of these processes require ATP. However, ATP level is decreased in PD; thus, it can be speculated that the dysfunction of these Ca^{2+}pump could be involved in PD pathogenesis. Moreover, decreases in ATP levels could decrease ER calcium levels that could stimulate store-operated Ca^{2+} entry which is dependent on TRPC and Orai channels. Thus, it is critical to establish their physiological functions as discussed in this chapter.

8.3.1 TRPC1

TRPC1, the founding member of TRPCs, is ubiquitously expressed in high levels in all neuronal tissues [2, 62]. Alterations in intracellular Ca^{2+}, especially in the storage organelles such as the ER and mitochondria, have also been shown to affect neuronal survival and are associated with PD [28]. Although several Ca^{2+} channels have been identified that bring Ca^{2+} into cells, Ca^{2+} entry through the store-operated calcium channels (SOC) could be the most important. Moreover, as Ca^{2+} entry through SOC has been shown to be essential for maintaining intracellular ER Ca^{2+} stores, along with regulating many cellular functions, they could play an important role in neuronal survival [8, 58, 64]. TRPC1 has been shown to be essential for the formation of the functional SOC as it is activated by store depletion per se in DA neurons [8, 58, 59, 64, 68].

One of the possible mechanisms that lead to neurodegeneration in PD could be the initiation of ER stress or the unfolded protein response (UPR) [22, 23]. Ca^{2+} in the ER is known to be important for protein synthesis and folding; thus, loss of this vital function could induce abnormal protein aggregation as well as ER stress that could activate cell death cascades [52, 73]. Ca^{2+} influx through the SOCE mechanism is essential for the refilling of the ER Ca^{2+} stores, which could prevent abnormal protein aggregation and ER stress. Recently published work [8, 58, 59] suggests that in DA neurons TRPC1 functions as the endogenous plasma membrane SOC Ca^{2+} channel. Importantly, the expression of TRPC1, but not that of other TRPCs as well as ORAI1 or SOCE modulator stromal interaction molecule 1 (STIM1), was decreased in cell and animal models as well as in PD patients. Interestingly, Ca^{2+} entry into DA neurons was also inhibited by neurotoxin that induced PD-like symptoms and decreased ER Ca^{2+} levels. Consistent with these results, overexpression of TRPC1 reduced neurotoxicity induced by MPP^{+} or salsolinol [8, 58]. In contrast, knockdown of TRPC1 or by addition of TRPC channel blockers inhibited DA neuron survival, indicating that the neuron protection role of TRPC1 might result from its Ca^{2+} influx ability. Similar results are also found in in vivo model and in tissue samples from human patients. Loss of TRPC1 shows a decrease in ER Ca^{2+} level and initiates the unfolded protein response. Moreover, overexpression of functional TRPC1 is protected against neurotoxin-induced loss of SOCE and the resultant ER stress response. In contrast, silencing of TRPC1 or its modulator STIM1 increased the UPR, which suggests that ER stress is induced in PD and TRPC1 attenuates ER stress [58] (Fig. 8.2). Furthermore, Ca^{2+} entry via TRPC1 also activates AKT/mTOR signaling and contributes to neuronal survival [58].

Mitochondrial dysfunction is another mechanism that has been demonstrated to have a role in the PD pathogenesis [41]. Importantly, calcium entry via SOC has been shown to modulate mitochondrial Ca^{2+} levels, and alterations in these Ca^{2+} levels could also lead to neurodegeneration. In addition, a strong correlation between SOCE and apoptosis has also been proposed, indicating that lack of Ca^{2+} entry would contribute to apoptosis [27]. Neurotoxins induce neuronal loss by decreasing TRPC1 levels, which could decrease mitochondrial Ca^{2+} levels necessary for ATP synthesis followed by disrupting mitochondrial membrane potential and initiation of apoptosis. Consistent with these results, activation of TRPC1 maintains mitochondrial membrane potential and inhibits Bax translocation to the mitochondria to prevent cytochrome c release and mitochondrial-mediated apoptosis. These results suggest that TRPC1 could prevent neurotoxin-induced cellular death by maintaining mitochondrial membrane potential, which prevents neurons from apoptosis [8, 10, 59]. Moreover, another study shows that downregulation of STIM1 expression, which is known to activate TRPC channels, inhibits cell apoptosis and reduces intracellular ROS production in PC12 cells by 6-hydroxydopamine [38].

8.3.2 TRPC2

Among the seven members of the TRPC, the TRPC2 channel, being a pseudogene in human, is the least investigated [33, 67]. Therefore, little is known about its physiological function. However, in rodents TRPC2 is highly expressed to the dendritic tip of the vomeronasal sensory neurons [30, 72] and plays an important role in pheromone sensing [18, 29], while its role in PD is not yet identified.

8.3.3 TRPC3

TRPC3 is highly expressed in the brain and oxidative stress has been shown to activate TRPC3 channels. The oxidant tertiary butyl hydroperoxide completely depolarized endothelial cells by activating TRPC3 [57], suggesting that TRPC3 determines endothelial redox sensitivity. In addition, overexpression of TRPC3 in HEK293T cells shows an increase in basal membrane conductance upon tertiary butyl hydroperoxide treatment, which is mainly due to the influx of Na^+ [51]. In another study in primary rat cortical neurons and astrocytes, TRPC3 levels and TRPC3-mediated Ca^{2+} flux are dose-dependently decreased upon treatment with oxidative stressors [56]. In addition, in murine striatal astrocytes, additions of neurotoxins which mimic PD decrease ATP level and OAG-induced Ca^{2+} transients, [60]. Importantly, a slight increase in TRPC3 expression is observed in PD condition as well as neurotoxin models of PD [58]. Disruption of Ca^{2+} signaling especially in astrocytes significantly impairs neuronal function and survival in neurological injury and in disease conditions such as PD. Together, these studies suggest that TRPC3 dysfunction is involved in Ca^{2+} dyshomeostasis and oxidative stress signaling observed in PD.

Parkinsonian movement disorders are also associated with abnormalities in SN pars reticulata (SNr) [34, 45, 54, 69]. TRPC3 channels are expressed in SNr GABA projection neurons, where TRPC3 channels are tonically active and mediate a voltage-independent inward current, leading to a substantial depolarization in these neurons [76]. Inhibition of TRPC3 channels induces hyperpolarization, decreases firing frequency, and increases firing irregularity, suggesting that TRPC3 channels play critical roles in maintaining the depolarized membrane potential, high firing frequency, and firing regularity in these basal ganglia output neurons crucial to Parkinsonian movement disorders [76]. In addition, dopamine released via the dopamine receptors from the dendrites activates TRPC3 channels in SNr GABA neurons and mediates an inward, Na^+-dependent current, leading to a substantial depolarization and ensuring appropriate firing intensity and pattern in SNr GABA projection neurons [75]. TRPC3 channels have also been shown to modulate motor coordination [5, 19]. In an ataxic mouse mutant (moonwalker, Mwk mice) that displays motor and coordination defects, a gain-of-function mutation (T635A) in TRPC3 channels is observed. Sustained activation of TRPC3 channels is observed to be associated with diminished dendritic arborization and progressive loss of Purkinje neurons. Similarly, another study also shows that loss of TRPC3 exhibits atrophy and progressive paralysis [55].

8.3.4 TRPC4

TRPC4α/β isoforms are the most abundantly expressed and functionally characterized in brain. TRPC4 and TRPC5 are the major TRPC subtypes in the adult rat brain because both are expressed highly in the pyramidal cell layer of the hippocampus, frontal cortex, and dentate gyrus [14, 61]. TRPC4 is specifically detected throughout the layers (2–6) of the prefrontal cortex or the motor cortex [25, 42]. Although the role of TRPC4 in PD is not yet defined, its role in axonal regeneration in adult rat dorsal root ganglia (DRG) has been reported [70]. The expression of TRPC4 is enhanced, whereas TRPC1, TRPC3, TRPC6, and TRPC7 expression remains unchanged after nerve injury persuaded by either sciatic nerve transection or intra-ganglionic microinjection of dibutyryl cAMP [57, 70]. In addition, TRPC4 expression in various neuronal cells has been shown to be increased upon addition of NGF and dibutyryl cAMP that induced differentiation [57, 70]. Suppression of TRPC4 by specific small interfering RNA significantly reduced the length of neuritis in cultured DRG neurons [70]. Taken together, these results suggest that TRPC4 contributes to axonal regeneration especially after nerve injury. If these findings have generality, TRPC4 could be an important molecular target for potential regeneration therapies in patients suffering from neuronal injury. In contrast, by using whole-genome sequencing, a recent report showed that gain-of-function mutations in TRPC4 gene induce cell death in DA neurons through a defined, calcium-related downstream pathway [43]. High expression of TRPC4 is also found in the PM of soma and proximal dendrites of lateral septal neurons, colocalizing with mGlu receptors, which could also contribute to cell death. Studies also show that TRPC4 is expressed in cells in the ventral tegmental area, a region with extensive inputs from dopamine neurons which are important in regulating the animal behavior [24]. Importantly, a recent report has shown that self-administration of cocaine was significantly less in the TRPC4 KO group than WT controls [31]. Also, spontaneous DA neuronal activity in the ventral tegmental area revealed fewer cells with high-frequency firing rate in rats that lack TRPC4. Together these studies show the roles of TRPC4 channels in various functions of the CNS making them a potential target, especially for neurological diseases associated with excitotoxicity like PD or drug addiction that are also dependent on DA neurons.

8.4 Conclusion

Plenty of research in the past decades has explored the role of TRPC channels in neuronal survival, differentiation, and neurodegeneration as observed in Parkinson's disease. It could be suggested that Ca^{2+} influx via the TRPC channels has an important role in the neurodegenerative diseases such as Parkinson's, although further studies are needed. Therefore, targeting Ca^{2+} entry (both inhibition and activation) through TRPC channels could be critical for maintaining normal physiological function in dopaminergic neurons. Moreover, recent findings have also implicated STIM1 as a regulator for SOCE, making STIM1 a potential target, as they may be involved in neurological diseases such as Parkinson's.

Acknowledgments We thank the grant support from the NIH (DE017102; DE024300; GM113123) awarded to B.B.S, and the assistance of John Lee, School of Medicine and Health Sciences, in making the figures.

References

1. Albers DS, Beal MF (2000) Mitochondrial dysfunction and oxidative stress in aging and neurodegenerative disease. J Neural Transm Suppl 59:133–154
2. Ambudkar IS, Bandyopadhyay BC, Liu X, Lockwich TP, Paria B, Ong HL (2006) Functional organization of TRPC-Ca2+ channels and regulation of calcium microdomains. Cell Calcium 40(5–6):495–504. doi:10.1016/j.ceca.2006.08.011
3. Arundine M, Tymianski M (2003) Molecular mechanisms of calcium-dependent neurodegeneration in excitotoxicity. Cell Calcium 34(4–5):325–337
4. Badger JL, Cordero-Llana O, Hartfield EM, Wade-Martins R (2014) Parkinson's disease in a dish – using stem cells as a molecular tool. Neuropharmacology 76(Pt A):88–96. doi:10.1016/j.neuropharm.2013.08.035
5. Becker EB, Oliver PL, Glitsch MD, Banks GT, Achilli F, Hardy A, Nolan PM, Fisher EM, Davies KE (2009) A point mutation in TRPC3 causes abnormal Purkinje cell development and cerebellar ataxia in moonwalker mice. Proc Natl Acad Sci U S A 106(16):6706–6711. doi:10.1073/pnas.0810599106
6. Berridge MJ, Bootman MD, Roderick HL (2003) Calcium signalling: dynamics, homeostasis and remodelling. Nat Rev Mol Cell Biol 4(7):517–529. doi:10.1038/nrm1155
7. Berridge MJ, Lipp P, Bootman MD (2000) The versatility and universality of calcium signalling. Nat Rev Mol Cell Biol 1(1):11–21. doi:10.1038/35036035. 35036035 [pii]
8. Bollimuntha S, Ebadi M, Singh BB (2006) TRPC1 protects human SH-SY5Y cells against salsolinol-induced cytotoxicity by inhibiting apoptosis. Brain Res 1099(1):141–149. doi:10.1016/j.brainres.2006.04.104
9. Bollimuntha S, Selvaraj S, Singh BB (2011) Emerging roles of canonical TRP channels in neuronal function. Adv Exp Med Biol 704:573–593. doi:10.1007/978-94-007-0265-3_31

10. Bollimuntha S, Singh BB, Shavali S, Sharma SK, Ebadi M (2005) TRPC1-mediated inhibition of 1-methyl-4-phenylpyridinium ion neurotoxicity in human SH-SY5Y neuroblastoma cells. J Biol Chem 280(3):2132–2140. doi:10.1074/jbc.M407384200
11. Cali T, Ottolini D, Brini M (2011) Mitochondria, calcium, and endoplasmic reticulum stress in Parkinson's disease. Biofactors 37(3):228–240. doi:10.1002/biof.159
12. Chen BT, Rice ME (2001) Novel Ca2+ dependence and time course of somatodendritic dopamine release: substantia nigra versus striatum. J Neurosci 21(19):7841–7847
13. Choi DW (1988) Calcium-mediated neurotoxicity: relationship to specific channel types and role in ischemic damage. Trends Neurosci 11(10):465–469
14. Chung YH, Sun Ahn H, Kim D, Hoon Shin D, Su Kim S, Yong Kim K, Bok Lee W, Ik Cha C (2006) Immunohistochemical study on the distribution of TRPC channels in the rat hippocampus. Brain Res 1085(1):132–137. doi:10.1016/j.brainres.2006.02.087
15. Dhar M, Wayman GA, Zhu M, Lambert TJ, Davare MA, Appleyard SM (2014) Leptin-induced spine formation requires TrpC channels and the CaM kinase cascade in the hippocampus. J Neurosci 34(30):10022–10033. doi:10.1523/JNEUROSCI.2868-13.2014
16. Fendt M, Schmid S, Thakker DR, Jacobson LH, Yamamoto R, Mitsukawa K, Maier R, Natt F, Husken D, Kelly PH, McAllister KH, Hoyer D, van der Putten H, Cryan JF, Flor PJ (2008) mGluR7 facilitates extinction of aversive memories and controls amygdala plasticity. Mol Psychiatry 13(10):970–979. doi:10.1038/sj.mp.4002073
17. Fiorio Pla A, Maric D, Brazer S-C, Giacobini P, Liu X, Chang YH, Ambudkar IS, Barker JL (2005) Canonical transient receptor potential 1 plays a role in basic fibroblast growth factor (bFGF)/FGF receptor-1-induced Ca2+ entry and embryonic rat neural stem cell proliferation. J Neurosci 25(10):2687–2701. doi:10.1523/JNEUROSCI.0951-04.2005
18. Flanagan KA, Webb W, Stowers L (2011) Analysis of male pheromones that accelerate female reproductive organ development. PLoS ONE 6(2):e16660. doi:10.1371/journal.pone.0016660
19. Hartmann J, Dragicevic E, Adelsberger H, Henning HA, Sumser M, Abramowitz J, Blum R, Dietrich A, Freichel M, Flockerzi V, Birnbaumer L, Konnerth A (2008) TRPC3 channels are required for synaptic transmission and motor coordination. Neuron 59(3):392–398. doi:10.1016/j.neuron.2008.06.009
20. Herman B, Gores GJ, Nieminen AL, Kawanishi T, Harman A, Lemasters JJ (1990) Calcium and pH in anoxic and toxic injury. Crit Rev Toxicol 21(2):127–148. doi:10.3109/10408449009089876
21. Hirschler-Laszkiewicz I, Tong Q, Conrad K, Zhang W, Flint WW, Barber AJ, Barber DL, Cheung JY, Miller BA (2009) TRPC3 activation by erythropoietin is modulated by TRPC6. J Biol Chem 284(7):4567–4581. doi:10.1074/jbc.M804734200
22. Hoozemans JJ, van Haastert ES, Eikelenboom P, de Vos RA, Rozemuller JM, Scheper W (2007) Activation of the unfolded protein response in Parkinson's disease. Biochem Biophys Res Commun 354(3):707–711. doi:10.1016/j.bbrc.2007.01.043
23. Hoozemans JJ, van Haastert ES, Nijholt DA, Rozemuller AJ, Scheper W (2012) Activation of the unfolded protein response is an early event in Alzheimer's and Parkinson's disease. Neurodegener Dis 10(1–4):212–215. doi:10.1159/000334536
24. Huang J, Du W, Yao H, Wang Y (2011) TRPC channels in neuronal survival. In: Zhu MX (ed) TRP channels. CRC Press, Boca Raton
25. Huang WC, Young JS, Glitsch MD (2007) Changes in TRPC channel expression during postnatal development of cerebellar neurons. Cell Calcium 42(1):1–10. doi:10.1016/j.ceca.2006.11.002
26. Javitch JA, D'Amato RJ, Strittmatter SM, Snyder SH (1985) Parkinsonism-inducing neurotoxin, N-methyl-4-phenyl-1,2,3,6 -tetrahydropyridine: uptake of the metabolite N-methyl-4-phenylpyridine by dopamine neurons explains selective toxicity. Proc Natl Acad Sci U S A 82(7):2173–2177
27. Jayadev S, Petranka JG, Cheran SK, Biermann JA, Barrett JC, Murphy E (1999) Reduced capacitative calcium entry correlates with vesicle accumulation and apoptosis. J Biol Chem 274(12): 8261–8268
28. Kazuno AA, Munakata K, Nagai T, Shimozono S, Tanaka M, Yoneda M, Kato N, Miyawaki A, Kato T (2006) Identification of mitochondrial DNA polymorphisms that alter mitochondrial matrix pH and intracellular calcium dynamics. PLoS Genet 2(8):e128. doi:10.1371/journal.pgen.0020128
29. Kimchi T, Xu J, Dulac C (2007) A functional circuit underlying male sexual behaviour in the female mouse brain. Nature 448(7157):1009–1014. doi:10.1038/nature06089
30. Kiselyov K, van Rossum DB, Patterson RL (2010) TRPC channels in pheromone sensing. Vitam Horm 83:197–213. doi:10.1016/S0083-6729(10)83008-0
31. Klipec WD, Burrow KR, O'Neill C, Cao JL, Lawyer CR et al (2016) Loss of the trpc4 gene is associated with a reduction in cocaine self-administration and reduced spontaneous ventral tegmental area dopamine neuronal activity, without deficits in learning for natural rewards. Behav Brain Res 306:117–127
32. Komuro H, Kumada T (2005) Ca2+ transients control CNS neuronal migration. Cell Calcium 37:387–393
33. Löf C, Viitanen T, Sukumaran P, Törnquist K (2011) TRPC2: of mice but not men. Adv Exp Med Biol 704:125–134. doi:10.1007/978-94-007-0265-3_6
34. Lee JI, Verhagen Metman L, Ohara S, Dougherty PM, Kim JH, Lenz FA (2007) Internal pallidal neuronal activity during mild drug-related dyskinesias in Parkinson's disease: decreased firing rates and altered firing patterns. J Neurophysiol 97(4):2627–2641. doi:10.1152/jn.00443.2006

35. Li H, Huang J, Du W, Jia C, Yao H, Wang Y (2012) TRPC6 inhibited NMDA receptor activities and protected neurons from ischemic excitotoxicity. J Neurochem 123(6):1010–1018. doi:10.1111/jnc.12045
36. Li HS, Xu XZ, Montell C (1999) Activation of a TRPC3-dependent cation current through the neurotrophin BDNF. Neuron 24(1):261–273
37. Li M, Chen C, Zhou Z, Xu S, Yu Z (2012) A TRPC1-mediated increase in store-operated Ca2+ entry is required for the proliferation of adult hippocampal neural progenitor cells. Cell Calcium 51(6):486–496. doi:10.1016/j.ceca.2012.04.014
38. Li Y, Guo Y, Tang J, Jiang J, Chen Z (2014) New insights into the roles of CHOP-induced apoptosis in ER stress. Acta Biochim Biophys Sin 46:629–640
39. Mattson MP, Guthrie PB, Kater SB (1989) Intrinsic factors in the selective vulnerability of hippocampal pyramidal neurons. Prog Clin Biol Res 317:333–351
40. Montecinos-Oliva C, Schuller A, Parodi J, Melo F, Inestrosa NC (2014) Effects of tetrahydrohyperforin in mouse hippocampal slices: neuroprotection, long-term potentiation and TRPC channels. Curr Med Chem 21(30):3494–3506
41. Moon HE, Paek SH (2015) Mitochondrial dysfunction in Parkinson's disease. Exp Neurobiol 24(2):103–116. doi:10.5607/en.2015.24.2.103
42. Morelli MB, Amantini C, Liberati S, Santoni M, Nabissi M (2013) TRP channels: new potential therapeutic approaches in CNS neuropathies. CNS Neurol Disord Drug Targets 12(2):274–293
43. Nagarajan A, Ning Y, Reisner K, Buraei Z, Larsen JP et al (2014) Progressive degeneration of dopaminergic neurons through TRP channel-induced cell death. J Neurosci 34:5738–5746
44. Naziroglu M (2011) TRPM2 cation channels, oxidative stress and neurological diseases: where are we now? Neurochem Res 36(3):355–366. doi:10.1007/s11064-010-0347-4
45. Nevet A, Morris G, Saban G, Fainstein N, Bergman H (2004) Discharge rate of substantia nigra pars reticulata neurons is reduced in non-parkinsonian monkeys with apomorphine-induced orofacial dyskinesia. J Neurophysiol 92(4):1973–1981. doi:10.1152/jn.01036.2003
46. O'Bryant SE, Hobson V, Hall JR, Waring SC, Chan W, Massman P, Lacritz L, Cullum CM, Diaz-Arrastia R (2009) Brain-derived neurotrophic factor levels in Alzheimer's disease. J Alzheimers Dis 17(2): 337–341. doi:10.3233/JAD-2009-1051doi:052G70T078857553 [pii]
47. Ovey IS, Naziroglu M (2015) Homocysteine and cytosolic GSH depletion induce apoptosis and oxidative toxicity through cytosolic calcium overload in the hippocampus of aged mice: involvement of TRPM2 and TRPV1 channels. Neuroscience 284:225–233. doi:10.1016/j.neuroscience.2014.09.078
48. Paschen W (2000) Role of calcium in neuronal cell injury: which subcellular compartment is involved? Brain Res Bull 53(4):409–413
49. Patel JC, Witkovsky P, Avshalumov MV, Rice ME (2009) Mobilization of calcium from intracellular stores facilitates somatodendritic dopamine release. J Neurosci 29(20):6568–6579. doi:10.1523/JNEUROSCI.0181-09.2009
50. Phelan KD, Shwe UT, Abramowitz J, Wu H, Rhee SW, Howell MD, Gottschall PE, Freichel M, Flockerzi V, Birnbaumer L, Zheng F (2013) Canonical transient receptor channel 5 (TRPC5) and TRPC1/4 contribute to seizure and excitotoxicity by distinct cellular mechanisms. Mol Pharmacol 83(2):429–438. doi:10.1124/mol.112.082271
51. Poteser M, Graziani A, Rosker C, Eder P, Derler I, Kahr H, Zhu MX, Romanin C, Groschner K (2006) TRPC3 and TRPC4 associate to form a redox-sensitive cation channel. Evidence for expression of native TRPC3-TRPC4 heteromeric channels in endothelial cells. J Biol Chem 281(19):13588–13595. doi:10.1074/jbc.M512205200
52. Putney JW (2005) Physiological mechanisms of TRPC activation. Pflugers Arch Eur J Physiol 451(1):29–34. doi:10.1007/s00424-005-1416-4
53. Putney JW Jr (2003) Capacitative calcium entry in the nervous system. Cell Calcium 34(4–5):339–344. doi:S014341600300143X [pii]
54. Rivlin-Etzion M, Marmor O, Heimer G, Raz A, Nini A, Bergman H (2006) Basal ganglia oscillations and pathophysiology of movement disorders. Curr Opin Neurobiol 16(6):629–637. doi:10.1016/j.conb.2006.10.002
55. Rodriguez-Santiago M, Mendoza-Torres M, Jimenez-Bremont JF, Lopez-Revilla R (2007) Knockout of the trcp3 gene causes a recessive neuromotor disease in mice. Biochem Biophys Res Commun 360(4):874–879. doi:10.1016/j.bbrc.2007.06.150
56. Roedding AS, Tong SY, Au-Yeung W, Li PP, Warsh JJ (2013) Chronic oxidative stress modulates TRPC3 and TRPM2 channel expression and function in rat primary cortical neurons: relevance to the pathophysiology of bipolar disorder. Brain Res 1517:16–27. doi:10.1016/j.brainres.2013.04.025
57. Selvaraj S, Sun Y, Singh BB (2010) TRPC channels and their implication in neurological diseases. CNS Neurol Disord Drug Targets 9(1):94–104
58. Selvaraj S, Sun Y, Watt JA, Wang S, Lei S, Birnbaumer L, Singh BB (2012) Neurotoxin-induced ER stress in mouse dopaminergic neurons involves downregulation of TRPC1 and inhibition of AKT/mTOR signaling. J Clin Invest 122(4):1354–1367. doi:10.1172/JCI61332
59. Selvaraj S, Watt JA, Singh BB (2009) TRPC1 inhibits apoptotic cell degeneration induced by dopaminergic neurotoxin MPTP/MPP(+). Cell Calcium 46(3):209–218. doi:10.1016/j.ceca.2009.07.008
60. Streifel KM, Gonzales AL, De Miranda B, Mouneimne R, Earley S, Tjalkens R (2014) Dopaminergic neurotoxicants cause biphasic inhibition of purinergic calcium signaling in astrocytes. PLoS ONE 9:e110996

61. Strübing C, Krapivinsky G, Krapivinsky L, Clapham DE (2001) TRPC1 and TRPC5 form a novel cation channel in mammalian brain. Neuron 29(3):645–655
62. Strübing C, Krapivinsky G, Krapivinsky L, Clapham DE (2003) Formation of novel TRPC channels by complex subunit interactions in embryonic brain. J Biol Chem 278(40):39014–39019. doi:10.1074/jbc.M306705200
63. Su B, Ji Y-S, X-l S, Liu X-H, Chen Z-Y (2014) Brain-derived neurotrophic factor (BDNF)-induced mitochondrial motility arrest and presynaptic docking contribute to BDNF-enhanced synaptic transmission. J Biol Chem 289(3):1213–1226. doi:10.1074/jbc.M113.526129
64. Sun Y, Sukumaran P, Bandyopadhyay BC, Singh BB (2014) Physiological function and characterization of TRPCs in neurons. Cell 3(2):455–475. doi:10.3390/cells3020455
65. Sun Y, Sukumaran P, Schaar A, Singh BB (2015) TRPM7 and its role in neurodegenerative diseases. Channels (Austin) 9(5):253–261. doi:10.1080/19336950.2015.1075675
66. Tong Q, Hirschler-Laszkiewicz I, Zhang W, Conrad K, Neagley DW, Barber DL, Cheung JY, Miller BA (2008) TRPC3 is the erythropoietin-regulated calcium channel in human erythroid cells. J Biol Chem 283(16):10385–10395. doi:10.1074/jbc.M710231200
67. Tornquist K, Sukumaran P, Kemppainen K, Lof C, Viitanen T (2014) Canonical transient receptor potential channel 2 (TRPC2): old name-new games. Importance in regulating of rat thyroid cell physiology. Pflugers Arch Eur J Physiol 466(11):2025–2034. doi:10.1007/s00424-014-1509-z
68. Wang GX, Poo MM (2005) Requirement of TRPC channels in netrin-1-induced chemotropic turning of nerve growth cones. Nature 434(7035):898–904. doi:10.1038/nature03478
69. Wichmann T, Soares J (2006) Neuronal firing before and after burst discharges in the monkey basal ganglia is predictably patterned in the normal state and altered in parkinsonism. J Neurophysiol 95(4):2120–2133. doi:10.1152/jn.01013.2005
70. Wu D, Huang W, Richardson PM, Priestley JV, Liu M (2008) TRPC4 in rat dorsal root ganglion neurons is increased after nerve injury and is necessary for neurite outgrowth. J Biol Chem 283(1):416–426. doi:10.1074/jbc.M703177200
71. Wu X, Zagranichnaya TK, Gurda GT, Eves EM, Villereal ML (2004) A TRPC1/TRPC3-mediated increase in store-operated calcium entry is required for differentiation of H19-7 hippocampal neuronal cells. J Biol Chem 279(42):43392–43402. doi:10.1074/jbc.M408959200
72. Yildirim E, Birnbaumer L (2007) TRPC2: molecular biology and functional importance. Handb Exp Pharmacol 179:53–75. doi:10.1007/978-3-540-34891-7_3
73. Yoshida I, Monji A, Tashiro K, Nakamura K, Inoue R, Kanba S (2006) Depletion of intracellular Ca2+ store itself may be a major factor in thapsigargin-induced ER stress and apoptosis in PC12 cells. Neurochem Int 48(8):696–702. doi:10.1016/j.neuint.2005.12.012
74. Yuruker V, Naziroglu M, Senol N (2015) Reduction in traumatic brain injury-induced oxidative stress, apoptosis, and calcium entry in rat hippocampus by melatonin: possible involvement of TRPM2 channels. Metab Brain Dis 30(1):223–231. doi:10.1007/s11011-014-9623-3
75. Zhou FM (2010) A transient receptor potential channel regulates basal ganglia output. Rev Neurosci 21:95–118
76. Zhou FW, Matta SG, Zhou FM (2008) Constitutively active TRPC3 channels regulate basal ganglia output neurons. J Neurosci 28(2):473–482. doi:10.1523/JNEUROSCI.3978-07.2008
77. Zuccato C, Cattaneo E (2007) Role of brain-derived neurotrophic factor in Huntington's disease. Prog Neurobiol 81(5–6):294–330 . doi:10.1016/j.pneurobio.2007.01.003doi:S0301-0082(07)00015-9 [pii]
78. Zuccato C, Cattaneo E (2009) Brain-derived neurotrophic factor in neurodegenerative diseases. Nat Rev Neurol 5(6):311–322 . doi:10.1038/nrneurol.2009.54doi:nrneurol.2009.54 [pii]

TRPC Channels and Neuron Development, Plasticity, and Activities

9

Yilin Tai and Yichang Jia

Abstract

In this chapter, we mainly focus on the functions of TRPC channels in brain development, including neural progenitor proliferation, neurogenesis, neuron survival, axon guidance, dendritic morphology, synaptogenesis, and neural plasticity. We also notice emerging advances in understanding the functions of TRPC channels in periphery, especially their functions in sensation and nociception in dorsal root ganglion (DRG). Because TRPC channels are expressed in all major types of glial cells, which account for at least half of total cells in the brain, TRPC channels may act as modulators for glial functions as well. The future challenges for studying these channels could be (1) the detailed protein structures of these channels, (2) their cell type-specific functions, (3) requirement for their specific blockers or activators, and (4) change in the channel conformation in the brain.

Keywords

TRPC channels • Neuron • Development • Plasticity

9.1 Introduction

Neurons are born to be connected, but the way to precise connection is a long journey. TRPC channels are faithful companions along the journey. In general, most of neurons are bipolar. However, after a neuron is born from a neural progenitor cell, it undergoes multipolar stage when the neuron extends several neurites without defined polarity. Affected by both extrinsic environmental cues and intrinsic genetic programs, the neuron changes from multipolar to bipolar stage

Y. Tai
Cold Spring Harbor Laboratory,
1 Bungtown Road, Cold Spring Harbor, NY 11724, USA

Y. Jia (✉)
School of Medicine, Tsinghua University,
Medical Science Building, Room D204, Beijing 100084, China

Peking-Tsinghua Joint Center for Life Sciences,
Beijing 100084, China

IDG/McGovern Institute for Brain Research at Tsinghua, Beijing 100084, China
e-mail: yichangjia@tsinghua.edu.cn

Y. Wang (ed.), *Transient Receptor Potential Canonical Channels and Brain Diseases*,
Advances in Experimental Medicine and Biology 976, DOI 10.1007/978-94-024-1088-4_9

when it has a leading process toward the direction of migration and a trailing process on the opposite. Once the polarity is established, the immature neuron starts to migrate along the radial glial fiber to its final residence where it starts to differentiate. The leading process branches out to become dendrites where the neuron receives inputs from other neurons. The trailing process extends into a long axon to send information to other neurons. With a growth cone that actively senses environmental cue, an axon can be guided to specific targets that in many cases locate very far away from the cell body. During the process, neurons also experience harsh survival selections. In the brain, a plethora of neurons is produced during early brain development. They compete with each other for limited trophic factors. Those who triumphed thrive. Synapses are then formed between axons (presynaptic) and dendrites (postsynaptic) of the survived neurons. Electrical or chemical signals can pass from the axons of presynaptic output neurons to the dendrites of postsynaptic input neurons through synapses. The dendrites and the cell body integrate the signals received from different sources and send it out through axon toward its own targets. In this way, complex neural circuits are formed between and within brain regions, which is the biological basis of animal behaviors. Once the neural connections are formed, they can undergo plastic changes in response to experience, so that the nervous system can better adapt to the environment.

TRPC channels, as discussed in Chaps. 2 and 3, are nonselective cation channels that are permeable to Ca^{2+}. The established modes of gating of TRPC channels are mainly divided into store operated and receptor operated [69]. Through activation of phospholipase C (PLC), G protein-coupled receptors (GPCRs) and receptor tyrosine kinases (RTKs) can modulate TRPC channel activity by hydrolysis of phosphatidylinositol bisphosphate (PIP2) to diacylglycerol (DAG) and inositol triphosphate (IP3). The former is required for receptor-operated activation of TRPC3/6/7, while the latter causes Ca^{2+} release from internal Ca^{2+} store (endoplasmic reticulum, ER), a process that triggers store-operated channel activation [10, 48, 63, 92]. This unique gating mechanism allows TRPC channels to act as downstreams of many GPCRs and RTKs, transducing extracellular stimuli into intracellular Ca^{2+} signals [94]. Ca^{2+} signals have indispensable roles throughout brain development [47], not only because it is an important second messenger but also because it is critical to control neuronal excitability and plasticity [8].

The expression of TRPC channels is ubiquitous in the nervous system, with different combinations in different cell types, and is strictly regulated in different developmental stages, suggesting their unique roles in each context. In this chapter, the functions of TRPC channels will be overviewed in the background of both central nervous system and peripheral nervous system development (Fig. 9.1). And, how TRPC channels affect the neuronal activity and plasticity will also be discussed.

9.2 TRPC Channels in the Central Nervous System

9.2.1 Neuroepithelial Proliferation and Neurogenesis

Ca^{2+} is a ubiquitous second messenger which is critical for proliferation and differentiation through its regulation of cytoskeleton dynamics and gene expression. The development of the central nervous system involves extensive proliferation of uncommitted precursor cells which divide to self-renew committed neural progenitor cells, which in turn differentiate into neurons. Steady-state levels of intracellular Ca^{2+} concentration in proliferating neuroepithelial cells are largely dependent on Ca^{2+} entry [55]. TRPC1, 3, 4, 5, and 6 are highly expressed in cultured embryonic neural stem cells, suggesting developmental roles of TRPC channels. Freshly sorted neural precursor cells from mouse E13 telencephalon have higher expression of TRPC1 and 4 than TRPC3, 5, and 6. TRPC1 is coimmunoprecipitated with FGFR-1, which is a receptor of fibroblast growth factor (FGF). FGF is essential for neuroepithelial cell proliferation. Knockdown

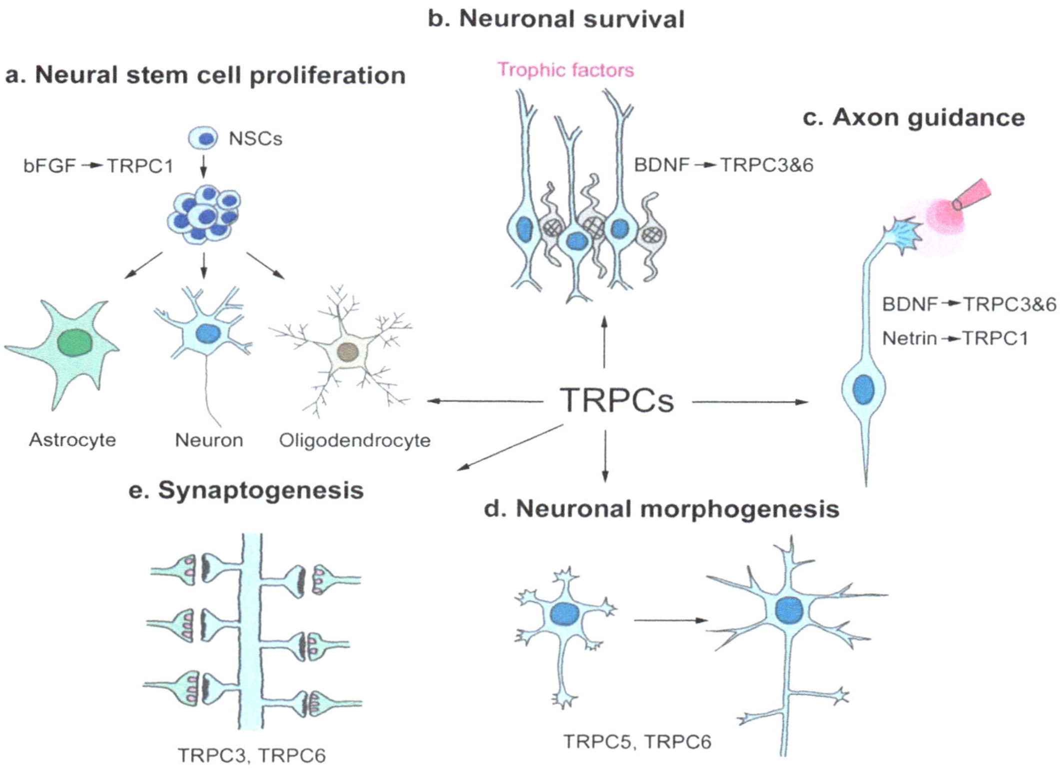

Fig. 9.1 (**a**) TRPC1 channels promote bFGF-induced neural stem cell proliferation. (**b**) TRPC3 and TRPC6 channels promote cerebellum granule cell survival in response to BDNF. (**c**) BDNF-induced chemoattraction is mediated by TRPC3 and TRPC6 channels. TRPC1 channels are required for growth cone turning in response to microscopic gradients of netrin-1. (**d**) TRPC5 negatively regulates dendritic/neurite outgrowth, whereas TRPC6 promotes dendritic growth. (**e**) TRPC6 channel is required for excitatory synaptogenesis. TRPC3 and TRPC6 channels mediate BDNF-induced spine formation [106]

of TRPC1 attenuates FGF-induced Ca^{2+} influx, which is involved in neural precursor cell proliferation [24]. Platelet-derived growth factor-BB (PDGF-BB) has been reported to provide tropic support for neurons in the CNS. It regulates neurogenesis in the context of HIV-associated neurological disorder and drug abuse. Pretreatment of rat hippocampal neuronal progenitor cells (NPCs) with PDGF-BB restored proliferation that had been impaired by the co-application of HIV transactivating protein (Tat) together with cocaine. The PDGF-BB-induced Ca^{2+} influx is mediated by TRPC1, and downregulation of TRPC1 attenuates PDGF-BB-induced NPC proliferation [103].

Adult hippocampal neurogenesis is a unique form of neurogenesis, and the adult neural progenitor cells (aNPCs) generate new neurons in dentate gyrus throughout life [23]. Increasing evidence supports the idea that neurogenesis is involved in normal function of adult hippocampus and impairment of this process will lead to neurological diseases [11, 76, 77]. Adult NPCs share many features with embryonic neural progenitors. Similarly, TRPC1 channels are highly expressed in aNPCs, and inhibition of TRPC1 channel activity arrests the cell cycle at G0/G1 phase and reduces the aNPC proliferation [52].

The function of TRPC1 in neural stem cell is conserved across species. In human telencephalic neuroepithelial cells and post-mitotic neurons, the spontaneous Ca^{2+} transients are sensitive to SKF96365, the antagonist to TRPC channels, and shRNA targeting TRPC1. Knockdown of

TRPC1 in dividing human neuroepithelial cells leads to a significant reduction in their proliferation [99].

9.2.2 Neuronal Survival During Development

Strictly controlled cell death is essential for normal development in multicellular organisms [95]. In the developing brain, neurons compete with each other for limited trophic factors. Those that thrive in the competition survive [46]. The intracellular Ca^{2+} level acts as a double-edged sword in this process. Excessive intracellular Ca^{2+} is a trigger of cell death. However, a modest increase in intracellular Ca^{2+} concentration promotes survival through a pathway that requires calmodulin [102]. Interestingly, many trophic factors, such as brain-derived trophic factor (BDNF), have been reported to be able to trigger a slow but sustained intracellular Ca^{2+} elevation [57]. In fact, BDNF is one of the pro-survival cues during the development of the cerebellum [59]. In addition, BDNF-induced Ca^{2+} influx is partially mediated by TRPC3 and TRPC6 channels [3, 50]. The peak expression of TRPC3 and 6 in the developing cerebellum is observed between postnatal days 10 and 17 [39]. During this period, there is a proliferation of granule cell progenitors in the external granule cell layer, followed by maturation of granule cells into the internal granule cell layer and apoptosis of about 30% of a subpopulation of these cells [84]. Downregulation of TRPC3 or TRPC6 blocks BDNF-induced intracellular Ca^{2+} elevation and subsequent pro-survival effect. Blocking either calmodulin-dependent protein kinase (CaMK) or mitogen-activated protein kinase (MAPK) inhibits the protective effect of TRPC channels. Moreover, the protective effect of TRPC3 and 6 is suppressed by a dominant negative form of CREB. Interestingly, the other pro-survival signaling, AKT signaling, activated by BDNF is not affected by TRPC3 and 6, suggesting the unique downstream of TRPC3/6 activation followed by BDNF bound to its high-affinity receptor TrkB [39]. Activity-dependent process is also important for neuronal survival via activation of the CREB pathway [79]. It is known that Ca^{2+} entry through L-type voltage-dependent Ca^{2+} channels (VDCCs) can lead to the Ca^{2+}-dependent activation of ERK and CaMKIV [18, 29]. Both kinases contribute to CREB phosphorylation and subsequent neuronal survival. These intracellular pro-survival signaling pathways are quite similar to those induced by TRPC channels. However, TRPC channels are not activated by depolarization. It thus appears that CREB is a key point on which both activity-dependent survival signaling and TRPC-induced survival signaling converge.

9.2.3 Neuronal Navigation

Once the neurons are born, they either undergo radial migration along the radial glial fiber or tangential migration directed by the environmental cues [82]. During this process, growth cone, a specialized structure at the tips of the neuritis and palm-shaped filopodia, constantly extends or retracts its protrusion in response to various environmental cues. The movements of these protrusions are believed to be driven by Ca^{2+} influx, which regulates growth cone morphology [31], the cytoskeleton dynamics [16], and positioning of precursor vesicles [12]. In the growth cone, the remnant Ca^{2+} current in the presence of blockers for VDCCs is abolished by SKF96365. Therefore, both TRPC channels and VDCCs are responsible for elevating the Ca^{2+} level in the growth cone. BDNF, also a chemoattractor, can affect growth cone turning. In cultured rat cerebellar granule cells, elevation in $[Ca^{2+}]_i$ and growth cone steering induced by BDNF are abolished when TRPC3 and 6 are inhibited by downregulation or dominant negative mutants of TRPC3 and 6. However, downregulating TRPC1 did not affect the elevation in $[Ca^{2+}]_i$ and growth cone turning induced by BDNF. In contrast, inhibition of XTRPC1, a homologue of TRPC1 in *Xenopus*, makes the growth cone insensitive to the attraction induced by netrin-1 in the cultured *Xenopus* spinal neurons [80, 97]. Meanwhile, knocking down XTRPC1 affects the myelin-associated glycoprotein-induced repulsion, but leaves semaphorin 3A-induced repulsion intact. Injection of XTRPC1 morpholinos into the *Xenopus* blastomere leads to a 65% reduction in the midline crossing of the ventrally projecting commissural axons. These discrepancies could be the result of differential distribution, concentration, and the combination of channel subunits in different cell types and species. Furthermore, stimulation of the same TRPC channel via different agonists could have a different response that could be attributed to the spatial temporal resolution of Ca^{2+} signaling.

9.2.4 Neuronal Morphogenesis

Development of neuronal circuits requires the establishment of a proper neuronal morphology in which Ca^{2+} can have profound effects on this process [72]. Activity-induced Ca^{2+} influx through VDCCs and N-methyl-D-aspartate (NMDA)-type glutamate receptors controls neurite morphogenesis [1, 73]. However, in the early developmental stages, especially in embryonic days and early postnatal days, neuronal activity in the brain is relatively low, which makes it unlikely that VDCCs and NMDARs are the major components for Ca^{2+} influx. The activation of TRPC channels is independent of membrane depolarization [96], and emerging evidence has shown that Ca^{2+} influx through TRPC channels plays a critical role in neurite morphogenesis during early stages of brain development.

The expression of TRPC4, 5, and 6 in the hippocampus is detectable as early as in embryonic days [90]. TRPC5 is delivered to growth cones of young cultured hippocampal neurons by the growth cone protein Stathmin2. Once in the growth cone, TRPC5 inhibits neurite growth, a developmental process that requires proper intracellular calcium range [33]. However, in the process of hippocampal neuronal axon formation, Ca^{2+} influx through TRPC5 seems to be critical for axon formation and is responsible for activating Ca^{2+}/calmodulin kinase kinase (CaMKK) signaling cascades [14]. According to Ca^{2+} set-point hypothesis, growth cone motility and neurite extension depend on an optimal range of $[Ca^{2+}]_i$. Altered Ca^{2+} homeostasis will retard the growth cone protrusion [32]. On the contrary, growth cone steering induced by various guidance cues requires localized Ca^{2+} influx. The different roles of TRPC5 in growth cone formation and morphogenesis suggest different Ca^{2+} requirements for different developmental stages of axon.

TRPC channel also participates in the dendritic development. TRPC6 channel promotes dendritic growth of the hippocampal neurons both in cultures and in mouse brain [90]. Interestingly, TRPC5 inhibits hippocampal neuron dendritic growth [37]. It is intriguing that members in the TRPC family have opposing roles in regulating dendritic growth. During development, environmental cues either promote or inhibit dendritic growth to shape the proper morphology of neurons. Many of the environmental cues are ligands for RTKs or GPCRs, which can activate TRPC channels. It is possible that growth-promoting cues selectively activate TRPC6, whereas inhibitory cues selectively activate TRPC5. It is indeed the case that BDNF, which is the pro-growth signal for dendrites, can activate TRPC6 through TrkB receptor, whereas neurotrophin-3 (NT-3), which is known to inhibit dendritic growth, activates the TRPC5 [37]. Why does TRPC5 or TRPC6 affect dendritic growth in the opposite way? One possibility is that they achieve opposite functions through the activation of different Ca^{2+}-sensing molecules. TRPC6 promotes dendritic growth through activation of CaMKIV-CREB pathway, whereas TRPC5 inhibits dendritic growth via activating CaMKIIα. In cerebellum granule neurons, TRPC5 inhibits dendritic growth through the activation of CaMKIIβ [68]. It has also been shown that TRPC5 and TRPC6 antagonize each other to regulate actin dynamics and cell mobility. At molecular level, TRPC5 is in a protein complex with Rac1, whereas TRPC6 is in the RhoA complex. The Ca^{2+} influx through TRPC5 activates Rac1 and promotes cell migration. On the other hand, the Ca^{2+} influx through TRPC6 activates RhoA and inhibits cell migration [93]. In the context of dendritic development, TRPC5 and TRPC6 might also regulate cytoskeleton dynamics in the opposite way by acting the Rac1 and

RhoA pathway differently. Thus, establishment of proper dendritic arborization may be regulated by multiple TRPC channels through which signals of different environmental cues can be translated into distinct intracellular Ca^{2+} signal cascades.

9.2.5 Synaptogenesis

Synapse formation is fundamental to the establishment of neuronal circuits. During development, synaptic formation is tightly coupled to neuronal differentiation. Shortly after neurons extend axonal and dendritic processes, vesicles carrying pre- and postsynaptic protein complexes are delivered and accumulated in the synaptic areas [56]. Dendritic spines serve as preferential sites of excitatory synaptic connections. A dendritic spine is a small membranous protrusion from a neuron's dendrite that typically receives input from a single axon. The function of dendritic spines appears to insulate signaling molecules to a particular postsynaptic density (PSD); certainly, all important Ca^{2+} signal is limited to the individual dendritic spines. The dynamics of dendritic spines depends on actin cytoskeleton and the proteins that regulate cytoskeleton. Some of these proteins are sensitive to changes in Ca^{2+} concentration [78]. In the subcellular fraction of synaptosome from rat cerebellum or cortex, all TRPCs are detectable [30]. TRPC1, 4, and 5 have been shown to interact with Homer1 or Homer3, molecules known as postsynaptic density scaffolding proteins [100]. Homer1 also regulates the gating property of TRPC1 [40]. Although TRPC6 does not interact with Homer, it localizes to both pre- and postsynaptic area [106]. Overexpression of TRPC6 induces excitatory synapse formation both in vitro and in vivo. TRPC6 transgenic mice also show better performance in water maze test than wild-type mice, indicating a better spatial learning and memory in these mice [106]. TRPC3, which forms heterotetromer with TRPC6 in synaptic regions, can be indirectly activated by BDNF and is necessary for the BDNF-induced increase in dendritic spine density [3].

TRPC channels may affect synaptogenesis through several ways. First, Ca^{2+} influx through TRPC channels may directly affect the activity of Rho GTPase to rearrange cytoskeleton to change spine motility [104]. This is an immediate effect of TRPC channels on spine formation. TRPC6 can regulate RhoA activity in human pulmonary arterial endothelial cells [81]. Whether RhoA activity is regulated by TRPC channels in neurons needs further exploration. Second, Ca^{2+} influx can activate pathways that cause a delayed change in spine density. In cultured hippocampal neurons, TRPC6 is able to activate CaMKIV-CREB transcriptional pathway to increase the spine number [106]. It has also been shown that TRPC1 and 3 mediate the leptin-induced spine formation through the activation of CaMKIγ in hippocampal neurons [17]. Furthermore, by promoting translocation of AMPA receptor to PSD, TRPC channels mediate the BDNF-induced maturation of silent synapse [62]. Much less is known, however, about the effect of TRPC channels on inhibitory synapse formation.

9.2.6 Neuronal Activity and Plasticity

Given the fact that TRPC channels play important roles in neuronal morphogenesis and spine formation, it is not surprising that they also regulate neuronal activity and plasticity.

Glutamate is the most predominant neurotransmitter used by excitatory synapses in the mammalian brain. Glutamate binds to two major classes of receptors, the ionotropic glutamate receptors (iGluRs) and the metabotropic glutamate receptors (mGluRs). TRPC channels have been reported to have cross talk with both groups. TRPC3 channels are most abundantly expressed in cerebellar Purkinje cells, the principle cells and the sole output of the cerebellar cortex. Interestingly, the highest expression of mGluR1, one of the metabotropic glutamate receptors, is also in Purkinje cells. Binding of glutamate to mGluR1 activates the PLC, followed by IP3 production and the subsequent ER Ca^{2+} release. The induction of long-term depression (LTD) in

Purkinje cells, which is presumed cellular basis of motor learning in the cerebellum [38], requires conjunctive stimulation of parallel and climbing fiber input [27]. During LTD induction, parallel fiber firing activates mGluR1-dependent Ca^{2+} release from internal stores, whereas climbing fiber activity strongly depolarizes the Purkinje cells and induces Ca^{2+} influx through voltage-gated Ca^{2+} channels [44]. Repetitive stimulation of parallel fiber inputs evokes a slow EPSC characterized by a time course that starts with a latency of 100–200 ms after stimulation and lasts for about 1 s [6]. Pharmacological experiments first suggest that mGluR1-dependent slow EPSC is mediated by TRPC1 [42]. But later studies investigated in more details suggest TRPC3 as the postsynaptic channel that mediates the slow EPSC [36, 41]. The mGluR1-dependent slow EPSC is gone in TRPC3 null mice, and the mice show a movement deficit of their hind paws that leads to an ataxic wide-based gait and poor performance when they walk on a horizontal ladder or an elevated beam [36]. The role of TRPC3 in mGluR1-mediated signaling is further proved in a study that used mice with a gain-of-function mutation (T635A) in the *trpc3*gene. These *moonwalker* (*Mwk*) mice have increased mGluR1-dependent inward currents, and their motor coordination is heavily impaired [7]. Increased TRPC3-mediated signaling is accompanied by a reduced dendritic arborization of Purkinje cells and cell loss after 4 months of age.

Excessive neuronal activity usually leads to neuronal death. Different TRPC family members play opposing roles in regulating neuronal excitotoxicity by regulating the activities of glutamate receptors. The first hint that TRPC channels might be involved in the NMDA receptor-mediated signaling came from the study that TRPC6 has protective roles in focal cerebral ischemia [19]. In this study, the authors found that TRPC6 undergoes calpain-dependent proteolytic degradation in response to Ca^{2+} entry through NMDA receptors in a rat model of ischemia. Blockade of TRPC6 proteolytic degradation preserves neuronal survival and prevents ischemic brain damage. The mechanisms through which TRPC6 channels protect neurons from ischemic death can be explained by the activation of CREB pro-survival signaling or inhibition of NMDA receptor-induced Ca^{2+} overload [51]. However, how TRPC6 regulates NMDA current needs further investigation. In physiological conditions, TRPC channels have been shown to be activated downstream of NMDA receptors. In olfactory granule cells, TRPC1 and TRPC4 are activated downstream of NMDA receptor activation and contribute to slow synaptic transmission in the olfactory bulb, including the Ca^{2+} dynamics required for asynchronous release from the granule cell spines [88]. During early developmental stages, before functional NMDA receptors are expressed, neurons transiently express Ca^{2+}-permeable AMPA receptors (CP-AMPARs). The transient nature of the CP-AMPARs is thought to play an important role in Ca^{2+} during synapse maturation. The expression of CP-AMPARs is elevated in response to BDNF. BDNF-induced CP-AMPAR translation and membrane insertion are dependent on CaMKK activation followed by TRPC5 and TRPC6 activation [25].

Seizures arise from the synchronized firing of a large group of cortical neurons. Synchronized firing will cause overactivation of glutamate receptors which leads to excessive Ca^{2+} entry and then excitotoxicity. The NMDA receptors are thought to play key roles, but targeted treatment by using NMDA antagonist shows limited effect [15]. The group I mGluRs have also been implicated in seizure and excitotoxicity. The mGluR selective agonists have been shown to be able to induce limbic seizure and neuronal degeneration [5, 75]. The highest expression level of TRPC4 is in the lateral septum where mGluR agonist can induce an "epileptiform" burst firing with a large depolarizing plateau potential. The plateau potential is altered in TRPC1 knockout mice and is completely absent in TRPC4 knockout mice and TRPC1/4 double knockout mice but is normal in TRPC3, TRPC5, TRPC6, and TRPC7 knockout mice [65]. Heteromeric TRPC1/4 channels are thought to be activated through the mGluR-PLC signaling cascade which leads to the plateau potential underlying epileptiform burst. In the hippocampus, TRPC1/4 channels play a critical role in the epileptiform burst firing in CA1 pyramidal neurons paralleling the functional role of these channels in lateral septum neurons [66].

The cholinergic input, which is another type of excitatory input to the cerebral cortex other than glutamate input, plays an important role in supporting processes such as arousal, attention, memory, and learning [67]. This effect is mediated in part by the ability of muscarinic receptors to modulate multiple ion currents and thus sculpture how neurons would fire [53]. In this context, one of the most intriguing electrophysiological properties of muscarinic receptors is their ability to induce the appearance of a slow afterdepolarization (sADP) in pyramidal neurons which may serve as a mechanism for the transient storage of memory traces within neuronal networks [21]. Several lines of evidence have pointed TRPC channels as potential candidates to mediate the muscarinic receptor-induced sADP. Yan et al. [101] first found that in cultured cortical pyramidal neurons, genetic expression of a TRPC dominant negative subunit inhibits, while overexpression of TRPC5 or TRPC6 subunits enhances, the amplitude of cholinergic sADPs. Tai et al. [89] further explored more mechanistically. They found that insertion of TRPC5 to the cell membrane followed by muscarinic stimulation is important for the sADP genesis. Blockade of TRPC5 membrane insertion or applying pharmacological antagonist of TRPC channels, such as SKF-96365, 2-APB, and flufenamic acid (FFA), blocks the generation of sADP in cultured hippocampal neurons. These findings are further confirmed by another group showing that applying a peptide that can disrupt TRPC4/5 subunit interaction caused inhibition of plateau potentials and persistent firing of the layer V pyramidal neurons in the entorhinal cortex [105]. Despite all these lines of evidence, in which most of the conclusions are drawn by the pharmacological studies and overexpression systems, loss-of-function evidence is missing. In fact, the genetic deletion of several TRPC subunits, including TRPC1, TRPC5, and TRPC6 (single knockout), or both TRPC5 and TRPC6 together (double knockout) in mouse, failed to reduce the amplitude of sADP of the layer V pyramidal neurons in the medial prefrontal cortex (mPFC) [13]. These discrepancies could be due to the channel formation from preferential subunits in different brain regions and cell types. Furthermore, since TRPC family members have similar activation mechanisms, one may anticipate that TRPC homologues belonging to the same subfamily could compensate for each other's function.

There are very few studies on the functions of TRPC channels in inhibitory synapse development or inhibitory synaptic transmission. There is one report showing that gamma-aminobutyric acid (GABA) release from thalamic interneurons by the activation of 5-hydroxytryptamine type 2 receptors requires Ca^{2+} entry that is critically dependent on TRPC4 channel. TRPC4 knockout mice showed reduced GABA release from the thalamic interneurons upon 5-hydroxytryptamine stimulation, but not by acetyl-β-methylcholine [61]. These results suggest that perhaps the F2 terminals are still functional in TRPC4 knockout mice and the regulation of GABA release via different mechanisms may be critical for specific functions, such as the sleep-wake cycle and processing of visual information.

9.3 TRPC Channels in the Peripheral Nervous System

9.3.1 Pheromone Transduction

TRPC2 is a unique member of the TRPC family since its expression is lost in higher mammals, evidenced by that TRPC2 is a pseudogene in human beings. However, TRPC2 seems to be essential for small rodents, for example, in mice, TRPC2 is exclusively expressed in the vomeronasal sensory system (VNO) and is localized to the dendritic tip of VNO neurons. Mouse VNO neurons are responsible for pheromone sensory transduction, especially pheromone-mediated sexual behaviors [20]. The TRPC2 knockout male mice fail to display male-male aggression and show sexual and courtship behaviors toward both males and females. The TRPC2 knockout mice also show the decreased action potential of sensory responses induced by pheromone application in the VNO neurons [49]. Furthermore, the knockout males increase their efforts to mount other males [43]. Although humans do not express TRPC2 channels and do not have similar VNO neurons, they have a functional olfactory system, which is believed to be also involved in pheromone-mediated social behaviors [35]. One may speculate that other TRPC paralogs could decode olfactory cues and could be responsible for the functions in human.

9.3.2 Sensation and Nociception

Sensory neurons receive information from our external and internal environment by nerves terminating in specialized structures termed sensory receptors located in the skin, muscles, and organs of the body. Different sensory neurons are molecularly distinct from each other and are specialized to respond to different types of stimuli, with the receptive properties defined by the molecular repertoire of sensory receptive ion channels or receptors expressed by the sensory neurons. TRPC1, 3, and 6 channels are most abundantly expressed in the dorsal root ganglion (DRG), a cluster of sensory neurons that bring sensory information from peripheral to the spinal cord and then sensory cortex [22].

TRPC1 is expressed in the NF-200-positive subclass of neurons with a large size of the cell body, while TRPC3, which stains up to 35% of DRG neurons, is almost exclusively present in non-peptidergic isolectin B4 (IB4)-positive small-size neurons that are largely TRPV1 negative. These expression patterns suggest important roles of TRPCs in sensory physiology in both nociceptive as well as non-nociceptive neurons. Pharmacological studies by using putative antagonist (SKF96365) and agonist (OAG) for TRPC channels proved functional TRPC channels expressed in DRG neurons [45]. Indeed, knockdown of TRPC1 in cultured DRG neurons shows 65% reduction of neurons with stretch-activated responses, implicating the involvement of TRPC1 in mechanotransduction and mechanosensitivity in DRG neurons [86]. TRPC1 null animals show a decrease in sensitivity to innocuous mechanical stimuli, evidenced by a 50% reduction in slowly adapting Aβ-fiber and in rapidly adapting Aδ-Down-hair afferent fibers in response to innocuous mechanical stimulation [28].

Loss of TRPC3 expression causes a shift of rapidly adapting mechanosensitive currents to intermediate-adapting currents in DRG neurons. Deletion of both TRPC3 and TRPC6 channels causes deficits in light touch and silences half of small-size sensory neurons expressing mechanically activated rapidly adapting currents [71]. TRPC6 is also activated by membrane stretch, while both TRPC5 and TRPC6 activities are blocked by a tarantula toxin known to inhibit mechanosensitive channels [85].

In addition to DRG neurons, TRPC channels are ubiquitously expressed in the inner ear structures including the organ of Corti and the spiral and vestibular ganglia [91]. TRPC3/6 double knockout mice show hearing impairment, vestibular deficits, and defective auditory brain stem responses to high-frequency sounds. Basal, but not apical, cochlear outer hair cells lost more than 75% of their responses to mechanical stimulation [71].

The first line of evidence indicating that TRPC channels might be involved in nociception is that TRPC1 and TRPC6 are often co-expressed with TRPV4, which plays a major role in mechanical hyperalgesia associated with pronociceptive inflammatory mediators and small fiber painful peripheral neuropathies [2]. Spinal intrathecal administration of antisense to TRPC1 and TRPC6 reversed the hyperalgesia to mechanical and hypotonic stimuli induced by inflammation. However, antisense to TRPC6, but not to TRPC1, reversed the mechanical hyperalgesia induced by a thermal injury or the TRPV4 agonist. These results indicate that TRPC1 and TRPC6 channels cooperate with TRPV4 channels to mediate mechanical hyperalgesia and primary afferent nociceptor sensitization, although they may have distinctive roles.

Another line of evidence points to TRPC3 as a key player in chronic pain. Chronic pain may accompany immune-related disorders with an elevated level of serum IgG immune complex (IgG-IC). IgG-IC activates a subpopulation of dorsal root ganglion (DRG) neurons through the neuronal Fc-gamma receptor I (FcϒRI). The FcϒRI and TRPC3 expression are found in the same DRG neuron. Knockdown of TRPC3 attenuates IgG-IC-induced nonselective cation currents, indicating that TRPC3 is a key molecule target for the excitatory effect of IgG-IC on DRG neurons [70].

Other than mechanosensation and nociception, TRPC5 channel has been shown to be a molecular component for detection and regional adaptation to cold temperature in the peripheral nervous system, which is distinct from noxious cold sensing [107]. The TRPC5 is present in mouse and human DRG neurons, and the gating of the channel is highly temperature sensitive. The TRPC5 current increases as temperature cooled below 37 °C, peaking around 25 °C as negative membrane potentials. Although TRPC5 knockout mice show no temperature-sensitive behavioral changes, TRPM8 and/or other menthol-sensitive channels appear to underpin a much larger component of noxious cold sensing after *trpc5* deletion and a shift in mechanosensitive C-fiber subtypes [107].

9.3.3 Neuromuscular Junction Synapses

The neuromuscular junction is a type of very specialized synapse that is formed by the contact between a motor neuron and a muscle fiber. Muscles require innervations to function, and they adapt to different patterns of motor nerve activity by alterations in gene expression that match specialized properties of contraction, metabolism, and muscle mass to changing work demands. Evidence has shown that TRPC3 is highly expressed in skeletal muscles. Calcium influx through TRPC3 can activate the nuclear factor of activated T cells (NFAT) and regulate gene expression in response to neuromuscular activity [74].

BDNF has been previously shown to induce Ca^{2+} elevation in the presynaptic neuronal terminals at the developing neuromuscular junction [87]. Given the fact that BDNF-induced Ca^{2+} influx is mediated by TRPC channels in other cell types, it is not surprising that the neuromuscular junctions adapt the same mechanism. Coordinated activation of presynaptic TrkB and postsynaptic p75NTR is required for full potentiation of neuromuscular junction synapses in response to BDNF. While inhibition of TrkB specifically in neurons completely abolished the potentiation, blocking TRPC channels in postsynaptic myocytes only blocked potentiation at the late phase, but not the initial phase, suggesting that TrkB receptor is activated first and necessary for BDNF-induced potentiation before the contribution of postsynaptic p75NTR and TRPC channels [58].

9.4 TRPC Channels in Glial Cells

Glial cells make up almost 50% of cells of a brain [4]. Emerging lines of evidence have shown that glial cells do not simply insulate neurons or provide support and nutrition to neurons. They can also regulate neuronal activity and synaptic transmissions. There are three major types of glial cells: astrocyte, oligodendrocyte, and microglial. The expression of TRPC channels has been found in all three types of glial cells, suggesting their potential roles in regulating glial cell functions.

Astrocytes, the largest population of glial cells, are fundamentally important for rapid regulation of extracellular ions and neurotransmitters, which, to a large extent, shapes neuronal excitability and synaptic transmission. Embryonic cultured astrocytes express mRNAs for TRPC1–6, which produce Ca^{2+} fluxes in response to OAG, an analog of DAG [34]. At protein level, relatively high expression of TRPC1 channels is detected in embryonic astrocyte cultures. Co-immunoprecipitation of TRPC1, $InsP_3$ receptors type II, and Homer is found in astrocyte cultures [98]. Similarly, co-localization of TRPC4 with ZO-1 scaffolding proteins is detected in cultured fetal human astrocytes [83]. Activation of

TRPC1 channels in astrocytes has been observed in various physiological and pathological contexts. The TRPC1 channel contributes to Ca^{2+} transients induced by stimulation of purinergic and glutamatergic metabotropic receptors [54]. Treatment of astrocytes with anti-TRPC antibody substantially reduces the plateau phase of the Ca^{2+} transients. TRPC1 channel is also instrumental for astroglial Ca^{2+} signaling following mechanical stimulation. The TRPC6 channel is claimed to contribute to Ca^{2+} entry following stimulation of interleukin-1β (IL-1β) receptors in embryonic astrocytes [9].

The functions of TRPC channels in the other two major glial cell types are relatively less clear. The expression of TRPC1 and TRPC3 is found in oligodendrocyte precursor cells (OPCs) and oligodendrocytes, respectively [26, 64]. TRPC1 channel is the store-operated calcium channel that mediates the Ca^{2+} influx induced by golli, myelin basic proteins essential for normal myelination during OPC development. TRPC1 also is required for the effects of golli on OPC proliferation [64]. TRPC3 expression is also found in microglials. Surface expression of TRPC3 channels in microglials is regulated by BDNF. Application of BDNF can rapidly induce TRPC3 membrane insertion, which can produce a sustained intracellular Ca^{2+} elevation. Pretreatment with BDNF suppresses the production of NO induced by tumor necrosis factor α (TNFα), which is prevented by coadministration of a selective TRPC3 inhibitor. These results suggest that TRPC3 channels play important roles in microglial functions, which might be important for the regulation of inflammatory responses [60].

9.5 Perspectives

TRPC channels are ubiquitously expressed in almost all the cell types inside the brain. Their gating mechanisms place them in a unique position to translate extracellular stimuli into intracellular Ca^{2+} signaling and its downstream cascades. In the past few years, there have been intensive studies on the various functions of TRPC channels in brain development. The specific expression of individual TRPC channels in the subpopulation of DRG neurons and in different glial cells makes their functions greatly extended. The considerable discrepancies regarding the functions of individuals may be due to the expression pattern diversity in different tissues, cell types, and channel conformation difference among different TRPC members. Although almost all the TRPC knockout mice are available, they all appear viable and fertile, suggesting their modulator roles in the brain functions. The future outstanding questions include but not limited to (1) crystal structures of TRPC channels, (2) the cell type-specific functions of TRPCs by using the combination of various TRPC conditional knockout alleles, (3) identification and characterization of the more specific antagonists or agonists for the channels, and (4) endogenous channel conformation and channel gating mechanisms. We believe that the more we gain insights into the physiological regulation of these channels, the better we can design the therapeutic intervention against the related diseases associated with the TRPC channels.

Acknowledgment Dr. Yichang Jia acknowledges the funding support from "1000-talents Plan" for young researchers, the Chinese Central Government; from Peking-Tsinghua Joint Center for Life Sciences and IDG/McGovern Institute for Brain Research at Tsinghua; from National Science Foundation of China (31571097, 81371361); from ALS association (16-IIP-284); and from NIH Pathway to Independence Award (K99/R00, NS079476).

References

1. Aizawa H, Hu SC, Bobb K, Balakrishnan K, Ince G, Gurevich I, Cowan M, Ghosh A (2004) Dendrite development regulated by CREST, a calcium-regulated transcriptional activator. Science 303 (5655):197–202. doi:10.1126/science.1089845303/5655/197 [pii]
2. Alessandri-Haber N, Dina OA, Chen X, Levine JD (2009) TRPC1 and TRPC6 channels cooperate with TRPV4 to mediate mechanical hyperalgesia and nociceptor sensitization. J Neurosci 29(19):6217–6228. doi:10.1523/jneurosci.0893-09.2009
3. Amaral MD, Pozzo-Miller L (2007) TRPC3 channels are necessary for brain-derived

neurotrophic factor to activate a nonselective cationic current and to induce dendritic apine formation. J Neurosci 27(19):5179–5189. doi:10.1523/jneurosci.5499-06.2007

4. Azevedo FA, Carvalho LR, Grinberg LT, Farfel JM, Ferretti RE, Leite RE, Jacob Filho W, Lent R, Herculano-Houzel S (2009) Equal numbers of neuronal and nonneuronal cells make the human brain an isometrically scaled-up primate brain. J Comp Neurol 513(5):532–541. doi:10.1002/cne.21974
5. Barton ME, Shannon HE (2005) Behavioral and convulsant effects of the (S) enantiomer of the group I metabotropic glutamate receptor agonist 3,5-DHPG in mice. Neuropharmacology 48 (6):779–787. doi:S0028-3908(05)00035-3 [pii]10.1016/j.neuropharm.2005.01.017
6. Batchelor AM, Madge DJ, Garthwaite J (1994) Synaptic activation of metabotropic glutamate receptors in the parallel fibre-Purkinje cell pathway in rat cerebellar slices. Neuroscience 63 (4):911–915. doi:0306-4522(94)90558-4 [pii]
7. Becker EBE, Oliver PL, Glitsch MD, Banks GT, Achilli F, Hardy A, Nolan PM, Fisher EMC, Davies KE (2009) A point mutation in TRPC3 causes abnormal Purkinje cell development and cerebellar ataxia in moonwalker mice. Proc Natl Acad Sci U S A 106(16):6706–6711. doi:10.1073/pnas.0810599106
8. Berridge MJ (1998) Neuronal calcium signaling. Neuron 21(1):13–26
9. Beskina O, Miller A, Mazzocco-Spezzia A, Pulina MV, Golovina VA (2007) Mechanisms of interleukin-1beta-induced Ca2+ signals in mouse cortical astrocytes: roles of store- and receptor-operated Ca2+ entry. Am J Phys Cell Phys 293 (3):C1103–C1111. doi:00249.2007 [pii]10.1152/ajpcell.00249.2007
10. Clapham DE (1995) Calcium signaling. Cell 80 (2):259–268. doi:0092-8674(95)90408-5 [pii]
11. Clelland CD, Choi M, Romberg C, Clemenson GD, Jr., Fragniere A, Tyers P, Jessberger S, Saksida LM, Barker RA, Gage FH, Bussey TJ (2009) A functional role for adult hippocampal neurogenesis in spatial pattern separation. Science 325 (5937):210–213. doi:10.1126/science.1173215325/5937/210 [pii]
12. Craig AM, Wyborski RJ, Banker G (1995) Preferential addition of newly synthesized membrane protein at axonal growth cones. Nature 375(6532):592–594. doi:10.1038/375592a0
13. Dasari S, Abramowitz J, Birnbaumer L, Gulledge AT (2013) Do canonical transient receptor potential channels mediate cholinergic excitation of cortical pyramidal neurons? Neuroreport 24(10):550–554. doi:10.1097/WNR.0b013e3283621344
14. Davare MA, Fortin DA, Saneyoshi T, Nygaard S, Kaech S, Banker G, Soderling TR, Wayman GA (2009) Transient receptor potential canonical 5 channels activate Ca2+/Calmodulin Kinase Iγ to promote axon formation in hippocampal neurons. J Neurosci 29(31):9794–9808. doi:10.1523/jneurosci.1544-09.2009
15. Davis SM, Lees KR, Albers GW, Diener HC, Markabi S, Karlsson G, Norris J (2000) Selfotel in acute ischemic stroke: possible neurotoxic effects of an NMDA antagonist. Stroke 31(2):347–354
16. Dent EW, Gertler FB (2003) Cytoskeletal dynamics and transport in growth cone motility and axon guidance. Neuron 40(2):209–227
17. Dhar M, Wayman GA, Zhu M, Lambert TJ, Davare MA, Appleyard SM (2014) Leptin-induced spine formation requires TrpC channels and the CaM Kinase Cascade in the hippocampus. J Neurosci 34(30):10022–10033. doi:10.1523/jneurosci.2868-13.2014
18. Dolmetsch RE, Pajvani U, Fife K, Spotts JM, Greenberg ME (2001) Signaling to the nucleus by an L-type calcium channel-calmodulin complex through the MAP kinase pathway. Science 294 (5541):333–339. doi:10.1126/science.1063395294/5541/333 [pii]
19. Du W, Huang J, Yao H, Zhou K, Duan B, Wang Y (2010) Inhibition of TRPC6 degradation suppresses ischemic brain damage in rats. J Clin Invest 120(10):3480–3492. doi:10.1172/jci43165
20. Dulac C, Axel R (1998) Expression of candidate pheromone receptor genes in vomeronasal neurons. Chem Senses 23(4):467–475
21. Egorov AV, Hamam BN, Fransen E, Hasselmo ME, Alonso AA (2002) Graded persistent activity in entorhinal cortex neurons. Nature 420 (6912):173–178. doi:10.1038/nature01171nature01171 [pii]
22. Elg S, Marmigere F, Mattsson JP, Ernfors P (2007) Cellular subtype distribution and developmental regulation of TRPC channel members in the mouse dorsal root ganglion. J Comp Neurol 503(1):35–46. doi:10.1002/cne.21351
23. Eriksson PS, Perfilieva E, Bjork-Eriksson T, Alborn AM, Nordborg C, Peterson DA, Gage FH (1998) Neurogenesis in the adult human hippocampus. Nat Med 4(11):1313–1317. doi:10.1038/3305
24. Fiorio Pla A, Maric D, Brazer SC, Giacobini P, Liu X, Chang YH, Ambudkar IS, Barker JL (2005) Canonical transient receptor potential 1 plays a role in basic fibroblast growth factor (bFGF)/FGF receptor-1-induced Ca2+ entry and embryonic rat neural stem cell proliferation. J Neurosci 25 (10):2687–2701. doi:25/10/2687 [pii]10.1523/JNEUROSCI.0951-04.2005
25. Fortin DA, Srivastava T, Dwarakanath D, Pierre P, Nygaard S, Derkach VA, Soderling TR (2012) Brain-derived neurotrophic factor activation of CaM-kinase kinase via transient receptor potential canonical channels induces the translation and synaptic incorporation of GluA1-containing calcium-permeable AMPA receptors. J Neurosci 32 (24):8127–8137. doi:10.1523/JNEUROSCI.6034-11.201232/24/8127 [pii]
26. Fusco FR, Martorana A, Giampà C, March ZD, Vacca F, Tozzi A, Longone P, Piccirilli S, Paolucci S, Sancesario G, Mercuri NB, Bernardi G (2004) Cellular localization of TRPC3 channel in rat brain: preferential distribution to oligodendrocytes.

Neurosci Lett 365(2):137–142. doi:http://dx.doi.org/10.1016/j.neulet.2004.04.070
27. Gao W, Dunbar RL, Chen G, Reinert KC, Oberdick J, Ebner TJ (2003) Optical imaging of long-term depression in the mouse cerebellar cortex in vivo. J Neurosci 23 (5):1859–1866. doi:23/5/1859 [pii]
28. Garrison SR, Dietrich A, Stucky CL (2012) TRPC1 contributes to light-touch sensation and mechanical responses in low-threshold cutaneous sensory neurons. J Neurophysiol 107(3):913–922
29. Ghosh A, Carnahan J, Greenberg ME (1994) Requirement for BDNF in activity-dependent survival of cortical neurons. Science 263(5153): 1618–1623
30. Goel M, Sinkins WG, Schilling WP (2002) Selective association of TRPC channel subunits in rat brain synaptosomes. J Biol Chem 277 (50):48303–48310. doi:10.1074/jbc.M207882200M207882200 [pii]
31. Gomez TM, Robles E, Poo M, Spitzer NC (2001) Filopodial calcium transients promote substrate-dependent growth cone turning. Science 291 (5510): 1983–1987. doi:10.1126/science.1056490291/5510/1983 [pii]
32. Gomez TM, Zheng JQ (2006) The molecular basis for calcium-dependent axon pathfinding. Nat Rev Neurosci 7 (2):115–125. doi:nrn1844 [pii]10.1038/nrn1844
33. Greka A, Navarro B, Oancea E, Duggan A, Clapham DE (2003) TRPC5 is a regulator of hippocampal neurite length and growth cone morphology. Nat Neurosci 6 (8):837–845. doi:10.1038/nn1092nn1092 [pii]
34. Grimaldi M, Maratos M, Verma A (2003) Transient receptor potential channel activation causes a novel form of [Ca 2+]I oscillations and is not involved in capacitative Ca 2+ entry in glial cells. J Neurosci 23 (11):4737–4745. doi:23/11/4737 [pii]
35. Grosser BI, Monti-Bloch L, Jennings-White C, Berliner DL (2000) Behavioral and electrophysiological effects of androstadienone, a human pheromone. Psychoneuroendocrinology 25(3):289–299
36. Hartmann J, Dragicevic E, Adelsberger H, Henning HA, Sumser M, Abramowitz J, Blum R, Dietrich A, Freichel M, Flockerzi V, Birnbaumer L, Konnerth A (2008) TRPC3 channels are required for synaptic transmission and motor coordination. Neuron 59(3):392–398. doi:http://dx.doi.org/10.1016/j.neuron.2008.06.009
37. He Z, Jia C, Feng S, Zhou K, Tai Y, Bai X, Wang Y (2012) TRPC5 channel is the mediator of neurotrophin-3 in regulating dendritic growth via CaMKIIα in rat hippocampal neurons. J Neurosci 32(27): 9383–9395. doi:10.1523/jneurosci.6363-11.2012
38. Ito M (2000) Mechanisms of motor learning in the cerebellum. Brain Res 886(1–2):237–245
39. Jia Y, Zhou J, Tai Y, Wang Y (2007) TRPC channels promote cerebellar granule neuron survival. Nat Neurosci 10 (5):559–567. doi:nn1870 [pii]10.1038/nn1870
40. Kim JY, Zeng W, Kiselyov K, Yuan JP, Dehoff MH, Mikoshiba K, Worley PF, Muallem S (2006) Homer 1 mediates store- and inositol 1,4,5-trisphosphate receptor-dependent translocation and retrieval of TRPC3 to the plasma membrane. J Biol Chem 281(43):32540–32549. doi:10.1074/jbc.M602496200
41. Kim SJ (2013) TRPC3 channel underlies cerebellar long-term depression. Cerebellum 12(3):334–337. doi:10.1007/s12311-013-0455-1
42. Kim SJ, Kim YS, Yuan JP, Petralia RS, Worley PF, Linden DJ (2003) Activation of the TRPC1 cation channel by metabotropic glutamate receptor mGluR1. Nature 426 (6964):285–291. doi:http://www.nature.com/nature/journal/v426/n6964/suppinfo/nature02162_S1.html
43. Kimchi T, Xu J, Dulac C (2007) A functional circuit underlying male sexual behaviour in the female mouse brain. Nature 448 (7157):1009–1014. doi:nature06089 [pii]10.1038/nature06089
44. Knopfel T, Anchisi D, Alojado ME, Tempia F, Strata P (2000) Elevation of intradendritic sodium concentration mediated by synaptic activation of metabotropic glutamate receptors in cerebellar Purkinje cells. Eur J Neurosci 12(6):2199–2204
45. Kress M, Karasek J, Ferrer-Montiel AV, Scherbakov N, Haberberger RV (2008) TRPC channels and diacylglycerol dependent calcium signaling in rat sensory neurons. Histochem Cell Biol 130(4):655–667. doi:10.1007/s00418-008-0477-9
46. Kuan CY, Roth KA, Flavell RA, Rakic P (2000) Mechanisms of programmed cell death in the developing brain. Trends Neurosci 23(7):291–297
47. Leclerc C, Néant I, Moreau M (2011) Early neural development in vertebrates is also a matter of calcium. Biochimie 93 (12):2102–2111. doi:http://dx.doi.org/10.1016/j.biochi.2011.06.032
48. Lemonnier L, Trebak M, Putney JW, Jr. (2008) Complex regulation of the TRPC3, 6 and 7 channel subfamily by diacylglycerol and phosphatidylinositol-4,5-bisphosphate. Cell Calcium 43 (5):506–514. doi:S0143–4160(07)00166–2 [pii]10.1016/j.ceca.2007.09.001
49. Leypold BG, Yu CR, Leinders-Zufall T, Kim MM, Zufall F, Axel R (2002) Altered sexual and social behaviors in trp2 mutant mice. Proc Natl Acad Sci U S A 99 (9):6376–6381. doi:10.1073/pnas.082127599082127599 [pii]
50. Li H-S, Xu X-ZS, Montell C (1999) Activation of a TRPC3-dependent cation current through the neurotrophin BDNF. Neuron 24(1):261–273. doi:http://dx.doi.org/10.1016/S0896-6273(00)80838-7
51. Li H, Huang J, Du W, Jia C, Yao H, Wang Y (2012) TRPC6 inhibited NMDA receptor activities and protected neurons from ischemic excitotoxicity. J Neurochem 123(6):1010–1018. doi:10.1111/jnc.12045
52. Li M, Chen C, Zhou Z, Xu S, Yu Z (2012) A TRPC1-mediated increase in store-operated Ca2+ entry is required for the proliferation of adult hippocampal

neural progenitor cells. Cell Calcium 51(6):486–496. doi:http://dx.doi.org/10.1016/j.ceca.2012.04.014
53. Lucas-Meunier E, Fossier P, Baux G, Amar M (2003) Cholinergic modulation of the cortical neuronal network. Pflugers Arch 446(1):17–29. doi:10.1007/s00424-002-0999-2
54. Malarkey EB, Ni Y, Parpura V (2008) Ca2+ entry through TRPC1 channels contributes to intracellular Ca2+ dynamics and consequent glutamate release from rat astrocytes. Glia 56(8):821–835. doi:10.1002/glia.20656
55. Maric D, Maric I, Barker JL (2000) Developmental changes in cell calcium homeostasis during neurogenesis of the embryonic rat cerebral cortex. Cereb Cortex 10(6):561–573
56. McAllister AK (2007) Dynamic aspects of CNS synapse formation. Annu Rev Neurosci 30:425–450. doi:10.1146/annurev.neuro.29.051605.112830
57. McCutchen ME, Bramham CR, Pozzo-Miller LD (2002) Modulation of neuronal calcium signaling by neurotrophic factors. Int J Dev Neurosci 20(3–5):199–207
58. McGurk JS, Shim S, Kim JY, Wen Z, Song H, G-l M (2011) Postsynaptic TRPC1 function contributes to BDNF-induced synaptic potentiation at the developing neuromuscular junction. J Neurosci 31(41):14754–14762. doi:10.1523/jneurosci.3599-11.2011
59. Minichiello L, Klein R (1996) TrkB and TrkC neurotrophin receptors cooperate in promoting survival of hippocampal and cerebellar granule neurons. Genes Dev 10(22):2849–2858
60. Mizoguchi Y, Kato TA, Seki Y, Ohgidani M, Sagata N, Horikawa H, Yamauchi Y, Sato-Kasai M, Hayakawa K, Inoue R, Kanba S, Monji A (2014) Brain-derived neurotrophic factor (BDNF) induces sustained intracellular Ca2+ elevation through the up-regulation of surface transient receptor potential 3 (TRPC3) channels in rodent microglia. J Biol Chem 289 (26):18549–18555. doi:10.1074/jbc.M114.555334M114.555334 [pii]
61. Munsch T, Freichel M, Flockerzi V, Pape HC (2003) Contribution of transient receptor potential channels to the control of GABA release from dendrites. Proc Natl Acad Sci U S A 100 (26):16065–16070. doi:10.1073/pnas.25353111002535311100 [pii]
62. Nakata H, Nakamura S (2007) Brain-derived neurotrophic factor regulates AMPA receptor trafficking to post-synaptic densities via IP3R and TRPC calcium signaling. FEBS Lett 581 (10):2047–2054. doi:S0014–5793(07)00429–2 [pii]10.1016/j.febslet.2007.04.041
63. Nilius B (2003) From TRPs to SOCs, CCEs, and CRACs: consensus and controversies. Cell Calcium 33(5–6):293–298
64. Paez PM, Fulton D, Spreuer V, Handley V, Campagnoni AT (2011) Modulation of canonical transient receptor potential channel 1 in the proliferation of oligodendrocyte precursor cells by the golli products of the myelin basic protein gene. J Neurosci 31 (10):3625–3637. doi:10.1523/JNEUROSCI.4424-10.201131/10/3625 [pii]
65. Phelan KD, Mock MM, Kretz O, Shwe UT, Kozhemyakin M, Greenfield LJ, Dietrich A, Birnbaumer L, Freichel M, Flockerzi V, Zheng F (2012) Heteromeric canonical transient receptor potential 1 and 4 channels play a critical role in epileptiform burst firing and seizure-induced neurodegeneration. Mol Pharmacol 81 (3):384–392. doi:10.1124/mol.111.075341mol.111.075341 [pii]
66. Phelan KD, Shwe UT, Abramowitz J, Wu H, Rhee SW, Howell MD, Gottschall PE, Freichel M, Flockerzi V, Birnbaumer L, Zheng F (2013) Canonical transient receptor channel 5 (TRPC5) and TRPC1/4 contribute to seizure and excitotoxicity by distinct cellular mechanisms. Mol Pharmacol 83(2):429–438. doi:10.1124/mol.112.082271
67. Phillis JW (2005) Acetylcholine release from the central nervous system: a 50-year retrospective. Crit Rev Neurobiol 17 (3–4):161–217. doi:0c6ebf4a55da92b6,7f45271578c3b24d [pii]
68. Puram SV, Riccio A, Koirala S, Ikeuchi Y, Kim AH, Corfas G, Bonni A (2011) A TRPC5-regulated calcium signaling pathway controls dendrite patterning in the mammalian brain. Genes Dev 25(24): 2659–2673. doi:10.1101/gad.174060.111
69. Putney JW (2005) Physiological mechanisms of TRPC activation. Pflugers Arch 451(1):29–34. doi:10.1007/s00424-005-1416-4
70. Qu L, Li Y, Pan X, Zhang P, LaMotte RH, Ma C (2012) Transient Receptor Potential Canonical 3 (TRPC3) is required for IgG immune complex-induced excitation of the rat dorsal root ganglion neurons. J Neurosci 32(28):9554–9562. doi:10.1523/jneurosci.6355-11.2012
71. Quick K, Zhao J, Eijkelkamp N, Linley JE, Rugiero F, Cox JJ, Raouf R, Gringhuis M, Sexton JE, Abramowitz J, Taylor R, Forge A, Ashmore J, Kirkwood N, Kros CJ, Richardson GP, Freichel M, Flockerzi V, Birnbaumer L, Wood JN (2012) TRPC3 and TRPC6 are essential for normal mechanotransduction in subsets of sensory neurons and cochlear hair cells. Open Biol 2(5)
72. Redmond L, Ghosh A (2005) Regulation of dendritic development by calcium signaling. Cell Calcium 37 (5):411–416. doi:S0143–4160(05)00025–4 [pii]10.1016/j.ceca.2005.01.009
73. Redmond L, Kashani AH, Ghosh A (2002) Calcium regulation of dendritic growth via CaM kinase IV and CREB-mediated transcription. Neuron 34(6):999–1010
74. Rosenberg P, Hawkins A, Stiber J, Shelton JM, Hutcheson K, Bassel-Duby R, Shin DM, Yan Z, Williams RS (2004) TRPC3 channels confer cellular memory of recent neuromuscular activity. Proc Natl Acad Sci U S A 101 (25):9387–9392. doi:10.1073/pnas.03081791010308179101 [pii]
75. Sacaan AI, Schoepp DD (1992) Activation of hippocampal metabotropic excitatory amino acid receptors leads to seizures and neuronal damage. Neurosci Lett 139 (1):77–82. doi:0304-3940(92)90862-2 [pii]

76. Santarelli L, Saxe M, Gross C, Surget A, Battaglia F, Dulawa S, Weisstaub N, Lee J, Duman R, Arancio O, Belzung C, Hen R (2003) Requirement of hippocampal neurogenesis for the behavioral effects of antidepressants. Science 301 (5634):805–809. doi:10.1126/science.1083328301/5634/805 [pii]
77. Schmidt-Hieber C, Jonas P, Bischofberger J (2004) Enhanced synaptic plasticity in newly generated granule cells of the adult hippocampus. Nature 429 (6988):184–187. doi:10.1038/nature02553nature02553 [pii]
78. Schubert V, Da Silva JS, Dotti CG (2006) Localized recruitment and activation of RhoA underlies dendritic spine morphology in a glutamate receptor-dependent manner. J Cell Biol 172 (3):453–467. doi:jcb.200506136 [pii]10.1083/jcb.200506136
79. See V, Boutillier AL, Bito H, Loeffler JP (2001) Calcium/calmodulin-dependent protein kinase type IV (CaMKIV) inhibits apoptosis induced by potassium deprivation in cerebellar granule neurons. FASEB J 15 (1):134–144. doi:10.1096/fj.00-0106com15/1/134 [pii]
80. Shim S, Goh EL, Ge S, Sailor K, Yuan JP, Roderick HL, Bootman MD, Worley PF, Song H, Ming GL (2005) XTRPC1-dependent chemotropic guidance of neuronal growth cones. Nat Neurosci 8 (6):730–735. doi:nn1459 [pii]10.1038/nn1459
81. Singh I, Knezevic N, Ahmmed GU, Kini V, Malik AB, Mehta D (2007) Galphaq-TRPC6-mediated Ca2+ entry induces RhoA activation and resultant endothelial cell shape change in response to thrombin. J Biol Chem 282 (11):7833–7843. doi:M608288200 [pii]10.1074/jbc.M608288200
82. Song H, Poo M (2001) The cell biology of neuronal navigation. Nat Cell Biol 3(3):E81–E88. doi:10.1038/35060164
83. Song X, Zhao Y, Narcisse L, Duffy H, Kress Y, Lee S, Brosnan CF (2005) Canonical transient receptor potential channel 4 (TRPC4) co-localizes with the scaffolding protein ZO-1 in human fetal astrocytes in culture. Glia 49(3):418–429. doi:10.1002/glia.20128
84. Sotelo C (2004) Cellular and genetic regulation of the development of the cerebellar system. Prog Neurobiol 72 (5):295–339. doi:10.1016/j.pneurobio.2004.03.004S0301008204000401 [pii]
85. Spassova MA, Hewavitharana T, Xu W, Soboloff J, Gill DL (2006) A common mechanism underlies stretch activation and receptor activation of TRPC6 channels. Proc Natl Acad Sci U S A 103 (44):16586–16591.doi:0606894103[pii]10.1073/pnas.0606894103
86. Staaf S, Maxvall I, Lind U, Husmark J, Mattsson JP, Ernfors P, Pierrou S (2009) Down regulation of TRPC1 by shRNA reduces mechanosensitivity in mouse dorsal root ganglion neurons in vitro. Neurosci Lett 457(1):3–7. doi:http://dx.doi.org/10.1016/j.neulet.2009.03.082
87. Stoop R, Poo MM (1996) Synaptic modulation by neurotrophic factors: differential and synergistic effects of brain-derived neurotrophic factor and ciliary neurotrophic factor. J Neurosci 16(10):3256–3264
88. Stroh O, Freichel M, Kretz O, Birnbaumer L, Hartmann J, Egger V (2012) NMDA receptor-dependent synaptic activation of TRPC channels in olfactory bulb granule cells. J Neurosci 32(17):5737–5746. doi:10.1523/JNEUROSCI.3753-11.201232/17/5737 [pii]
89. Tai C, Hines DJ, Choi HB, MacVicar BA (2011) Plasma membrane insertion of TRPC5 channels contributes to the cholinergic plateau potential in hippocampal CA1 pyramidal neurons. Hippocampus 21(9):958–967. doi:10.1002/hipo.20807
90. Tai Y, Feng S, Ge R, Du W, Zhang X, He Z, Wang Y (2008) TRPC6 channels promote dendritic growth via the CaMKIV-CREB pathway. J Cell Sci 121(14):2301–2307
91. Takumida M, Anniko M (2009) Expression of canonical transient receptor potential channel (TRPC) 1-7 in the mouse inner ear. Acta Otolaryngol 129 (12):1351–1358. doi:10.3109/00016480902798 35010.3109/00016480902798350 [pii]
92. Tang J, Lin Y, Zhang Z, Tikunova S, Birnbaumer L, Zhu MX (2001) Identification of common binding sites for calmodulin and inositol 1,4,5-trisphosphate receptors on the carboxyl termini of trp channels. J Biol Chem 276 (24):21303–21310. doi:10.1074/jbc.M102316200M102316200 [pii]
93. Tian D, Jacobo SMP, Billing D, Rozkalne A, Gage SD, Anagnostou T, Pavenstädt H, Hsu H-H, Schlondorff J, Ramos A, Greka A (2010) Antagonistic regulation of actin dynamics and cell motility by TRPC5 and TRPC6 channels. Sci Signal 3(145):ra77–ra77
94. Trebak M, Lemonnier L, Smyth JT, Vazquez G, Putney JW Jr (2007) Phospholipase C-coupled receptors and activation of TRPC channels. Handb Exp Pharmacol 179:593–614. doi:10.1007/978-3-540-34891-7_35
95. Vaux DL, Korsmeyer SJ (1999) Cell death in development. Cell 96(2):245–254
96. Venkatachalam K, Montell C (2007) TRP channels. Annu Rev Biochem 76:387–417. doi:10.1146/annurev.biochem.75.103004.142819
97. Wang GX, Poo MM (2005) Requirement of TRPC channels in netrin-1-induced chemotropic turning of nerve growth cones. Nature 434 (7035):898–904. doi:nature03478 [pii]10.1038/nature03478
98. Weerth SH, Holtzclaw LA, Russell JT (2007) Signaling proteins in raft-like microdomains are essential for Ca2+ wave propagation in glial cells. Cell Calcium 41 (2):155–167. doi:S0143-4160(06)00124-2 [pii]10.1016/j.ceca.2006.06.006
99. Weick JP, Austin Johnson M, Zhang SC (2009) Developmental regulation of human embryonic stem cell-derived neurons by calcium entry via transient receptor potential channels. Stem Cells 27(12):2906–2916. doi:10.1002/stem.212

100. Xiao B, Tu JC, Worley PF (2000) Homer: a link between neural activity and glutamate receptor function. Curr Opin Neurobiol 10(3):370–374
101. Yan HD, Villalobos C, Andrade R (2009) TRPC channels mediate a muscarinic receptor-induced afterdepolarization in cerebral cortex. J Neurosci 29 (32):10038–10046. doi:10.1523/JNEUROSCI.1042-09.200929/32/10038 [pii]
102. Yano S, Tokumitsu H, Soderling TR (1998) Calcium promotes cell survival through CaM-K kinase activation of the protein-kinase-B pathway. Nature 396(6711):584–587. doi:10.1038/25147
103. Yao H, Duan M, Yang L, Buch S (2012) Platelet-derived growth factor-BB restores human immunodeficiency virus Tat-cocaine-mediated impairment of neurogenesis: role of TRPC1 channels. J Neurosci 32 (29):9835–9847. doi:10.1523/JNEUROSCI.0638-12.201232/29/9835 [pii]
104. Yuste R, Bonhoeffer T (2001) Morphological changes in dendritic spines associated with long-term synaptic plasticity. Annu Rev Neurosci 24:1071–1089. doi:10.1146/annurev.neuro.24.1.107124/1/1071 [pii]
105. Zhang Z, Reboreda A, Alonso A, Barker PA, Seguela P (2011) TRPC channels underlie cholinergic plateau potentials and persistent activity in entorhinal cortex. Hippocampus 21(4):386–397. doi:10.1002/hipo.20755
106. Zhou J, Du W, Zhou K, Tai Y, Yao H, Jia Y, Ding Y, Wang Y (2008) Critical role of TRPC6 channels in the formation of excitatory synapses. Nat Neurosci 11(7):741–743. doi:http://www.nature.com/neuro/journal/v11/n7/suppinfo/nn.2127_S1.html
107. Zimmermann K, Lennerz JK, Hein A, Link AS, Kaczmarek JS, Delling M, Uysal S, Pfeifer JD, Riccio A, Clapham DE (2011) Transient receptor potential cation channel, subfamily C, member 5 (TRPC5) is a cold-transducer in the peripheral nervous system. Proc Natl Acad Sci U S A 108(44):18114–18119. doi:10.1073/pnas.1115387108

TRPC Channels and Brain Inflammation

10

Yoshito Mizoguchi and Akira Monji

Abstract

Nonresolving low-grade inflammation is supposed to underly the basis of chronic disorders including cardiovascular diseases, cancer, diabetes, obesity, and psychiatric disorders such as depression and Alzheimer's diseases. There is increasing evidence suggesting that pathophysiology of psychiatric disorders is related to the inflammatory responses mediated by microglial cells. Elevation of intracellular Ca^{2+} is important for the activation of microglial cell functions, including proliferation, release of NO, cytokines, and BDNF. It has been shown that alteration of intracellular Ca^{2+} signaling underlies the pathophysiology of psychiatric disorders, including depression. BDNF induces a sustained intracellular Ca^{2+} elevation through the upregulation of the surface expression of TRPC3 channels in rodent microglial cells. Microglial cells are able to respond to BDNF, which may be important for the regulation of inflammatory responses and may also be involved in the pathophysiology and/or the treatment of psychiatric disorders. We also need to study the effect of proBDNF on microglial cells especially by focusing on the TRPC channels.

Keywords

Inflammation • Microglia • Depression • BDNF • TRPC3

10.1 Bullet Points

Nonresolving low-grade inflammation underlies the basis of chronic disorders including cardiovascular diseases, cancer, diabetes, obesity, and psychiatric disorders.

Brain inflammation and/or activated microglia contribute to pathologies such as stroke, trauma,

Y. Mizoguchi (✉) • A. Monji
Department of Psychiatry, Faculty of Medicine, Saga University, Saga, Japan
e-mail: ymizo@cc.saga-u.ac.jp

Y. Wang (ed.), *Transient Receptor Potential Canonical Channels and Brain Diseases*,
Advances in Experimental Medicine and Biology 976, DOI 10.1007/978-94-024-1088-4_10

psychiatric disorders including depression, and neurodegenerative diseases such as Alzheimer's disease.

For electrically non-excitable cells such as microglia, SOCE mediated by TRPC channels could play some important roles in inflammatory processes in the brain.

BDNF induces a sustained intracellular Ca^{2+} elevation through the upregulation of the surface expression of TRPC3 channels in rodent microglial cells. In addition, pretreatment with BDNF suppresses the production of NO induced by TNFα, which is prevented by coadministration of a selective TRPC3 inhibitor.

10.2 Introduction

There are many reports showing that brain inflammation and/or activated microglia contribute to the pathologies such as stroke, trauma, and meningitis; psychiatric disorders including depression, schizophrenia, and autism; and neurodegenerative diseases such as multiple sclerosis, Alzheimer's disease, Parkinson's disease, and motor neuron disease [11, 34]. In addition, peripheral, chronic, and nonresolving inflammation can readily spread to the brain. Numerous environmental and biological factors including early life or psychosocial stress, obesity, sleep disorders, dietary intake, and the bacterial composition of the gut microbiota can contribute to chronic inflammation. Both epidemiologic and mechanistic studies show that the chronic inflammation, in turn, is now recognized to be a common pathway underlying pathology including cardiovascular disease, cancer, diabetes, obesity, depression, and Alzheimer's disease. Thus, to prevent and treat many diseases mediated by brain inflammation and/or activated microglia, we need to target many contributors to chronic inflammation, such as macrophage, T cell, and/or endothelial cell dysfunction (Fig. 10.1; [14]).

Microglia are immune cells derived from progenitors that have migrated from the periphery and are from mesodermal/mesenchymal origin [21]. After invading the brain parenchyma, microglia transform into the "resting" ramified phenotype and are distributed in the whole brain. However, microglia revert to an ameboid appearance when they are activated when with infection, trauma, ischemia, neurodegenerative diseases, or disruption of brain homeostasis [1, 8]. Recent in vivo imaging has shown that microglial cells actively scan their environment with motile protrusions even in their resting state and are ready to transform to "activated" state in responses to injury, ischemia, or autoimmune challenges in the brain [62]. Microglia are the most active cytokine-producing cells in the brain and can release many factors including pro-inflammatory cytokines (such as TNFα, IL-6), nitric oxide (NO), and neurotrophic factors (such as BDNF) when they are activated in response to immunological stimuli [21, 57]. In addition, microglia are shown to be involved in the development of neural circuits or synaptic plasticity thereby maintaining the brain homeostasis [49].

There is increasing evidence suggesting that pathophysiology of depression is related to the inflammatory responses mediated by microglia [23]. Microglial activation can be estimated by positron-emission tomography (PET) using radiopharmaceuticals. For example, a quantitative [18F]FEPPA PET scan shows that activated microglia, translocator protein density measured by distribution volume (TSPO VT), is elevated in the prefrontal cortex, anterior cingulate cortex (ACC), hippocampus, and insula in patients suffering from major depression, and the correlation between higher brain TSPO VT and severity of depression is consistent with the concept that brain inflammation and/or activated microglia contribute to pathologies including depression [52].

Psychological stress is also shown to induce alterations in microglial phenotype and prime monocytes to lead to aberrant peripheral and central inflammation. Repeated social defeat (RSD) stress leads to microglia activation with increased pro-inflammatory cytokine and chemokine production that contributes to the development of reactive endothelium. Vascular endothelial cells increase cell adhesion molecule (CAM) expression that facilitates the adherence and extravasation of peripherally derived monocytes that

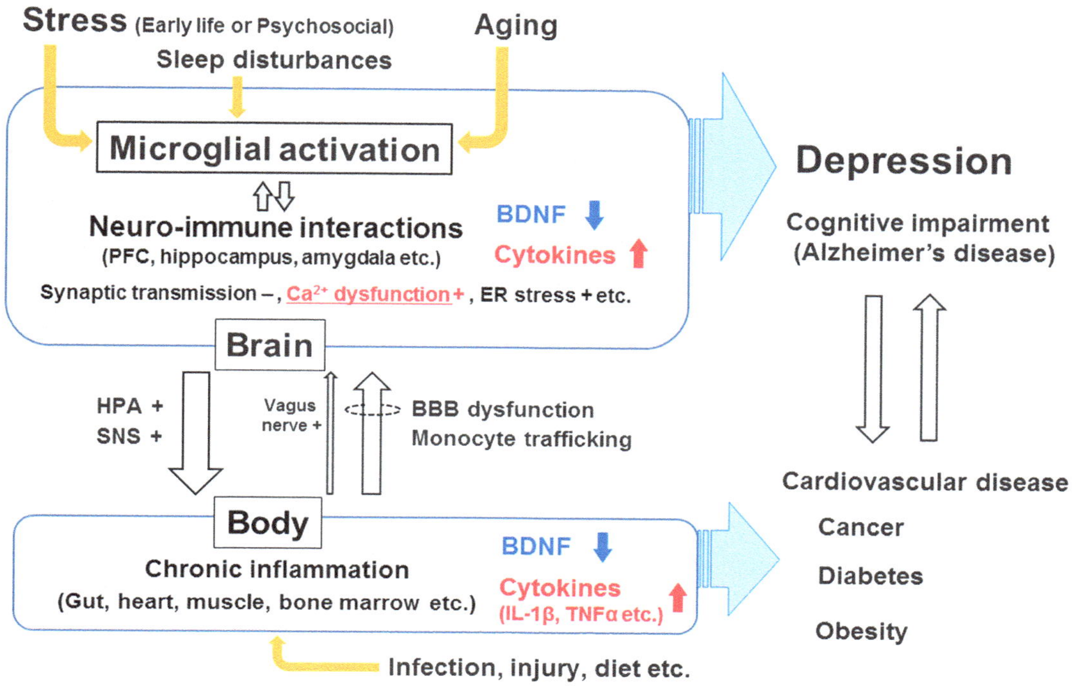

Fig. 10.1 Brain inflammation and/or activated microglia contribute to acute pathologies such as stroke, psychiatric disorders including depression, and neurodegenerative diseases such as Alzheimer's disease. Microglia are primed for activation by psychosocial stress, sleep disturbances, or aging. In addition, peripheral, chronic, and nonresolving inflammation can readily spread to the brain. Numerous factors including early life or psychosocial stress, obesity, sleep disorders, dietary intake, and the bacterial composition of the gut microbiota can contribute to chronic inflammation. The chronic inflammation, in turn, is now recognized to be a common pathway underlying pathology including cardiovascular disease, cancer, diabetes, obesity, depression, and Alzheimer's disease. Thus, to prevent and treat the diseases mediated by brain inflammation and/or activated microglia, we need to target many contributors to chronic inflammation, such as macrophage, T cell, and/or endothelial cell dysfunction. BDNF has major roles not only in the brain but also in cellular function in peripheral organs. The BDNF levels are found to be low in cardiovascular disease, diabetes, obesity, and also in depression and Alzheimer's disease. BDNF is expressed in non-neurogenic tissues, including skeletal muscle or endothelial cells. Thus, BDNF functions at the crossroads of two major routes, providing insight into psychiatric disorders, namely, synaptic plasticity and neuroinflammation. *BBB*; Blood Brain Barrier, *SNS*; Sympathetic nervous systems, *HPA*; Hypothalamic-pituitary-adrenal axis

differentiate into perivascular and parenchymal macrophages. Accumulation of macrophages in the brain converges with activated microglia and then amplifies brain inflammatory signaling. Elevated pro-inflammatory cytokine levels caused by activated microglia together with recruitment of monocytes to the brain contribute to the persistent anxiety-like behaviors [65].

In addition, we and others have reported that pretreatment with antidepressants significantly inhibits the release of pro-inflammatory cytokines and/or NO from activated microglial cells [17, 20]. Interestingly, pretreatment with antidepressant selective serotonin reuptake inhibitors (SSRIs), paroxetine or sertraline, significantly inhibits the generation of NO and tumor necrosis factor (TNF)-α from interferon (IFN)-γ-activated 6–3 microglial cells. In addition, pretreatment with paroxetine or sertraline suppresses the elevation of intracellular Ca^{2+} concentration ($[Ca^{2+}]_i$) induced by IFN-γ in rodent microglial cells. Moreover, we have previously reported that treatment of BAPTA-AM, a membrane-permeable intracellular Ca^{2+} chelator, significantly decreases the amount of NO released from IFN-γ-activated microglial cells [19], and the production of NO and TNF-α from IFN-γ-activated microglia is significantly inhibited by specific inhibitors of PKC, p38 MAPK, and ERK 1/2, all of which appear to be activated by elevation of $[Ca^{2+}]_i$ [19]. In the same study, JAK inhibitor-1 also significantly inhibits production of NO and TNFα from IFN-γ-activated microglia. The biological effects of IFN-γ are elicited through the activation of intracellular signaling pathways, including JAK-STAT pathway [18]. In this pathway, the phosphorylated STAT1 homodimer translocates to the nucleus where it initiates gene transcription, and the influx of Ca^{2+} induced by IFN-γ is required for the ser-727 phosphorylation of STAT1 in NIH 3 T3 cells [37]. Based on these results, we speculate that paroxetine and sertraline may inhibit signaling pathways including PKC, p38 MAPK, ERK 1/2, and JAK-STAT pathways in microglia by suppressing the IFN-γ-induced elevation of $[Ca^{2+}]_i$ that ultimately results in reduced production of NO and TNFα and suggest the importance of microglial intracellular Ca^{2+} signaling as a target of antidepressant for the treatment of depression [17].

The electrical activity of neurons (i.e., excitable cells) depends on a number of different types of voltage- or ligand-gated ion channels that are permeable to inorganic ions such as sodium, potassium, chloride, and calcium. While the former three ions predominantly support the electrogenic roles, Ca^{2+} are different in that they cannot only alter the membrane potential but also serve as important intracellular signaling entities by themselves. In the CNS, intracellular Ca^{2+} signaling regulates many different neuronal functions, such as cell proliferation, gene transcription, and exocytosis at synapses [3]. In neurons, because the prolonged elevation of $[Ca^{2+}]_i$ is cytotoxic, $[Ca^{2+}]_i$ is tightly regulated by intrinsic gating processes mediated by voltage-gated calcium channels and NMDA receptors (NMDARs; [55]). In addition, dysregulation of neuronal Ca^{2+} signaling has been linked to various psychiatric disorders including bipolar disorders (BD; [61]). Mood change can be caused by hyperactivity of phosphoinositide signaling with enhanced activity of the Ins(1,4,5)P3/Ca^{2+} pathway. The central feature of the inositol depletion hypothesis of BD is that mood stabilizers, Li^+ and valproate, which are used for the treatment of BD, act to inhibit the supply of inositol required to maintain the inositol lipid-signaling pathway. If this pathway is a target for Li^+, it suggests that BD may be caused by hyperactivity of phosphoinositide signaling with enhanced activity of the inositol 1,4,5-trisphosphate, Ins(1,4,5)P3/Ca^{2+}, and DAG/protein kinase C pathways. The abnormal elevations in Ca^{2+} will enhance membrane excitability, and this may distort the neural components of the circuits that control mood [4].

Elevation of intracellular Ca^{2+} is also important for the activation of microglia, including proliferation, migration, ramification, de-ramification, and release of NO, pro-inflammatory cytokines, and BDNF [21]. However, in microglial cells, an application of high $[K^+]_{out}$ or glutamate does not elevate $[Ca^{2+}]_i$. This observation is supported by the fact that both voltage-gated Ca^{2+} channels and NMDARs are not expressed in microglia [21]. For electrically non-excitable cells including microglia, the primary source of intracellular Ca^{2+} is the release from intracellular Ca^{2+} stores and the entry through the ligand-gated and/or store-operated Ca^{2+} channels [33]. Microglia contain at least two types of intracellular Ca^{2+} stores: the ER and mitochondria. The main route for the generation of intracellular Ca^{2+} signaling is associated with inositol 1,4,5-trisphosphate (InsP3) receptors on the ER membrane. Stimulation of G-protein-coupled metabotropic or tyrosine kinase receptors results in the activation of the phospholipase C (PLC), production of 2 s messengers including the diacylglycerol (DAG) and the InsP3, and the release of Ca^{2+} from the ER. Importantly, the depletion of ER activates the store-operated Ca^{2+} entry (SOCE), known as a capacitative Ca^{2+} influx, mediated by plasmalemmal channels such as calcium release-activated Ca^{2+} (CRAC) channels and/or transient receptor potential (TRP) channels [41]. In addition, STIM1, one of ER membrane proteins, senses the filling state of ER Ca^{2+}

and delivers the ER to the plasma membrane where it directly activates Orai1/CRAC channels, thereby facilitating the reuptake of Ca^{2+} to ER through the sarco(endo)plasmic reticulum Ca^{2+}-ATPases (SERCA). The concentration of Ca^{2+} in the ER is precisely controlled by SERCA. The influx of Ca^{2+} through the TRP channels could play some important roles in many inflammatory processes including the activation of microglia [39]. Some chronic pathological conditions may also affect the status of the ER Ca^{2+} store. In particular, the InsP3-mediated Ca^{2+} release following the stimulation of purinoceptors was reduced by more than 50% in microglial cells obtained from the brains of Alzheimer's disease patients, possibly indicating chronic depletion of the ER stores [25]. Disruption of microglial Ca^{2+} homeostasis triggers activation of death programs, which are regulated by the microglia activation status. Treatment of primary cultured microglial cells with thapsigargin or ionomycin induced apoptosis, whereas the same agents applied to LPS-activated microglia resulted in necrotic cell death [36]. Both apoptotic and necrotic pathways were regulated by $[Ca^{2+}]_i$ because treatment of cultures with BAPTA-AM reduced microglial cell death [36].

There are seven transient receptor potential canonical (TRPC) channels in mammalian species. Among them, TRPC2 is a pseudogene in humans. The remaining members of the TRPC subfamily are classified into three groups according to sequence homology, TRPC1, TRPC3/C6/C7, and TRPC4/C5. Quantitative comparisons of mRNA expression using real-time RT-PCR shows that TRPM7>TRPC6>TRPM2>TRPC1>TRPC3≥TRPC4>TRPC7>TRPC5>TRPC2, where ">" denotes a significant difference from the preceding gene, and "≥" indicates a nonsignificant difference, in microglial cells cultured from rats [40]. Again, for electrically non-excitable cells such as microglia, SOCE mediated by TRPC channels could play some important roles in inflammatory processes in the brain.

10.3 BDNF Induces Sustained Intracellular Ca^{2+} Elevation Through the Upregulation of Surface TRPC3 Channels in Rodent Microglia

Brain-derived neurotrophic factor (BDNF), one of the neurotrophins, has various important roles in cell survival, neurite outgrowth, neuronal differentiation, and gene expression in the brain [42, 59]. BDNF binds to the TrkB receptor and induces the activation of intracellular signaling pathways, including PLC-γ, phosphatidylinositol 3-kinase (PI3K), and mitogen-activated protein kinase-1/2 (MAPK-1/2; [44]). BDNF rapidly activates the PLC pathway, leading to the generation of inositol trisphosphate (IP3) and the mobilization of intracellular Ca^{2+} from the endoplasmic reticulum [26, 27]. BDNF is most abundantly expressed in the hippocampus and cerebral cortex and is also involved in the pathophysiology of psychiatric disorders such as depression [29, 50]. The BDNF hypothesis of depression was originally based on the findings obtained from rodents indicating that acute or chronic stress decreases the expression of BDNF in the hippocampus, and the treatment of diverse classes of antidepressant prevents the action of stress [5]. Because the injection of BDNF into the rodent hippocampus exerts antidepressant-like effects in the forced swim and learned helplessness tests, activation of BDNF signaling in the hippocampus is thought to have antidepressant effects. There are many reports showing that serum BDNF levels are significantly low in patients suffering from major depression and that BDNF levels are elevated following a course of antidepressant treatment [32].

In the rodent brain, microglial cells express BDNF mRNA [12] and secrete BDNF following stimulation with lipopolysaccharide (LPS) [38]. BDNF released from activated microglia then induces the sprouting of nigrostriatal dopaminergic neurons [2], causing a shift in the neuronal anion gradient [7], or promotes the proliferation and survival of microglia themselves [69]. In addition, Parkhurst et al. showed that the Cre-dependent removal of BDNF from microglia induces deficits in multiple learning tasks mediated by reduction in learning-dependent spine elimination/formation. These results suggest that microglia serve important physiological functions in learning and memory by promoting learning-related synapse formation through the BDNF signaling [43].

We have reported that BDNF induces a sustained increase in $[Ca^{2+}]_i$ through binding with the truncated tropomyosin-related kinase B receptor (TrkB-T1), resulting in activation of the PLC pathway and store-operated calcium entry (SOCE) in rodent microglial cells. Sustained activation of SOCE occurs after a brief BDNF application and contributes to the maintenance of sustained $[Ca^{2+}]_i$ elevation. Pretreatment with BDNF significantly suppresses the release of NO from activated microglia. Additionally, pretreatment of BDNF suppresses the IFN-γ-induced increase in $[Ca^{2+}]_i$, along with a rise in basal levels of $[Ca^{2+}]_i$ in rodent microglial cells [28]. Thereafter, we observe that canonical transient receptor potential 3 (TRPC3) channels contribute to the maintenance of BDNF-induced sustained intracellular Ca^{2+} elevation. Immunocytochemical technique and flow cytometry also reveal that BDNF rapidly upregulated the surface expression of TRPC3 channels in rodent microglial cells. BDNF-induced upregulation of surface expression of TRPC3 channels also depends on activation of the PLC pathway, as previously shown by others [60]. In addition, pretreatment with BDNF suppresses the production of NO induced by TNFα, which is prevented by coadministration of a selective TRPC3 inhibitor, Pyr3. These results suggest that BDNF induces sustained intracellular Ca^{2+} elevation through the upregulation of surface TRPC3 channels, and TRPC3 channels could be important for the BDNF-induced suppression of the NO production in activated microglia. We show direct evidence that rodent microglial cells are able to respond to BDNF, and TRPC3 channels could also play important roles in microglial functions. Hall et al. have previously demonstrated the implication of the basal level of $[Ca^{2+}]_i$ in the activation of rodent microglia, including NO production [13]. BDNF-induced elevation of basal levels of $[Ca^{2+}]_i$ could regulate the microglial intracellular signal transduction to suppress the release of NO induced by IFN-γ [15, 28]. We observe that pretreatment with BDNF also suppresses the production of NO in murine microglial cells activated by TNFα, which is prevented by coadministration of Pyr3. We also find that pretreatment with both BDNF and Pyr3 do not elevate the basal $[Ca^{2+}]_i$ in rodent microglial cells. These results suggest that BDNF-induced elevation of basal levels of $[Ca^{2+}]_i$ mediated by TRPC3 channels is important for the BDNF-induced suppression of NO production in rodent microglial cells. TNFα plays a key role in the induction of sickness behaviors [54] and also in the development of depressive symptoms [9]. In rodent model of chronic unpredictable stress (CUS)-induced depression, Qin et al. recently reported that the TRPC3 but not TRPC5 expression is lower in the hippocampus of depression model group compared with that of control group [47]. Thus, these results suggest that BDNF might have an anti-inflammatory effect through the inhibition of microglial activation, and TRPC3 could be an important target for the treatment of psychiatric disorders including depression. We need to further examine the mechanism underlying the upregulation of surface TRPC3 channels induced by BDNF in rodent microglial cells [30, 31].

Although the mechanism underlying the activation of TRPCs via PLC stimulation is still not completely resolved, TRPC3, like TRPC6 and TRPC7, can be activated directly by diacylglycerol (DAG). The trafficking of TRPC3 channels to the plasma membrane depends on interactions with Cav-1, Homer1, PLC-γ, VAMP2, and RFN24 [10]. The TRPC3 signaling is also regulated by cytoskeleton status. Enhancement and stabilization of the cortical actin layer by jasplakinolide or calyculin A treatments promote internalization of TRPC3-signaling complex and reduce TRPC3 activity [24]. TRPC3 can form a heteromeric channel with TRPC1 and be activated when TRPC1 binds to and is gated by STIM1. This interaction with TRPC1 confers store-dependent regulation on TRPC3 channels [22]. Additionally, other proteins such as clathrin, dynamin, AP-2, syntaxin, synaptotagmin-1,

MxA, and RACK1 are supposed to interact with TRPC3 and to, possibly, regulate trafficking and activity of the channels [10]. We need additional studies to identify the molecular mechanisms that determine the trafficking and activity of TRPC3 channels and what underlies the BDNF-induced upregulation of surface TRPC3 channels in these mechanisms.

10.4 TRPC Channels and Brain Inflammation

In the brain, modulation of inflammatory processes is also mediated by astrocytes. In intracerebral hemorrhage (ICH), astrocytes accumulate in the perihematomal region and induce toxic edema, provoke inflammation, release cytotoxins, and form scars [63]. Moreover, neutrophils, macrophages, and microglia are supposed to provoke secondary brain damage after ICH. Shirakawa et al. [53] have previously shown that thrombin, a predominant blood-derived factor, induces functional activation of astrocytes via opening of the TRPC3 channels in human astrocytoma cell lines and cerebral astrocytes cultured from rats. In addition, Pyr3, a selective TRPC3 inhibitor, ameliorated reactive astrogliosis and neurological deficits resulting from ICH. They suggest that the Pyr3-mediated suppression of astrocytes contribute to the prevention of the activation of microglia/macrophages after ICH [35]. In astrocytes, TRPC1 channels are also reported to be an important target for mood stabilizers. The mRNA expression of TRPC1 in primary cultures of astrocytes was reduced to a similar degree after treatment with lithium (Li^+), carbamazepine (CBZ), or valproic acid (VPA; [66]).

The TRPC5 channels are stimulated by nitric oxide (NO) through a mechanism that requires oxidation (S-nitrosylation) at extracellular cysteine [68]. Although TRPC5 channels can form homomeric channels or assemble with TRPC1/4, under pathological conditions such as in Huntington's disease, the suppression of TRPC1 occurred, and oxidative stress leads to the elevation of TRPC5 S-glutathionylation at Cys176/Cys178 and also induces the formation of homomeric TRPC5 channels. This GSSG-activated TRPC5 current results in a sustained increase in cytosolic Ca^{2+}. The abnormal increase in a sustained cytosolic Ca^{2+} due to TRPC5 overactivation is supposed to cause neuronal damage in the striatum in Huntington's disease [16].

Systemic lupus erythematosus (SLE) is a prototypic autoimmune disease with protean clinical manifestations, including neurological and psychiatric disorders (NPSLE). NPSLE could affect 21–95% of patients with SLE and impact significantly patients' survival and quality of life. Ramirez et al. reported the association of a single nucleotide polymorphic variant of the TRPC6 gene with protection from the development of NPSLE [48]. TRPC6 has been repeatedly shown to exert neuroprotection, in particular, after ischemic injury [64, 70]. These reports suggest that TRPC channels other than TRPC3 also have important roles to modulate the inflammatory processes in the brain.

Recent reports show that BDNF has major roles not only in the brain but also in regulating energy metabolism or cellular function in peripheral organs. The concentrations of BDNF in serum and/or plasma are found to be lower in patients suffering from cardiovascular disease, diabetes, obesity, and also Alzheimer's disease. BDNF is expressed in non-neurogenic tissues, including skeletal muscle or endothelial cells, and exercise is shown to potentiate the production of BDNF not only in the brain but also in skeletal muscle or plasma as well [46]. The TRPC3 channels are highly expressed in endothelial [51], smooth, and cardiac muscle cells. In T lymphocytes, TRPC3 is activated by T-cell receptor-triggered immune responses and is probably directly coupled to PLC-γ, which mediates translocation of the channel [45]. TRPC3 is inhibited by cGMP/protein kinase G (PKG) and provides a negative feedback mechanism for the generation of nitric oxide (NO) in endothelial cells [67]. In endothelial cells, vascular cell adhesion molecule-1 (VCAM-1) expression and monocyte adhesion also depend on TRPC3 channels [56]. TRPC3 channels also contribute to the constitutive Ca^{2+} influx in macrophages, which is required for the survival of macrophage [39, 58].

Systemic inflammation can damage the blood-brain barrier (BBB) and allow to recruit lymphocytes, monocytes, and neutrophils into the brain. Especially, brain microvascular endothelial cells, which constitute the neurovascular unit of the BBB, play an essential role in the regulation of interplay between systemic and brain inflammation [6]. Thus, we also need to examine the effect of BDNF on TRPC3 functions in peripheral cells including lymphocytes, monocytes, neutrophils, and endothelial cells.

10.5 Conclusions

Nonresolving low-grade inflammation is supposed to underlie the basis of chronic disorders including cardiovascular diseases, cancer, diabetes, obesity, and psychiatric disorders such as depression or Alzheimer's diseases. There is increasing evidence suggesting that pathophysiology of psychiatric disorders is related to the inflammatory responses mediated by microglial cells. In addition, chronic disorders are associated with low circulating levels of BDNF. Elevation of intracellular Ca^{2+} is important for the activation of microglial cell functions, including proliferation, release of NO, cytokines, and BDNF. It has been shown that alteration of intracellular Ca^{2+} signaling underlies the pathophysiology of psychiatric disorders, including depression and bipolar disorders.

BDNF induces a sustained intracellular Ca^{2+} elevation through the upregulation of the surface expression of TRPC3 channels in rodent microglial cells (Fig. 10.2). Microglial cells are able to respond to BDNF, which is important for the regulation of inflammatory responses, and also involved in the pathophysiology and/or the treatment of psychiatric disorders. BDNF is firstly synthesized as proBDNF protein. ProBDNF is then either proteolytically cleaved intracellularly or by extracellular proteases, such as metalloproteinases and plasmin, to mature BDNF. Interestingly, interaction of mature neurotrophins with Trk receptors leads to cell survival, whereas binding of proBDNF to p75NTR leads to apoptosis. In addition, mature BDNF and proBDNF facilitate long-term potentiation (LTP) and long-term depression (LTD) at the hippocampal CA1 synapses, respectively. Thus, Trk and p75NTR preferentially bind mature and pro-

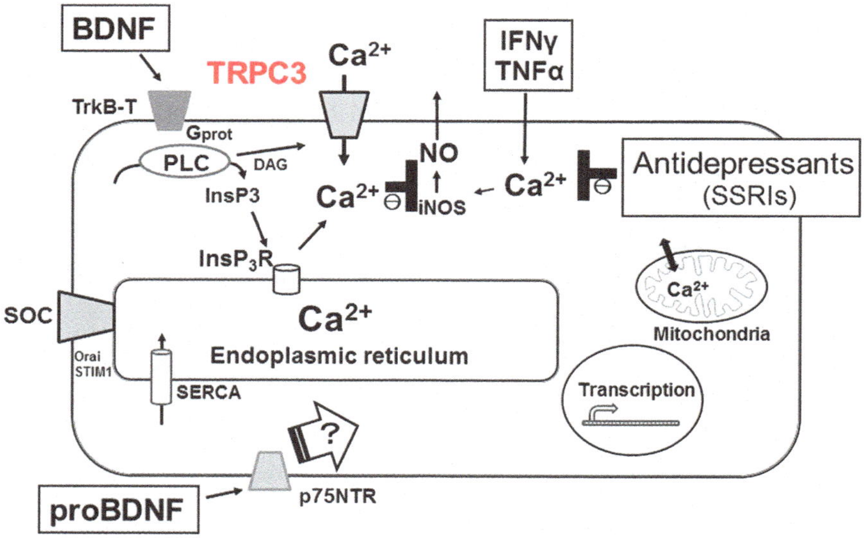

Fig. 10.2 Schematic illustration representing the effect of BDNF on microglial intracellular Ca^{2+} signaling mediated by TRPC3 channels. The effect of proBDNF on microglial intracellular Ca^{2+} signaling remains to be explored. *SERCA* sarco(endo)plasmic reticulum Ca^{2+}-ATPases, *SOC* store-operated calcium channel

neurotrophins, respectively, to elicit opposing biological responses in the CNS. However, the effects of proBDNF on microglial cells are not fully understood. Thus, it is of great importance to study the effect of proBDNF on microglial cells especially by focusing on the TRPC channels.

References

1. Aguzzi A, Barres BA, Bennett ML (2013) Microglia: scapegoat, saboteur, or something else? Science 339(6116):156–161
2. Batchelor PE, Liberatore GT, Wong JY, Porritt MJ, Frerichs F, Donnan GA, Howells DW (1999) Activated macrophages and microglia induce dopaminergic sprouting in the injured striatum and express brain-derived neurotrophic factor and glial cell line-derived neurotrophic factor. J Neurosci 19:1708–1716
3. Berridge MJ (1998) Neuronal calcium signaling. Neuron 21(1):13–26
4. Berridge MJ (2013) Dysregulation of neural calcium signaling in Alzheimer disease, bipolar disorder and schizophrenia. Prion 7(1):2–13
5. Berton O, Nestler EJ (2006) New approaches to antidepressant drug discovery: beyond monoamines. Nat Rev Neurosci 7:137–151
6. Cardoso FL, Brites D, Brito MA (2010) Looking at the blood-brain barrier: molecular anatomy and possible investigation approaches. Brain Res Rev 64(2):328–363
7. Coull JA, Beggs S, Boudreau D, Boivin D, Tsuda M, Inoue K, Gravel C, Salter MW, De Koninck Y (2005) BDNF from microglia causes the shift in neuronal anion gradient underlying neuropathic pain. Nature 438:1017–1021
8. Cunningham C (2013) Microglia and neurodegeneration: the role of systemic inflammation. Glia 61(1):71–90
9. Dantzer R, O'Connor JC, Freund GG, Johnson RW, Kelley KW (2008) From inflammation to sickness and depression: when the immune system subjugates the brain. Nat Rev Neurosci 9:46–56
10. de Souza LB, Ambudkar IS (2014) Trafficking mechanisms and regulation of TRPC channels. Cell Calcium 56(2):43–50
11. Dudvarski Stankovic N, Teodorczyk M, Ploen R, Zipp F, Schmidt MH (2016) Microglia-blood vessel interactions: a double-edged sword in brain pathologies. Acta Neuropathol 131(3):347–363
12. Elkabes S, DiCicco-Bloom EM, Black IB (1996) Brain microglia/macrophages express neurotrophins that selectively regulate microglial proliferation and function. J Neurosci 16:2508–2521
13. Hall AA, Herrera Y, Ajmo CT Jr, Cuevas J, Pennypacker KR (2009) Sigma receptors suppress multiple aspects of microglial activation. Glia 57(7):744–754
14. Haroon E, Raison CL, Miller AH (2012) Psychoneuroimmunology meets neuropsychopharmacology: translational implications of the impact of inflammation on behavior. Neuropsychopharmacology 37(1):137–162
15. Hoffmann A, Kann O, Ohlemeyer C, Hanisch UK, Kettenmann H (2003) Elevation of basal intracellular calcium as a central element in the activation of brain macrophages (microglia): suppression of receptor-evoked calcium signaling and control of release function. J Neurosci 23(11):4410–4419
16. Hong C, Seo H, Kwak M, Jeon J, Jang J, Jeong EM, Myeong J, Hwang YJ, Ha K, Kang MJ, Lee KP, Yi EC, Kim IG, Jeon JH, Ryu H, So I (2015) Increased TRPC5 glutathionylation contributes to striatal neuron loss in Huntington's disease. Brain 138(Pt 10):3030–3047
17. Horikawa H, Kato TA, Mizoguchi Y, Monji A, Seki Y, Ohkuri T, Gotoh L, Yonaha M, Ueda T, Hashioka S, Kanba S (2010) Inhibitory effects of SSRIs on IFN-γ induced microglial activation through the regulation of intracellular calcium. Prog Neuropsychopharmacol Biol. Psychiatry 34(7):1306–1316
18. Hu X, Ivashkiv LB (2009) Cross-regulation of signaling pathways by interferon-gamma: implications for immune responses and autoimmune diseases. Immunity 31(4):539–550
19. Kato T, Mizoguchi Y, Monji A, Horikawa H, Suzuki SO, Seki Y, Iwaki T, Hashioka S, Kanba S (2008) Inhibitory effects of aripiprazole on interferon-gamma-induced microglial activation via intracellular Ca2+ regulation in vitro. J Neurochem 106(2):815–825
20. Kato TA, Yamauchi Y, Horikawa H, Monji A, Mizoguchi Y, Seki Y, Hayakawa K, Utsumi H, Kanba S (2013) Neurotransmitters, psychotropic drugs and microglia: clinical implications for psychiatry. Curr Med Chem 20(3):331–344
21. Kettenmann H, Hanisch UK, Noda M, Verkhratsky A (2011) Physiology of microglia. Physiol Rev 91(2):461–553
22. Lee KP, Choi S, Hong JH, Ahuja M, Graham S, Ma R, So I, Shin DM, Muallem S, Yuan JP (2014) Molecular determinants mediating gating of Transient Receptor Potential Canonical (TRPC) channels by stromal interaction molecule 1 (STIM1). J Biol Chem 289(10):6372–6382
23. Leonard BE (2014) Impact of inflammation on neurotransmitter changes in major depression: an insight into the action of antidepressants. Prog Neuro-Psychopharmacol Biol Psychiatry 48:261–267
24. Lockwich T, Singh BB, Liu X, Ambudkar IS (2001) Stabilization of cortical actin induces internalization

of transient receptor potential 3 (Trp3)-associated caveolar Ca2+ signaling complex and loss of Ca2+ influx without disruption of Trp3-inositol trisphosphate receptor association. J Biol Chem 276(45):42401–42408
25. McLarnon JG, Choi HB, Lue LF, Walker DG, Kim SU (2005) Perturbations in calcium-mediated signal transduction in microglia from Alzheimer's disease patients. J Neurosci Res 81(3):426–435
26. Mizoguchi Y, Ishibashi H, Nabekura J (2003) The action of BDNF on GABA(A) currents changes from potentiating to suppressing during maturation of rat hippocampal CA1 pyramidal neurons. J Physiol 548(Pt 3):703–709
27. Mizoguchi Y, Kanematsu T, Hirata M, Nabekura J (2003) A rapid increase in the total number of cell surface functional GABAA receptors induced by brain-derived neurotrophic factor in rat visual cortex. J Biol Chem 278(45):44097–44102
28. Mizoguchi Y, Monji A, Kato T, Seki Y, Gotoh L, Horikawa H, Suzuki SO, Iwaki T, Yonaha M, Hashioka S, Kanba S (2009) Brain-derived neurotrophic factor induces sustained elevation of intracellular Ca2+ in rodent microglia. J Immunol 183(12):7778–7786
29. Mizoguchi Y, Monji A, Kato TA, Horikawa H, Seki Y, Kasai M, Kanba S, Yamada S (2011) Possible role of BDNF-induced microglial intracellular Ca(2+) elevation in the pathophysiology of neuropsychiatric disorders. Mini-Rev Med Chem 11(7):575–581
30. Mizoguchi Y, Kato TA, Seki Y, Ohgidani M, Sagata N, Horikawa H, Yamauchi Y, Sato-Kasai M, Hayakawa K, Inoue R, Kanba S, Monji A (2014) Brain-derived neurotrophic factor (BDNF) induces sustained intracellular Ca2+ elevation through the up-regulation of surface transient receptor potential 3 (TRPC3) channels in rodent microglia. J Biol Chem 289(26):18549–18555
31. Mizoguchi Y, Kato TA, Horikawa H, Monji A (2014) Microglial intracellular Ca(2+) signaling as a target of antipsychotic actions for the treatment of schizophrenia. Front Cell Neurosci 8:370
32. Molendijk ML, Spinhoven P, Polak M, Bus BA, Penninx BW, Elzinga BM (2014) Serum BDNF concentrations as peripheral manifestations of depression: evidence from a systematic review and meta-analyses on 179 associations (N=9484). Mol Psychiatry 19(7):791–800
33. Möller T (2002) Calcium signaling in microglial cells. Glia 40(2):184–194
34. Monji A, Kato TA, Mizoguchi Y, Horikawa H, Seki Y, Kasai M, Yamauchi Y, Yamada S, Kanba S (2013) Neuroinflammation in schizophrenia especially focused on the role of microglia. Prog Neuro-Psychopharmacol Biol Psychiatry 42:115–121
35. Munakata M, Shirakawa H, Nagayasu K, Miyanohara J, Miyake T, Nakagawa T, Katsuki H, Kaneko S (2013) Transient receptor potential canonical 3 inhibitor Pyr3 improves outcomes and attenuates astrogliosis after intracerebral hemorrhage in mice. Stroke 44(7):1981–1987
36. Nagano T, Kimura SH, Takai E, Matsuda T, Takemura M (2006) Lipopolysaccharide sensitizes microglia toward Ca(2+)-induced cell death: mode of cell death shifts from apoptosis to necrosis. Glia 53(1):67–73
37. Nair JS, DaFonseca CJ, Tjernberg A, Sun W, Darnell JE Jr, Chait BT, Zhang JJ (2002) Requirement of Ca2+ and CaMKII for Stat1 Ser-727 phosphorylation in response to IFN-gamma. Proc Natl Acad Sci U S A 99(9):5971–5976
38. Nakajima K, Honda S, Tohyama Y, Imai Y, Kohsaka S, Kurihara T (2001) Neurotrophin secretion from cultured microglia. J Neurosci Res 65:322–331
39. Nilius B, Szallasi A (2014a) Transient receptor potential channels as drug targets: from the science of basic research to the art of medicine. Pharmacol Rev 66(3):676–814
40. Ohana L, Newell EW, Stanley EF, Schlichter LC (2009) The Ca2+ release-activated Ca2+ current (I(CRAC)) mediates store-operated Ca2+ entry in rat microglia. Channels (Austin) 3(2):129–139
41. Parekh AB, Putney JW Jr (2005) Store-operated calcium channels. Physiol Rev 85(2):757–810
42. Park H, Poo MM (2013) Neurotrophin regulation of neural circuit development and function. Nat Rev Neurosci 14(1):7–23
43. Parkhurst CN, Yang G, Ninan I, Savas JN, Yates JR 3rd, Lafaille JJ, Hempstead BL, Littman DR, Gan WB (2013) Microglia promote learning-dependent synapse formation through brain-derived neurotrophic factor. Cell 155(7):1596–1609
44. Patapoutian A, Reichardt LF (2001) Trk receptors: mediators of neurotrophin action. Curr Opin Neurobiol 11:272–280
45. Philipp S, Strauss B, Hirnet D, Wissenbach U, Mery L, Flockerzi V, Hoth M (2003) TRPC3 mediates T-cell receptor-dependent calcium entry in human T-lymphocytes. J Biol Chem 278(29):26629–26638
46. Pedersen BK, Febbraio MA (2012 Apr 3) Muscles, exercise and obesity: skeletal muscle as a secretory organ. Nat Rev Endocrinol 8(8):457–465
47. Qin X, Liu Y, Zhu M, Yang Z (2015) The possible relationship between expressions of TRPC3/5 channels and cognitive changes in rat model of chronic unpredictable stress. Behav Brain Res 290:180–186
48. Ramirez GA, Lanzani C, Bozzolo EP, Citterio L, Zagato L, Casamassima N, Canti V, Sabbadini MG, Rovere-Querini P, Manunta P, Manfredi AA (2015) TRPC6 gene variants and neuropsychiatric lupus. J Neuroimmunol 288:21–24
49. Schwartz M, Kipnis J, Rivest S, Prat A (2013) How do immune cells support and shape the brain in health, disease, and aging? J Neurosci 33(45):17587–17596
50. Sen S, Duman R, Sanacora G (2008) Serum brain-derived neurotrophic factor, depression, and antidepressant medications: meta-analyses and implications. Biol Psychiatry 64:527–532

51. Senadheera S, Kim Y, Grayson TH, Toemoe S, Kochukov MY, Abramowitz J, Housley GD, Bertrand RL, Chadha PS, Bertrand PP, Murphy TV, Tare M, Birnbaumer L, Marrelli SP, Sandow SL (2012) Transient receptor potential canonical type 3 channels facilitate endothelium-derived hyperpolarization-mediated resistance artery vasodilator activity. Cardiovasc Res 95(4):439–447
52. Setiawan E, Wilson AA, Mizrahi R, Rusjan PM, Miler L, Rajkowska G, Suridjan I, Kennedy JL, Rekkas PV, Houle S, Meyer JH (2015) Role of translocator protein density, a marker of neuroinflammation, in the brain during major depressive episodes. JAMA Psychiatry 72(3):268–275
53. Shirakawa H, Sakimoto S, Nakao K, Sugishita A, Konno M, Iida S, Kusano A, Hashimoto E, Nakagawa T, Kaneko S (2010) Transient receptor potential canonical 3 (TRPC3) mediates thrombin-induced astrocyte activation and upregulates its own expression in cortical astrocytes. J Neurosci 30(39):13116–13129
54. Simen BB, Duman CH, Simen AA, Duman RS (2006) TNFα signaling in depression and anxiety: behavioral consequences of individual receptor targeting. Biol Psychiatry 59:775–785
55. Simms BA, Zamponi GW (2014) Neuronal voltage-gated calcium channels: structure, function, and dysfunction. Neuron 82(1):24–45
56. Smedlund KB, Birnbaumer L, Vazquez G (2015) Increased size and cellularity of advanced atherosclerotic lesions in mice with endothelial overexpression of the human TRPC3 channel. Proc Natl Acad Sci U S A 112(17):E2201–E2206
57. Smith AM, Dragunow M (2014) The human side of microglia. Trends Neurosci 37(3):125–135
58. Tano JY, Smedlund K, Lee R, Abramowitz J, Birnbaumer L, Vazquez G (2011) Impairment of survival signaling and efferocytosis in TRPC3-deficient macrophages. Biochem Biophys Res Commun 410(3):643–647
59. Thoenen H (1995) Neurotrophins and neuronal plasticity. Science 270(5236):593–598
60. van Rossum DB, Patterson RL, Sharma S, Barrow RK, Kornberg M, Gill DL, Snyder SH (2005) Phospholipase Cgamma1 controls surface expression of TRPC3 through an intermolecular PH domain. Nature 434(7029):99–104
61. Viswanath B, Jose SP, Squassina A, Thirthalli J, Purushottam M, Mukherjee O, Vladimirov V, Patrinos GP, Del Zompo M, Jain S (2015) Cellular models to study bipolar disorder: a systematic review. J Affect Disord 184:36–50
62. Wake H, Moorhouse AJ, Miyamoto A, Nabekura J (2013) Microglia: actively surveying and shaping neuronal circuit structure and function. Trends Neurosci 36(4):209–217
63. Wang J, Doré S (2007) Inflammation after intracerebral hemorrhage. J Cereb Blood Flow Metab 27:894–908
64. Weber EW, Han F, Tauseef M, Birnbaumer L, Mehta D, Muller WA (2015) TRPC6 is the endothelial calcium channel that regulates leukocyte transendothelial migration during the inflammatory response. J Exp Med 212(11):1883–1899
65. Wohleb ES, McKim DB, Sheridan JF, Godbout JP (2015) Monocyte trafficking to the brain with stress and inflammation: a novel axis of immune-to-brain communication that influences mood and behavior. Front Neurosci 8:447
66. Yan E, Li B, Gu L, Hertz L, Peng L (2013) Mechanisms for L-channel-mediated increase in [Ca(2+)]i and its reduction by anti-bipolar drugs in cultured astrocytes combined with its mRNA expression in freshly isolated cells support the importance of astrocytic L-channels. Cell Calcium 54(5):335–342
67. Yao X, Garland CJ (2005) Recent developments in vascular endothelial cell transient receptor potential channels. Circ Res 97(9):853–863
68. Yoshida T, Inoue R, Morii T, Takahashi N, Yamamoto S, Hara Y, Tominaga M, Shimizu S, Sato Y, Mori Y (2006) Nitric oxide activates TRP channels by cysteine S-nitrosylation. Nat Chem Biol 2(11):596–607
69. Zhang J, Geula C, Lu C, Koziel H, Hatcher LM, Roisen FJ (2003) Neurotrophins regulate proliferation and survival of two microglial cell lines in vitro. Exp Neurol 183:469–481
70. Zhang J, Mao X, Zhou T, Cheng X, Lin Y (2014) IL-17A contributes to brain ischemia reperfusion injury through calpain-TRPC6 pathway in mice. Neuroscience 274:419–428

11 TRPC Channels and Epilepsy

Fang Zheng

Abstract

Accumulating evidence suggest that TRPC channels play critical roles in various aspects of epileptogenesis. TRPC1/4 channels are major contributors to nonsynaptically derived epileptiform burst firing in the CA1 and the lateral septum. TRPC7 channels play a critical role in synaptically derived epileptiform burst firing. The reduction of spontaneous epileptiform bursting in the CA3 is correlated to a reduction in pilocarpine-induced SE in vivo in TRPC7 knockout mice. TRPC channels are also significant contributors to SE-induced neuronal cell death. Although the pilocarpine-induced SE itself is not significantly reduced, the SE-induced neuronal cell death is significantly reduced in the CA1 and the lateral septum, indicating that TRPC1/4 channels directly contribute to SE-induced neuronal cell death. Genetic ablation of TRPC5 also reduces SE-induced neuronal cell death in the CA1 and CA3 areas of the hippocampus.

Keywords

Seizures • Interictal spikes • Epileptiform bursting • Canonical transient receptor potential channels

Epilepsy is a group of neurological disorders characterized by recurrent, usually unprovoked seizures. Epilepsy is one of the most common brain disorders, affecting approximately 50 million people (i.e., 1% of the general population) globally [14]. In the United States, epilepsy is the fourth most common neurological diseases [24], and one in 26 people will develop epilepsy during their lifetime [23]. Although epilepsy can be successfully treated with antiepileptic drugs (AEDs), one third of patients suffer from refractory (i.e., drug-resistant) seizures. Therefore, there is a pressing need for the development of new anti-epileptic treatment options.

Historically, references to epileptic seizures can be found in many ancient cultures. The

F Zheng (✉)
Department of Pharmacology and Toxicology, College of Medicine, University of Arkansas for Medical Science, Little Rock, AR 72205, USA
e-mail: zhengfang@uams.edu

Y. Wang (ed.), *Transient Receptor Potential Canonical Channels and Brain Diseases*, Advances in Experimental Medicine and Biology 976, DOI 10.1007/978-94-024-1088-4_11

earliest description of epilepsy has been found in a Babylonian medical textbook, and it provides a remarkably accurate account of clinical manifestations of seizures [56]. Descriptions of seizure phenomena include generalized convulsions, partial motor seizures, and sensory symptoms indicative of an aura and repetitive occurrence as in *status epilepticus*. These clinical phenomena are today recognized as parts of a broad category of symptom complexes designated as "epileptic" first by Jackson in the nineteenth century [25]. The greatest contribution by Jackson is his conceptualization of a single mechanism capable of explaining the full spectrum of epileptic phenomena. According to Jackson, all the phenomena of epileptic fits may be explained by a discharge of gray matter: focal discharges of sensory areas may lead to aura, whereas focal discharges of motor areas cause contralateral convulsion; intensification and spread of focal discharges can lead to unconsciousness and convulsion. The understanding and treatment of seizures and epilepsy have advanced greatly in the twentieth century after the discovery of electroencephalography (EEG) and its application as a critical clinical tool. EEG findings have redefined epilepsy in modern terms [20]. There are distinct EEG ictal patterns for petit mal, grand mal, and psychomotor seizures. These human EEG findings form the foundation of modern understanding of the pathophysiology of epilepsy.

The rapid development of molecular biology and human genetics has revolutionized the epilepsy field in the last 20 years. In many cases, idiopathic seizures are "channelopathies" in nature, i.e., caused by mutations that cause malfunction of voltage-gated or ligand-gated ion channels [19, 34, 46]. The current list of voltage-gated ion channels associated with seizures includes sodium channels, calcium channels, potassium channels, and chloride channels. The current list of ligand-gated ion channels associated with seizures includes GABA-A receptors, nicotinic receptors, and NMDA receptors. And the list will continue to grow in the future. The number of ion channel families involved in idiopathic epilepsy syndromes indicates a complexity in the underlying pathogenic processes for epilepsy and presents a challenge for the rational design of the next generation of antiepileptic drugs. On the other hand, the identification of these ion channels also offers numerous clues for the understanding of the pathophysiology of epilepsy.

There are two essential epileptogenic factors that are required for the occurrence of a seizure [33, 34]. The first is the neuronal hyperexcitability that arises from cellular mechanisms that affect ion channels involved in the modulation of neuronal firing patterns. The second is a network abnormality that results in uncontrolled synchronization of large groups of neurons and propagation of the epileptic discharge along the neural pathways. Canonical transient receptor potential (TRPC) channels contribute to the occurrence of seizures by playing a role in both sets of disturbances. In this chapter, we will first review the current evidence regarding the role of various TRPC channels in the generation of epileptiform burst firing, followed by a discussion about the role of TRPC channels in synaptic plasticity and abnormal synchronization. Finally, the implication of these distinct mechanisms on seizure generation and propagation will be discussed.

11.1 Ictal vs Interictal Activities: From Clinics to Bench Side

The *International League Against Epilepsy* (ILAE) defined a seizure as "a transient occurrence of signs and/or symptoms due to abnormal excessive or synchronous neuronal activity in the brain" [17]. Clinically, physicians rely on EEG recording to identify, classify, and localize seizures. Abnormal EEG patterns are broadly defined as ictal (i.e., during seizures) or interictal (i.e., between seizures). However, the line between interictal and ictal epileptiform discharges can be ambiguous [16]. Practically, interictal epileptiform discharges are defined as paroxysmal activities lasting up to 200 ms, with those lasting 30–70 ms defined as "spikes" and those lasting 70–200 ms defined as "sharp waves." This classification based on duration is largely arbitrary and without clear clinical utility. Ictal epileptiform discharges are also complex and somewhat ambiguous. Absence seizures ("petit mal") are associated with 3 Hz rhythmic spike and wave patterns, whereas generalized tonic-clonic seizures ("grand mal") are characterized by 10 Hz spikes during the tonic phase and 10 Hz spikes mixed with slow waves during the clonic phase. Focal seizures are associated with localized ictal activities and interictal activities outside of the seizure foci.

Animal studies of seizures have been largely confined to generalize seizures because of the difficulty of detecting focal seizures either electrographically or behaviorally. The kindling model [35] utilizes repetitive electric stimulation of vulnerable brain structures to induce seizures, and the Racine scale [44] has been developed to describe the behavioral manifestation of seizures resulting from kindling of amygdala in rats. However, the kindling process does not result in spontaneous recurrent seizures. On the other hand, pilocarpine, a muscarinic agonist, induces *status epilepticus* (SE) acutely, and spontaneous recurrent seizures later on after the reorganization of neural network resulted from SE-induced neurodegeneration [10, 54]. Although intrahippocampal injection of kainic acid also can result in spontaneous recurrent seizures in about a third of mice, the pilocarpine model remains the mostly utilized animal model of epilepsy today. If the definition of seizures in the clinic is difficult and somewhat ambivalent, the definition of seizures in animal studies is even more challenging. The reason is simple: clinical diagnosis of epileptic seizures relies heavily on the report of symptoms and histories by patients, whereas it is difficult to discern and distinguish normal EEG patterns or behaviors from epileptic ones in animals. Although EEG recording technique has been used in animal epilepsy research, the Racine scale which describes a set of convulsive behaviors still plays a dominant role. This state of affairs has implications for the understanding of TRPC channel's role in seizure and epilepsy and will be discussed in more details later in this chapter.

The cellular correlates of interictal and ictal activities are first investigated by Matsumoto and Marsan in 1964 [30, 31]. The spikes, the simplest epileptiform interictal discharges in EEG, are correlated with a burst of action potentials with a large underlying depolarization called paroxysmal depolarization shift (PDS) [30]. Ayala et al. first proposed that interictal spikes are generated by altered network excitability and the PDS, i.e., a plateau potential, are synaptic in origin [4]. This view gained further support by the work of Johnston and Brown [26]. Later, a competing view that explains the PDS as synchronized firing of neurons with intrinsic bursting emerged [50]. Today, it is generally accepted that epileptiform bursting can be either synaptically derived or nonsynaptically derived, depending on the specific neuronal population and neural network properties [33].

11.2 TRPC Channel's Contribution to Intrinsic Epileptiform Burst Firing

Neuronal excitability is controlled by ion channels expressed in a given neuron. The information in the nervous system is encoded by action potentials. The firing pattern of a neuron is generally determined by its intrinsic properties, i.e., independent of synaptic inputs. An action potential is generated when the membrane potential is depolarized to a threshold point, when self-sustained activation of voltage-gated sodium channels is triggered. The rapid and large depolarization caused by this inward current carried by voltage-gated sodium channels then activates voltage-gated potassium channels, which repolarize the neuron quickly. In most cortical pyramidal neurons, the firing rate is rather limited. This is largely due to two factors: (a) the time needed to remove inactivation of voltage-gated sodium channels after each action potential and (b) the hyperpolarization before and after action potentials mediated by various types of voltage-gated potassium channels and non-voltage-gated potassium channels [9, 21]. The A-type potassium channels put break on the membrane depolarization, slowing down the triggering of the next action potential [12], whereas potassium channels responsible for the afterhyperpolarization delay the return to the resting membrane potential. One additional layer of control is the phenomenon known as "spike adaptation," i.e., a gradual decrease in action potential frequency during sustained membrane depolarization, which is mediated by M-type potassium channels [22] or the HERG-type potassium channels [55]. Dentate granule cells show similar AHP and spike adaptation as pyramidal neurons [47]. In contrast, interneurons have a higher firing frequency because they can recover more quickly from inactivation of voltage-gated sodium channels and they have less afterhyperpolarization. Thus, glutamatergic principal neurons typically fire at low frequency and exhibit strong spike adaptation, and there are built-in safeguards against hyper-excitability.

However, some pyramidal cells exhibit bursting behavior, i.e., a train of action potentials sustained by a slow membrane depolarization lasting 100–400 ms [28]. It should also be noted that the firing pattern of a pyramidal neurons is malleable and subjected to neuromodulation. For examples, activation of group I metabotropic glutamate receptors (mGluRs) can elicit burst firing in CA1 pyramidal neurons [11] or lateral septal neurons (which are GABAergic) [61]. Activation of muscarinic receptors by carbachol has similar effects in many cortical neurons. Synaptic plasticity also can alter the firing pattern of pyramidal neurons. The best known example is the CA3 pyramidal neuron, in which the activity-dependent long-term potentiation of recurrent collateral synapses leads to synchronized bursting of a large group of CA3 pyramidal neurons that resembles epileptiform discharges [2, 48].

The ion channels responsible for the slow depolarization underlying burst firing include low-threshold voltage-gated calcium channels (i.e., the T-type; $Ca_V3.1$–3.3) [3, 7] and hyperpolarization-activated cyclic nucleotide-gated (HCN) cation channels (i.e., the H-current or the Q-current; HCN1–4) [21, 27]. In addition to these voltage-gated ion channels, the so-called calcium-activated nonselective (CAN) cation channels are also known to contribute to the slow depolarization driving the bursting. The molecular identity of the CAN channels had been a mystery for many years, and recent studies revealed that they are largely heteromeric TRPC channels comprised of TRPC1 and TRPC4.

The activation of TRPC channels by mGluRs and its implication in the generation of epileptiform discharges has been thoroughly investigated in the lateral septal neurons. The lateral septal nucleus is an important relay nucleus in the limbic system, which is an integral part of the septo-hippocampal loop: (1) The lateral septal nucleus receives its excitatory input primarily from the CA3 region of the hippocampus; (2) lateral septal neurons project to the medial septal nucleus and the diagonal band of Broca; and (3) the neurons in the medial septum and the diagonal band of Broca project back to the hippocampus to complete the loop [18]. The lateral septal nucleus also receives prominent serotonin input from the dorsal raphe nucleus, norepinephrine input from the locus coeruleus, and dopamine input from the ventral tegmental area and has reciprocal connections with the hypothalamus. Thus, the firing of lateral septal neurons is modulated by essentially every neurotransmitter or neuromodulator in the central nervous system. Interestingly, the effects of most neurotransmitters or neuromodulators are inhibitory, i.e., causing membrane hyperpolarization and reduced firing. Only agonists for group I mGluRs and m1 muscarinic receptors are excitatory, i.e., causing membrane depolarization and bursting [18]. The ion channels responsible for epileptiform bursting induced by mGluR agonists are initially characterized as CAN channels because they are clearly calcium-dependent nonselective cation channels [61–63]. However, they exhibit a very strong negative slop region in their current-voltage relationship, a rather peculiar property for non-voltage-gated ion channels [45]. These channels have now been identified convincingly as TRPC1/4 channels because genetic ablation of either TRPC1 or TRPC4 is sufficient to abolish the plateau potential underlying epileptiform bursting induced by group I mGluR agonists [40]. Furthermore, genetic ablation of TRPC3, 5, 6, and 7 has no detectable effects [64]. It remains unclear whether the excitatory effects of carbachol in lateral septal neurons are also mediated by TRPC1/4 channels, since carbachol has yet to be tested in any TRPC knockout mice.

Activation of group I mGluRs also elicits epileptiform burst firing in CA1 pyramidal neurons, and these bursts are also dependent on heteromeric TRPC1/4 channels [42]. The epileptiform bursting in the CA1 pyramidal neurons can also be elicited by carbachol, a muscarinic agonist. The amplitude of the plateau underlying the epileptiform bursts in CA1 pyramidal neurons is significantly reduced in TRPC1KO and TRPC1/4DKO mice [42]. The number of spikes in each burst is also significantly reduced in TRPC1KO and TRPC1/4DKO mice [42]. These findings are similar to what are observed in the lateral septum, suggesting that the activation of heteromeric TRPC1/4 channels by group I mGluRs plays a critical role in the epileptiform bursting. TRPC1 and TRPC4 are not the only TRPC subfamily members expressed in CA1 pyramidal neurons. TRPC5 is also highly expressed in CA1 pyramidal neurons. A previous study [49] has reported that carbachol induces membrane insertion of TRPC5 channels in CA1 pyramidal neurons and postulated that TRPC5 channels are responsible for the epileptiform bursting induced by muscarinic agonists. Strikingly, genetic ablation of TRPC5 has no detectable effects on mGluR agonist-induced epileptiform bursting in CA1 pyramidal neurons [42]. This is really surprising given that TRPC4 and TRPC5 are highly homologous and are thought to be able to form heteromeric channels with each other and with TRPC1. These results indicate that TRPC4 and TRPC5 have distinct functional roles even though they may be expressed in the same neuron.

Due to a lack of specific antibodies for many members of the TRPC subfamily, it is unclear whether CA1 pyramidal neurons also express other TRPC subfamily members. A previous study reported that TRPC3 channels are required for BDNF-induced cation current and dendritic spine formation [1]. However, genetic ablation of TRPC3 channels has no detectable effects on mGluR agonist-induced epileptiform bursting in CA1 pyramidal neurons (personal observation; Phelan et al.). This is clearly opposite to the reported role of TRPC3 channels in immature cortex [65]. Epileptiform bursting induced by mGluR agonists is also unaltered in TRPC7KO mice [41]. Thus, the evidence so far indicates that participation in epileptiform bursting is a distinct role of heteromeric TRPC1/4 channels in at least two brain regions.

Can TRPC5 channels contribute to bursting at other cortical regions? Yan et al. reported that transfection of dominant negative TRPC5 into pyramidal neurons in cortical slice cultures abolishes carbachol-induced slow afterdepolarization, whereas overexpression of TRPC5 has the opposite effects [58]. These observations have been interpreted as TRPC channels being required for the carbachol-induced slow afterdepolarization in these neurons. However, the exact subunit composition remains unclear. Further studies are required to determine exactly which type of TRPC channels are responsible for the carbachol-induced slow afterdepolarization and burst firing in many cortical areas.

Although epileptiform bursting induced by mGluR agonists or muscarinic agonists are potential cellular correlates of seizure activities, it must be noted that permanent activation of TRPC channels in the absence of these agonists has to occur if they contribute to the generation of spontaneous recurrent seizures. There are reports of upregulation of some TRPC channels in human patients suffering from seizures [57, 60]. However, it remains unclear whether such changes occur in animal models of seizure and epilepsy. Furthermore, it remains unclear whether the upregulation of these TRPC channels is also associated with spontaneous epileptiform burst firing. These questions need to be answered by future studies.

11.3 Contribution to Synaptically Derived Epileptiform Burst Firing by TRPC Channels

In addition to changes of intrinsic membrane properties, changes of synaptic signaling strength can also contribute to the generation of epileptiform burst firing. Dingledine et al. reported that blocking GABAergic inhibition by bicuculline triggers epileptiform burst firing in CA1 pyramidal neurons, and these bursts (either evoked by stimulating Schaffer collateral or spontaneous) are reduced by AP5, a selective NMDA receptor antagonist [15]. However, it appears that the role of NMDA receptors is different in CA3 pyramidal neurons [2]. High-frequency stimulation of recurrent collaterals induces epileptiform bursting in CA3 pyramidal neurons, and AP5 can block it when applied before the high-frequency stimulation but is ineffective when applied after the induction of epileptiform bursting. Further studies showed that the epileptiform bursting in CA3 pyramidal neurons can also be elicited by bathing the slice in high K^+ extracellular solutions [5] or the GABA-A receptor antagonist bicuculline [48]. No matter how the bursting is elicited, it has been shown consistently that the bursting results from an activity-dependent long-lasting increase in synaptic strength at CA3 recurrent synapses, which recruits a group of CA3 pyramidal neurons to fire action potentials synchronously. This long-term potentiation of the CA3 recurrent synapses depends on the activation of both NMDA receptors and group I mGluRs. The most striking property of synaptically derived epileptiform bursting in CA3 pyramidal neurons is its self-sustaining nature. Once induced, these bursting can last for hours after the washout of high K^+ or bicuculline. This is in stark contrast to the epileptiform bursting in CA1 pyramidal neurons induced by muscarinic agonist or mGluR agonists, which is reversible upon washout of these agonists. Furthermore, the spontaneous epileptiform bursting in CA1 provides the drive for epileptiform bursting in entorhinal cortex because severing the connection between the two areas abolishes the bursting in entorhinal cortex but leaves the bursting in CA3 intact. For these reasons, the epileptiform bursting resulting from the LTP of RC synapses is the most convincing model of epileptogenesis in vitro.

Given the evidence that the involvement of mGluRs in the generation of epileptiform burst firing in CA3 pyramidal neurons, TRPC channels, which are coupled to mGluRs, are also likely involved. TRPC1 and TRPC4 are potential candidates because of the critical role of TRPC1/4 channels in the epileptiform bursting in CA1 pyramidal neurons. However, the spontaneous burst firing induced by bicuculline in CA3 pyramidal neurons is unaltered in TRPC1KO mice or TRPC1/4DKO mice, suggesting that TRPC1 and TRPC4 do not play a critical role in the generation epileptiform bursting in CA3 pyramidal neurons (personal communication, Phelan et al.). TRPC3 is another likely candidate because of its critical role in the BDNF-TrkB signaling pathway. Genetic ablation of TRPC3 also has no discernable effects on spontaneous epileptiform bursting in CA3 pyramidal neurons (personal communication, Phelan et al.). Surprisingly, TRPC7, a little known member of the TRPC subfamily, turns out to play a critical role in spontaneous epileptiform bursting in CA3 pyramidal neurons [41].

TRPC7 is the last member of the TRPC subfamily to be cloned and is highly expressed in the brain [38]. However, the distribution of TRPC7 mRNA is rather low in the mouse Allen Brain Atlas, with its highest expression level detected in the cortical subplate. The protein expression pattern of TRPC7 is unclear due to a lack of a specific antibody. TRPC7 shows a high degree of sequence homology to TRPC3 and TRPC6, and these three members of the TRPC subfamily belong to a subgroup that can be directly activated by diacylglycerol (DAG). Genetic ablation of TRPC7 significantly reduced spontaneous epileptiform bursting in CA3 pyramidal neurons. The duration of the self-sustained bursts (i.e., 20 min after washout of bicuculline) was reduced from 197 ms down to 97 ms, and the amplitude of the plateau underlying the burst was reduced from 36 mV down to 24 mV. The reduction of epileptiform bursting is associated with a reduction of LTP at CA3 recurrent synapses, whereas

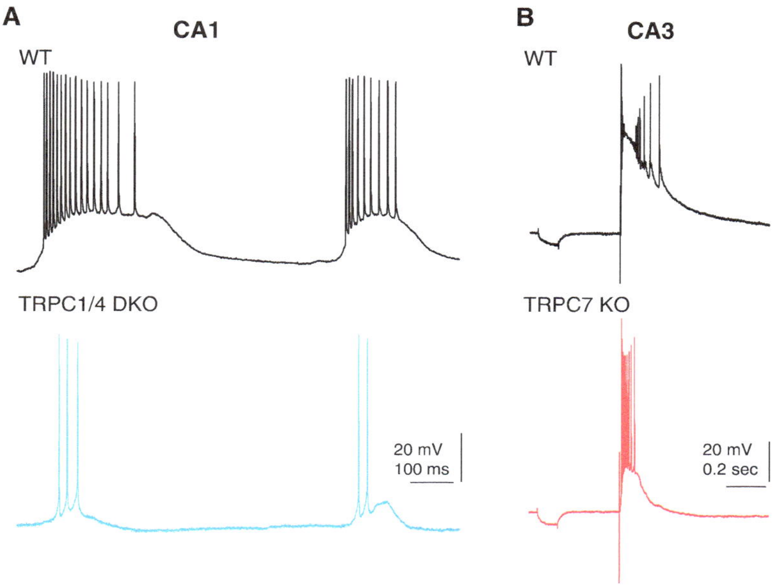

Fig. 11.1 Contribution of TRPC channels to epileptiform burst firing in the hippocampus. (**a**) The spontaneous epileptiform burst firing in CA1 pyramidal neurons shown is induced by 30 M 1S,3R-ACPD, a mGluR agonist. TRPC1/4 channels are required for this epileptiform bursting. (**b**) Epileptiform burst firing in CA3 pyramidal neurons is evoked by mossy fiber stimulation after bath application of 20 M bicuculline for 30 min. The persistent bursting results from activity-dependent strengthening of recurrent collaterals. TRPC7 channels play a critical role in this process (Adapted with permission from Refs. [41, 42])

the LTP at mossy fiber synapses is normal in TRPC7KO mice. Interestingly, the Schaffer collateral LTP in the CA1 is also reduced in TRPC7KO mice. Since Schaffer collaterals and recurrent collaterals share the same origin, these findings suggest that TRPC7 channels are likely located on the presynaptic terminals of both Schaffer collaterals and CA3 recurrent collaterals and are involved in the modulation of LTP through presynaptic mechanisms. What are the ion channels responsible for the plateau potential underlying spontaneous epileptiform bursting in CA3 pyramidal neurons? It has been reported that although the induction of bursting requires functional NMDA receptors, NMDA antagonists have no effects on the amplitude or the duration of the bursts. Therefore, NMDA receptors contribute little to the plateau potential underlying epileptiform bursting in CA3 pyramidal neurons. It is possible that TRPC channels contribute significantly to the plateau potential (Fig. 11.1), but this cannot be tested directly yet because of the lack of drugs that can selectively block specific subtypes of TRPC channels.

11.4 How Do the In Vitro Findings Correlate with Acute Seizures In Vivo?

The potential role of TRPC channels in in vivo seizures has just begun to be investigated using the pilocarpine model. The recent reports are limited to the role of TRPC channels in pilocarpine-induced SE and SE-induced neurodegeneration. The role of TRPC channels in the SE-induced spontaneous recurrent seizures has yet to be determined. Before a detailed discussion about the published data regarding any TRPC channel's role, it is critical to thoroughly discuss the pilocarpine model itself, the Racine scale, and other parameters frequently used to assess the role of a given molecular target.

Pilocarpine is an agonist for M1 subtype of muscarinic receptors, and administration of pilocarpine in rodents results in status epilepticus in a dose-dependent manner. This dose-response curve is a rather steep one [32, 37], and the lower dose of pilocarpine (40 mg/kg for mice) actually has anticonvulsant effects in the maximal electric shock (MES) model of seizures [59]. The initial site of action is the hippocampus, because the seizure activities in the hippocampus precede seizure activities in the cortex and the thalamus [54]. EEG recording in both rats and mice revealed a long latent period before the appearance of cortical seizures [43, 52]. This latent period is characterized by a suppression of normal cortical EEG activity, but a progression from immobility to forelimb clonus and Straub's tail [40, 43]. A burst of cortical seizures appears after a substantial delay (20–40 min following the administration of pilocarpine) and is followed by a postictal depression in the EEG signals but not behaviorally [43]. This process repeats several times with the length of the seizure increasing each time, until the SE state is finally reached [41, 43]. Most Racine stage 4 and 5 convulsive behaviors occur during this transition period, and once the SE is established, only stage 3 or lower convulsive behaviors remain during the remaining period of SE which lasts hours [43]. At lower doses, pilocarpine either only causes short bursts of epileptiform discharges in cortical EEG that can be characterized as interictal activities or only depression of normal EEG activities (personal communication, Zheng).

The Racine scale remains to be widely used in animal studies of seizure and epilepsy and is regarded by many as an indicator of seizure severity. This widely accepted notion has been recently challenged by quantitative EEG analysis of pilocarpine-induced seizures in mice [43]. With the exception of the transition period, the Racine scores do not correlate to the RMS power of cortical EEG signals. These results support previous reports that the Racine scale describes the involvement of distinct brain areas (or circuitry), rather than increasing intensity of seizures. Thus, a reduction of Racine scores does not necessarily indicate a reduction in cortical seizure severity or intensity. Another frequently used parameter in animal seizure research is the latency to the onset of seizures. The peculiar characteristic of the latency to pilocarpine-induced SE is that it is not dose-dependent (Personal communication, Zheng). The duration of the transition period is also fixed, with very little animal to animal variations. It is difficult at the moment to ascertain the pathophysiological implication of a change in the latency to SE.

Initial work on the role of TRPC channels in pilocarpine-induced acute seizures relied solely on the behavioral manifestation of seizures graded using the Racine scale [40, 42]. Using this approach, the pilocarpine-induced acute seizures in TRPC1 knockout mice or TRPC1/4 double knockout mice are comparable to the pilocarpine-induced acute seizures in wild-type mice. Since heteromeric TRPC1/4 channels are critical for epileptiform burst firings in the lateral septum and the CA1 area of the hippocampus, these results are puzzling. It is likely a reflection of the limited discriminating power of the behavioral approach. The EEG approach needs to be utilized to conclusively determine the role of TRPC1 and TRPC4 in pilocarpine-induced acute seizures.

Using the same behavioral approach, it has been reported that TRPC5 knockout mice exhibited reduced Racine scores during the late phase, but not the early phase of pilocarpine-induced SE [42]. One possible explanation for this observed change is the reduced Schaffer collateral LTP in TRPC5 knockout mice. In in vitro experiments in slices with intact hippocampal-entorhinal circuitry, the ictal activities are often initiated in the CA1 area, driven by epileptiform bursting in CA3 pyramidal neurons [6, 39, 51]. Therefore, a reduction of Schaffer collateral LTP is expected to hinder the hyper-excitability of the CA1 area. However, it remains unclear whether the reduced Racine score during the late phase of pilocarpine-induced SE truly reflects a reduction of cortical seizure intensity without confirmation from EEG analysis.

The TRPC channel associated with the most striking role in pilocarpine-induced SE is surprisingly the least studied TRPC7. Genetic ablation of TRPC7 drastically reduced the occurrence of pilocarpine-induced SE. Pilocarpine, at 280 mg/kg, induced SE in 10 out of 11 wild-type mice.

The same dosage of pilocarpine induced only a suppression of cortical EEG in a majority of TRPC7 knockout mice [41]. A detailed power spectral analysis indicates that there was a selective change in gamma wave activities in TRPC7 knockout mice [41]. Gamma wave activities normally increase significantly during the silent period after the administration of pilocarpine, and this increase precedes the occurrence of generalized cortical seizures that marks the beginning of the transition period and is correlated moderately to the Racine scores [43]. Genetic ablation of TRPC7 selectively abolishes this pilocarpine-induced increase in gamma wave activities during the latent period. These findings indicate that the pilocarpine-induced increase in gamma wave activities likely plays a critical role in the spread of pilocarpine-induced seizures from the hippocampus to the cortex. There are two possible gamma wave generators in the hippocampus: (1) the dentate gyrus and (2) the CA3 area [13]. There is a strong possibility that pilocarpine directly acts in the CA3 to produce convulsive behaviors and increased cortical gamma wave activities, because multiple studies have reported pilocarpine-induced interictal and ictal-like activities in the CA3 [6, 54].

11.5 The Contribution of TRPC Channels to SE-Induced Neurodegeneration

SE-induced neurodegeneration is a critical element in the pilocarpine model of epilepsy [10, 36, 53]. It is commonly accepted that SE-induced hippocampal neuronal cell death is a prerequisite to the epileptogenesis. In other words, it is believed that without SE-induced neurodegeneration, spontaneous recurrent seizures will not occur later on.

It has been reported that there appears to be a threshold for the "severity" of pilocarpine-induced seizures to induce neurodegeneration. Only mice that experienced seizures greater than Racine stage 3 show seizure-induced neuronal cell death [8, 40]. This threshold can be viewed simply as the presence of cortical SE, since it has been reported Racine stage 4 and 5 convulsions occur during the transition to SE [43]. There appears also to be a minimal duration of SE for resulting in neuronal cell death [32]. The general pattern of SE-induced neuronal cell death is also relatively consistent across multiple species [53]. Typical vulnerable brain regions include the hilar regions, CA1 and CA3 regions of the hippocampus, limbic regions such as amygdala and lateral septum, thalamus, hypothalamus, entorhinal cortex, and cingulate cortex. Interestingly, the dentate granule cells and CA2 hippocampal pyramidal cells are typically resistant to SE-induced cell death.

Changes in SE-induced neuronal cell death occur in various TRPC knockout mice. SE-induced neuronal cell death is reduced in TRPC1/4 double knockout mice but not in TRPC1 knockout mice or in TRPC5 knockout mice. The roles of TRPC3 and TRPC6 are suggested to be opposite: blocking TRPC3 channels by Pyr3 and activating TRPC6 channels by hyperforin both reduce pilocarpine-induced neuronal cell death [29]. However, these proposed roles are not supported by data from TRPC3 or TRPC6 knockout mice (Phelan et al., personal communication). The changes in SE-induced neuronal cell death by genetic ablation or pharmacological blockade of various TRPC channels raise several questions regarding the commonly accepted notions about the excitotoxicity and indicate that the underlying molecular mechanisms for seizure-induced neurodegeneration are not fully understood. On the one hand, neuronal cell death can be prevented without any significant changes in SE severity or duration. This occurs in TRPC1/4 double knockout mice. As discussed earlier, TRPC1/4 double knockout mice clearly exhibit unmitigated pilocarpine-induced SE but significantly reduced neuronal cell death in the CA1 area as well as lateral septum. This finding suggests that TRPC1/4 channels are directly involved in the signaling cascade leading to neuronal cell death (Fig. 11.2). However, the details of the signaling downstream of TRPC1/4 channel activation remain unexplored. On the other hand, unpublished data indicates that neuronal cell death can occur without

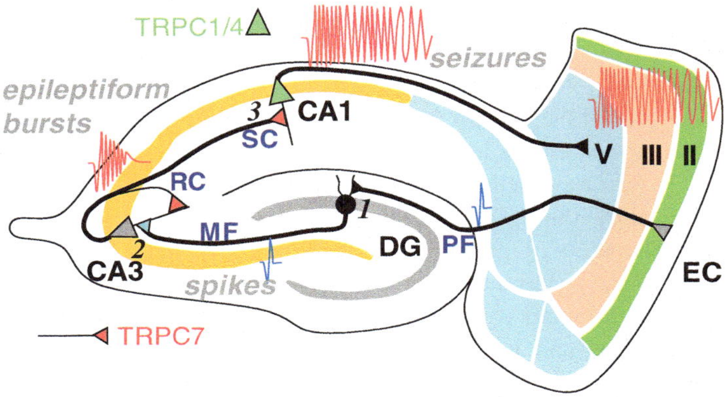

Fig. 11.2 The generation of seizures in the entorhinal-hippocampal loop. We propose that the generation of synaptically derived epileptiform bursting in CA3 requires TRPC7 channels. The epileptiform bursting in the CA3 drives the generation of ictal-like activities in the CA1, which then propagate to the entorhinal cortex. TRPC1/4 channels contribute to seizure-induced neuronal cell death in the CA1 (Adapted with permission from Ref. [41])

sustained SE in a TRPC knockout line (personal communication, Phelan et al.). This appears to support the notion that some TRPC subfamily members may play a neuroprotective role. Furthermore, the anatomical pattern of SE-induced neurodegeneration is also altered in some TRPC knockout lines. All these issues need to be taken into account for the future study of the TRPC channel's role in epileptogenesis using the pilocarpine model.

11.6 Perspectives

The critical roles of various TRPC channels in epileptiform bursting in brain slices suggest that these channels likely play a significant role in the generation of spontaneous recurrent seizures in vivo. Hopefully, future studies will provide direct evidence that will elucidate the precise role of each TRPC subfamily member in epileptogenesis. TRPC channels have great potentials as novel molecular targets for the development of the next generation of antiepileptic drugs.

Acknowledgments This work is supported in part by NINDS (NS050381), by NIGMS (GM103425), by the Fund to Cure Stroke Foundation, and by the University of Arkansas for Medical Sciences College of Medicine Pilot Award.

References

1. Amaral MD, Pozzo-Miller L (2007) TRPC3 channels are necessary for brain-derived neurotrophic factor to activate a nonselective cationic current and to induce dendritic spine formation. J Neurosci 27:5179–5189. doi:10.1523/JNEUROSCI.5499-06.2007
2. Anderson WW, Swartzwelder HS, Wilson WA (1987) The NMDA receptor antagonist 2-amino-5-phosphonovalerate blocks stimulus train-induced epileptogenesis but not epileptiform bursting in the rat hippocampal slice. J Neurophysiol 57:1–21
3. Astori S, Wimmer RD, Prosser HM, Corti C, Corsi M, Liaudet N, Volterra A, Franken P, Adelman JP, Lüthi A (2011) The Ca(V)3.3 calcium channel is the major sleep spindle pacemaker in thalamus. Proc Natl Acad Sci U S A 108:13823–13828. doi:10.1073/pnas.1105115108
4. Ayala GF, Dichter M, Gumnit RJ, Matsumoto H, Spencer WA (1973) Genesis of epileptic interictal spikes. New knowledge of cortical feedback systems suggests a neurophysiological explanation of brief paroxysms. Brain Res 52:1–17. doi:10.1016/0006-8993(73)90647-1
5. Bains JS, Longacher JM, Staley KJ (1999) Reciprocal interactions between CA3 network activity and strength of recurrent collateral synapses. Nat Neurosci 2:720–726. doi:10.1038/11184
6. Bear J, Lothman EW (1993) An in vitro study of focal epileptogenesis in combined hippocampal-parahippocampal slices. Epilepsy Res 14:183–193
7. Becker AJ, Pitsch J, Sochivko D, Opitz T, Staniek M, Chen C-C, Campbell KP, Schoch S, Yaari Y, Beck H (2008) Transcriptional upregulation of Cav3.2 mediates epileptogenesis in the pilocarpine model of epi-

lepsy. J Neurosci 28:13341–13353. doi:10.1523/JNEUROSCI.1421-08.2008
8. Borges K, McDermott DL, Dingledine R (2004) Reciprocal changes of CD44 and GAP-43 expression in the dentate gyrus inner molecular layer after status epilepticus in mice. Exp Neurol 188:1–10. doi:10.1016/j.expneurol.2004.03.019
9. Cai S-Q, Li W, Sesti F (2007) Multiple modes of a-type potassium current regulation. Curr Pharm Des 13:3178–3184
10. Cavalheiro EA, Santos NF, Priel MR (1996) The pilocarpine model of epilepsy in mice. Epilepsia 37:1015–1019
11. Chuang SC, Bianchi R, Kim D, Shin HS, Wong RK (2001) Group I metabotropic glutamate receptors elicit epileptiform discharges in the hippocampus through PLCbeta1 signaling. J Neurosci 21:6387–6394
12. Connor JA (1978) Slow repetitive activity from fast conductance changes in neurons. Fed Proc 37:2139–2145
13. Csicsvari J, Jamieson B, Wise KD, Buzsáki G (2003) Mechanisms of gamma oscillations in the hippocampus of the behaving rat. Neuron 37:311–322. doi:10.1016/S0896-6273(02)01169-8
14. de Boer HM, Mula M, Sander JW (2008) The global burden and stigma of epilepsy. Epilepsy Behav 12:540–546. doi:10.1016/j.yebeh.2007.12.019
15. Dingledine R, Hynes MA, King GL (1986) Involvement of N-methyl-D-aspartate receptors in epileptiform bursting in the rat hippocampal slice. J Physiol 380:175–189
16. Fisher RS, Scharfman HE, DeCurtis M (2014) How can we identify ictal and interictal abnormal activity? Adv Exp Med Biol 813:3–23. doi:10.1007/978-94-017-8914-1_1
17. Fisher RS, Van Emde Boas W, Blume W, Elger C, Genton P, Lee P, Engel J (2005) Epileptic seizures and epilepsy: definitions proposed by the International League Against Epilepsy (ILAE) and the International Bureau for Epilepsy (IBE). Epilepsia 46:470–472. doi:10.1111/j.0013-9580.2005.66104.x
18. Gallagher JP, Zheng F, Hasuo H, Shinnick-Gallagher P (1995) Activities of neurons within the rat dorsolateral septal nucleus (DLSN). Prog Neurobiol 45:373–395
19. George AL (2004) Inherited channelopathies associated with epilepsy. Epilepsy Curr 4:65–70. doi:10.1111/j.1535-7597.2004.42010.x
20. Gibbs FA, Davis H, Lennox WG (1935) The electroencephalogram in epilepsy and in conditions of impaired consciousness. Arch Neurol Psychiatr 34:1131–1148
21. Gu N, Vervaeke K, Hu H, Storm JF (2005) Kv7/KCNQ/M and HCN/h, but not KCa2/SK channels, contribute to the somatic medium after-hyperpolarization and excitability control in CA1 hippocampal pyramidal cells. J Physiol 566:689–715. doi:10.1113/jphysiol.2005.086835
22. Hernandez CC, Zaika O, Tolstykh GP, Shapiro MS (2008) Regulation of neural KCNQ channels: signalling pathways, structural motifs and functional implications. J Physiol 586:1811–1821. doi:10.1113/jphysiol.2007.148304
23. Hesdorffer DC, Logroscino G, Benn EKT, Katri N, Cascino G, Hauser WA (2011) Estimating risk for developing epilepsy: a population-based study in Rochester, Minnesota. Neurology 76:23–27. doi:10.1212/WNL.0b013e318204a36a
24. Hirtz D, Thurman DJ, Gwinn-Hardy K, Mohamed M, Chaudhuri AR, Zalutsky R (2007) How common are the "common" neurologic disorders? Neurology 68:326–337. doi:10.1212/01.wnl.0000252807.38124.a3
25. Jackson J (1958) Volume one: on epilepsy and epileptiform convulsions. In: Taylor J (ed) Selected writings of John Hughlings Jackson. Basic Books, New York, pp 1–486
26. Johnston D, Brown TH (1981) Giant synaptic potential hypothesis for epileptiform activity. Science 211:294–297
27. Jung S, Jones TD, Lugo JN, Sheerin AH, Miller JW, D'Ambrosio R, Anderson AE, Poolos NP (2007) Progressive dendritic HCN channelopathy during epileptogenesis in the rat pilocarpine model of epilepsy. J Neurosci 27:13012–13021. doi:10.1523/JNEUROSCI.3605-07.2007
28. Kandel ER, Spencer WA (1961) Electrophysiology of hippocampal neurons. II. After-potentials and repetitive firing. J Neurophysiol 24:243–259
29. Kim D-S, Ryu HJ, Kim J-E, Kang T-C (2013) The reverse roles of transient receptor potential canonical channel-3 and -6 in neuronal death following pilocarpine-induced status epilepticus. Cell Mol Neurobiol 33:99–109. doi:10.1007/s10571-012-9875-6
30. Matsumoto H, Marsan CA (1964) Cortical cellular phenomena in experimental epilepsy: interictal manifestations. Exp Neurol 9:286–304. doi:10.1016/0014-4886(64)90025-1
31. Matsumoto H, Marsan CA (1964) Cortical cellular phenomena in experimental epilepsy: ictal manifestations. Exp Neurol 9:305–326. doi:10.1016/0014-4886(64)90026-3
32. Mazzuferi M, Kumar G, Rospo C, Kaminski RM (2012) Rapid epileptogenesis in the mouse pilocarpine model: video-EEG, pharmacokinetic and histopathological characterization. Exp Neurol 238:156–167. doi:10.1016/j.expneurol.2012.08.022
33. McCormick DA, Contreras D (2001) On the cellular and network bases of epileptic seizures. Annu Rev Physiol 63:815–846. doi:10.1146/annurev.physiol.63.1.815
34. McNamara JO (1999) Emerging insights into the genesis of epilepsy. Nature 399:A15–A22. doi:10.1038/399a015
35. McNamara JO, Byrne MC, Dasheiff RM, Fitz JG (1980) The kindling model of epilepsy: a review.

Prog Neurobiol 15:139–159. doi:10.1016/0301-0082(80)90006-4
36. Morimoto K, Fahnestock M, Racine RJ (2004) Kindling and status epilepticus models of epilepsy: rewiring the brain. Prog Neurobiol 73:1–60. doi:10.1016/j.pneurobio.2004.03.009
37. Müller CJ, Gröticke I, Hoffmann K, Schughart K, Löscher W (2009) Differences in sensitivity to the convulsant pilocarpine in substrains and sublines of C57BL/6 mice. Genes Brain Behav 8:481–492. doi:10.1111/j.1601-183X.2009.00490.x
38. Nagamine K, Kudoh J, Minoshima S, Kawasaki K, Asakawa S, Ito F, Shimizu N (1998) Molecular cloning of a novel putative Ca2+channel protein (TRPC7) highly expressed in brain. Genomics 54:124–131. doi:10.1006/geno.1998.5551
39. Nagao T, Alonso A, Avoli M (1996) Epileptiform activity induced by pilocarpine in the rat hippocampal-entorhinal slice preparation. Neuroscience 72:399–408. doi:10.1016/0306-4522(95)00534-X
40. Phelan KD, Mock MM, Kretz O, Shwe UT, Kozhemyakin M, Greenfield LJ, Dietrich A, Birnbaumer L, Freichel M, Flockerzi V, Zheng F (2012) Heteromeric canonical transient receptor potential 1 and 4 channels play a critical role in epileptiform burst firing and seizure-induced neurodegeneration. Mol Pharmacol 81:384–392. doi:10.1124/mol.111.075341
41. Phelan KD, Shwe UT, Abramowitz J, Birnbaumer L, Zheng F (2014) Critical role of canonical transient receptor potential channel 7 in initiation of seizures. Proc Natl Acad Sci U S A 111:11533–11538. doi:10.1073/pnas.1411442111
42. Phelan KD, Shwe UT, Abramowitz J, Wu H, Rhee SW, Howell MD, Gottschall PE, Freichel M, Flockerzi V, Birnbaumer L, Zheng F (2013) Canonical transient receptor channel 5 (TRPC5) and TRPC1/4 contribute to seizure and excitotoxicity by distinct cellular mechanisms. Mol Pharmacol 83:429–438. doi:10.1124/mol.112.082271
43. Phelan KD, Shwe UT, Williams DK, Greenfield LJ, Zheng F (2015) Pilocarpine-induced status epilepticus in mice: a comparison of spectral analysis of electroencephalogram and behavioral grading using the Racine scale. Epilepsy Res 117:90–96. doi:10.1016/j.eplepsyres.2015.09.008
44. Racine RJ (1972) Modification of seizure activity by electrical stimulation. II. Motor seizure. Electroencephalogr Clin Neurophysiol 32:281–294
45. Raggenbass M, Pierson P, Metzger D, Alberi S (1997) Action of a metabotropic glutamate receptor agonist in rat lateral septum: induction of a sodium-dependent inward aftercurrent. Brain Res 776:75–87. doi:10.1016/S0006-8993(97)00945-1
46. Reid CA, Berkovic SF, Petrou S (2009) Mechanisms of human inherited epilepsies. Prog Neurobiol 87:41–57. doi:10.1016/j.pneurobio.2008.09.016
47. Scharfman HE (1992) Differentiation of rat dentate neurons by morphology and electrophysiology in hippocampal slices: granule cells, spiny hilar cells and aspiny "fast-spiking" cells. Epilepsy Res Suppl 7:93–109
48. Stoop R, Conquet F, Zuber B, Voronin LL, Pralong E (2003) Activation of metabotropic glutamate 5 and NMDA receptors underlies the induction of persistent bursting and associated long-lasting changes in CA3 recurrent connections. J Neurosci 23:5634–5644
49. Tai C, Hines DJ, Choi HB, MacVicar BA (2011) Plasma membrane insertion of TRPC5 channels contributes to the cholinergic plateau potential in hippocampal CA1 pyramidal neurons. Hippocampus 21:958–967. doi:10.1002/hipo.20807
50. Traub RD, Wong RK (1982) Cellular mechanism of neuronal synchronization in epilepsy. Science 216:745–747. doi:10.1126/science.7079735
51. Traynelis SF, Dingledine R (1988) Potassium-induced spontaneous electrographic seizures in the rat hippocampal slice. J Neurophysiol 59:259–276
52. Treiman DM, Walton NY, Kendrick C (1990) A progressive sequence of electroencephalographic changes during generalized convulsive status epilepticus. Epilepsy Res 5:49–60. doi:10.1016/0920-1211(90)90065-4
53. Turski L, Ikonomidou C, Turski WA, Bortolotto ZA, Cavalheiro EA (1989) Review: cholinergic mechanisms and epileptogenesis. The seizures induced by pilocarpine: a novel experimental model of intractable epilepsy. Synapse 3:154–171. doi:10.1002/syn.890030207
54. Turski WA, Cavalheiro EA, Bortolotto ZA, Mello LM, Schwarz M, Turski L (1984) Seizures produced by pilocarpine in mice: a behavioral, electroencephalographic and morphological analysis. Brain Res 321:237–253
55. Vandenberg JI, Perry MD, Perrin MJ, Mann SA, Ke Y, Hill AP (2012) hERG K(+) channels: structure, function, and clinical significance. Physiol Rev 92:1393–1478
56. Wilson JV, Reynolds EH (1990) Texts and documents. Translation and analysis of a cuneiform text forming part of a Babylonian treatise on epilepsy. Med Hist 34:185–198
57. Xu G-Z, Shu H, Yue H-Y, Zheng D-H, Guo W, Yang H (2015) Increased expression of TRPC5 in cortical lesions of the focal cortical dysplasia. J Mol Neurosci 55:561–569. doi:10.1007/s12031-014-0390-8
58. Yan H-D, Villalobos C, Andrade R (2009) TRPC channels mediate a muscarinic receptor-induced afterdepolarization in cerebral cortex. J Neurosci 29:10038–10046. doi:10.1523/JNEUROSCI.1042-09.2009
59. Zablocka B, Esplin DW (1963) Central excitatory and depressant effects of pilocarpine in rats and mice. J Pharmacol Exp Ther 140:162–169
60. Zeng C, Zhou P, Jiang T, Yuan C, Ma Y, Feng L, Liu R, Tang W, Long X, Xiao B, Tian F (2015) Upregulation and diverse roles of TRPC3 and TRPC6 in synaptic reorganization of the mossy fiber

pathway in temporal lobe epilepsy. Mol Neurobiol 52:562–572. doi:10.1007/s12035-014-8871-x
61. Zheng F, Gallagher JP (1991) Trans-ACPD (trans-D,L-1-amino-1,3-cyclopentanedicarboxylic acid) elicited oscillation of membrane potentials in rat dorsolateral septal nucleus neurons recorded intracellularly in vitro. Neurosci Lett 125:147–150. doi:10.1016/0304-3940(91)90013-J
62. Zheng F, Gallagher JP (1992) Burst firing of rat septal neurons induced by 1S,3R-ACPD requires influx of extracellular calcium. Eur J Pharmacol 211:281–282. doi:10.1016/0014-2999(92)90542-C
63. Zheng F, Gallagher JP, Connor JA (1996) Activation of a metabotropic excitatory amino acid receptor potentiates spike-driven calcium increases in neurons of the dorsolateral septum. J Neurosci 16:6079–6088
64. Zheng F, Phelan KD (2014) The role of canonical transient receptor potential channels in seizure and excitotoxicity. Cells 3:288–303. doi:10.3390/cells3020288
65. Zhou F-W, Roper SN (2014) TRPC3 mediates hyperexcitability and epileptiform activity in immature cortex and experimental cortical dysplasia. J Neurophysiol 111:1227–1237. doi:10.1152/jn.00607.2013

TRPC Channels and Mental Disorders

12

Karina Griesi-Oliveira, Angela May Suzuki, and Alysson Renato Muotri

Abstract

Transient receptor potential canonical (TRPC) channels mediate the influx of different types of cations through the cell membrane and are involved in many functions of the organism. Evidences of involvement of TRPC channels in neuronal development suggest that this family of proteins might play a role in certain neurological disorders. As reported, knockout mice for different TRPC channels show alterations in neuronal morphological and functional parameters, with behavioral abnormalities, such as in exploratory and social behaviors. Although mutations in TRPC channels could be related to mental/neurological disorders, there are only a few cases reported in literature, indicating that this correlation should be further explored. Nonetheless, other functional evidences support the implication of these channels in neurological diseases. In this chapter, we summarize the main findings relating TRPC channels to neurological disorders, such as autism spectrum disorders, bipolar disorder, and intellectual disability among others.

Keywords

TRPC • Autism • Hyperforin • Mental disorders

K. Griesi-Oliveira
Hospital Albert Einstein, São Paulo, SP, Brazil

A.M. Suzuki
Department of Genetics and Evolutionary Biology, Bioscience Institute, University of São Paulo, São Paulo, SP, Brazil

A.R. Muotri (✉)
Department of Pediatrics and Department of Cellular & Molecular Medicine, University of California San Diego School of Medicine, La Jolla, CA, USA

Rady Children's Hospital San Diego, San Diego, CA, USA

UCSD Stem Cell Program, Institute for Genomic Medicine, New York, NY, USA
e-mail: muotri@ucsd.edu

Y. Wang (ed.), *Transient Receptor Potential Canonical Channels and Brain Diseases*, Advances in Experimental Medicine and Biology 976, DOI 10.1007/978-94-024-1088-4_12

12.1 Introduction

Nervous system development is an intricate process that involves a series of steps, including proliferation of progenitor cells, cellular migration, neurite development and guidance, and establishment of synapses. All these steps have to be strictly regulated in order to enable the correct configuration of neuronal networks and adequate brain functioning. Disruption in such processes can lead to abnormal development or functioning of neuronal circuits, contributing to abnormal regulation of neuronal plasticity, which ultimately results in psychiatry/neurological diseases.

The TRPC channels have important roles in different cellular processes in several tissues, including nervous system. They are divided in three main groups according to their sequence similarity: TRPC1/4/5, TRPC3/6/7, and TRPC2 which, in humans, is actually a pseudogene. TRPC channels are highly expressed in the brain, and their roles on neuronal development have been intensively investigated. Some examples of TRPC channels functions are summarized below.

TRPC1 is reported to have an important role in neural progenitor cell proliferation [11, 23, 49, 62]. TRPC3 induces long-term depression in Purkinje cells [42] and participates in the regulation of hippocampal neuronal excitability, affecting the memory formation [59]. In conjunction with TRPC6, TRPC3 affects axonal path finding induced by BDNF [51] and has a protective effect in neuronal survival [34]. Moreover, TRPC channels are also important for synaptogenesis, for example, TRPC proteins could interact with postsynaptic membrane scaffolding proteins [87]. TRPC6 has a role in excitatory synapse formation [90], and both TRPC3 and TRPC6 are necessary for BDNF-induced increase in dendritic spine density [2].

Different TRPC channels even have opposite roles in the regulation of neurite outgrowth. In PC12 cells, TRPC1 induces neurite outgrowth through a mechanism that is independent of Ca^{2+} influx, while TRPC5 has an opposite effect [31]. Activation of TRPC4 by Gαi2 inhibits neurite growth and dendritic arborization of hippocampal neurons [33]. Neurotrophin-3 induces Ca^{2+} influx through TRPC5 and inhibits neurite outgrowth in rat hippocampal neurons, while neurotrophin-4 promotes neurite growth through TRPC6 activation [30]. Growth cone collapse induced by Semaphorin-3A is reduced in hippocampal neurons from TRPC5 knockout mice [35]. Finally, TRPC6 promotes dendritic growth via a Ca^{2+}-CaMKIV-dependent mechanism in rat neurons [79]. It is worthy to highlight that most of the function of TRPC in neurons are mediated by Ca^{2+} influx, although they are nonselective cation channels. In fact, Ca^{2+} signaling is a crucial mechanism for brain development (reviewed in [73]).

In summary, TRPC channels contribute to the regulation of neuronal development and function. Malfunction of these channels might have an impact on such processes and thus contribute to psychiatric/neurological disorders. Here we enumerate some human disorders for which there are evidences of involvement of TRPC channels in their pathophysiology. For some of them, direct disruption of the sequence of a *-trpc* gene was found in the patients, while for some other disorders, the relationship between TRPC channels and disease is based on indirect evidences.

12.2 TRPC6 and Autism

Autism spectrum disorders (ASD) are defined as a group of neurodevelopmental disorders characterized by repetitive behaviors as well as impairments in communicative and social interaction skills. Although environmental factors might play a role in ASD etiology, evidences support a major contribution of genetic factors to these disorders. However, these genetic factors are heterogeneous across autistic individuals, and in fact, dozens of genes have already been implicated in ASD.

trpc6 is one of the genes that have been recently implicated in ASD etiology. The relation between *trpc6* and ASD is originally based on the work of Griesi-Oliveira et al. [28] that has identified an autistic patient with a chromosomal translocation between chromosomes 3 and 11. Translocation mapping indicated that the sequence of genes *vprbp* on chromosome 3 and *trpc6* on chromosome 11 are disrupted, leading to haploinsufficiency. Previous studies have shown that TRPC6 overexpression in rat neurons leads to an enhancement on dendritic growth, dendritic spine density, and excitatory synapse formation, both in vitro and in vivo [79, 90]. Based on these evidences, the authors invested in the analysis of the consequences of such disruption in the function of TRPC6 and its downstream signaling pathway in the cells of the reported patient. The role of TRPC6 on neuronal development is mainly related to the calcium influx through the channel, leading to CREB phosphorylation [79], which in turn mediates the regulation of expression of many genes. Promoting the activation of the channel using hyperforin, a specific activator of TRPC6, will be discussed later, and authors demonstrated that CREB phosphorylation is reduced in patient's dental pulp cells compared to control cells. Accordingly, a global expression analysis suggests that a significant number of putative CREB target genes are dysregulated in patient's dental pulp cells.

Griesi-Oliveira et al. generated induced pluripotent stem cells (iPSC) from the patients and healthy controls through cell reprogramming to investigate *trpc6* disruption consequences on targeted cortical neuronal cells derived from the iPSC. Again, after inducing TRPC6 activation with hyperforin, the authors showed that calcium influx through neuronal progenitor cells is reduced in the patient's sample. In addition, patient's cortical neurons are found to have a reduced dendritic arborization, shorter neurites, and reduced density of dendritic spines and glutamatergic vesicles compared to controls (cortical neurons derived from non-ASD, healthy individuals). All these abnormalities are in accordance to previous findings on rat neurons overexpressing TRPC6, which have the opposite characteristics [79]. Moreover, the authors perform gain and loss of function experiments in neurons. Reducing TRPC6 expression in control cells leads to the same neuronal alterations seen in the patient's neurons. On the contrary, complementation of TRPC6 expression in patient's cells rescues the neuronal phenotypes. These results corroborate that the alterations in patient's neurons are related to *trpc6* disruption and consequent haploinsufficiency. Behavioral studies with TRPC6 knockout mice, although they do not indicate any alteration in social or repetitive behavior, have a reduced exploratory behavior, which is associated to clinical signs of ASD, including the autistic patients used by Griesi and colleagues [8, 28]. Finally, the patient's neuronal abnormalities is rescued by treating the neurons with hyperforin, raising the hypothesis that autistic patients with *trpc6* mutations, or other genes mutated in this pathway, could benefit from this chemical. However, the number of patients with truncating mutations in *trpc6* is only around 0.2%. Thus, an effective clinical trial would benefit from a previous stratification of the ASD individuals based on the genetic alterations in *trpc6* or its downstream genes. The idea to use genome sequencing and functional tests using cell reprogramming to stratify a heterogeneous population for clinical trials is recently proposed for ASDs [54].

12.3 TRPC Channels, BDNF, and Rett Syndrome

Rett syndrome is an X-linked neurodevelopmental disorder caused mostly by loss-of-function mutations in *mecp2* (methyl-CpG-binding protein) gene and affects primarily females. Male patients usually die early in development. The disease is characterized by a period of apparently normal development until 6–18 months of age followed by a progressive regression of developmental and motor skills, seizures, and hypotonia, and the majority of the patients can also present autistic features. MeCP2 protein influences neuronal development in several ways. MeCP2 presents a wide range of actions on transcriptional regulation, through its binding to CpG-methylated islands, where it recruits other transcriptional repressors and histones acetylases. In addition, MeCP2 has been shown to regulate TRPC channels expression in mouse brain, and Ca^{2+} signals elicited by TRPC3/6 activation are impaired in neurons of male *mecp2* mutant mice [10, 50]. Furthermore, MeCP2 is found in association with *trpc6* promoter region in human neurons, and in accordance, neurons expressing a loss-of-function allele of *mecp2* presented deregulation of *trpc6* expression [28]. Interestingly, neurons from RTT patients present a series of abnormalities, such as shorter neurites, reduced dendritic arborization, and lower spine density, which is also found in *trpc6* mutant autistic patient's neurons described in the previous section [53, 54]. In accordance, Nageshappa et al. [58] show opposite effects in neuronal cells from patients with *mecp2* duplication. MeCP2 also seems to exert an indirect action in regulating TRPC channels, through the control of the expression of another target, the *bdnf* gene [12, 55]. Involvement of BDNF on pathophysiology of Rett syndrome is widely documented (reviewed by Katz [36]). This gene codes for a neurotrophin that acts as a chemoattractant molecule for neuronal growing process and supports neuronal survival and maturation. Studies suggest that these roles of BDNF on synapse formation are mediated, at least in part, by TRPC channels. The first evidence comes from the observation that exogenous BDNF triggers membrane cation currents resembling those that occur through TRPC channels [48]. Later, it is demonstrated that Ca^{2+} currents induced by BDNF are in fact abolished by inhibition of TRPC channels via pharmacological inhibition or downregulation of their expression [51]. Such inhibition also blocks BDNF-induced cone growth. The protective effect of BDNF on neuronal survival as well as its induction of dendritic spine formation is also dependent on TRPC channel expression [2, 34]. Finally, Fortin et al. [24] show that pharmacological blocking of TRPC channel activity prevents BDNF-induced translation and incorporation of GluA1-containing AMPA receptors in synaptic membrane. Thus, it would be of interest to investigate whether Rett syndrome patients could also benefit from chemicals that can modulate TRPC channels activity.

12.4 TRPC5 and X-Linked Intellectual Disability

Intellectual disability (ID) is a clinical phenotype characterized by significant limitations in cognitive function and in adaptive behavior, i.e., social and practical skills needed for everyday life activities, such as communication, self-care, and socialization. ID is associated with many different genetic syndromes and genetic alterations and can also be caused by pre- or postnatal environmental factors, such as alcohol abuse by the mother, exposure to pathogens during pregnancy, problems at the time of birth, or diseases in early childhood such as meningitis. It is estimated that ID affects about 3% of the population [56]. Prevalence of ID is biased toward males, indicating that X-linked genes have an important contribution to cognitive development. In fact, many different X-linked genes were already identified as being the causative gene for ID, especially in pedigrees in which this phenotype segregates in an X-linked pattern [14, 65, 72]. Recently, Mignon-Ravix et al. [57] use a custom high-resolution CGH array to investigate the presence of copy number variations (CNV) in the X chromosome, in a cohort of 54 male individuals presenting ID. In this study, the group finds a patient with a 47 kb deletion in Xq23 region involving the first exon of *trpc5* gene. A CNV that is considered to be probably pathogenic since it has not been previously identified in any control sample and since *trpc5* is highly expressed in the brain. The deletion is also found in the mother, who is unaffected and presented a 50:50 ratio of X chromosome inactivation in her lymphocytes. The individual is reported to present autistic features such as repetitive and stereotyped behavior. TRPC5 is shown to be important for the control of neurite extension and growth cone morphology [27, 30, 31, 35]. Downregulation of TRPC5 in rat neuronal progenitor cells reduced the Ca^{2+} currents through the cells and blocked the neuronal differentiation [76]. Finally, TRPC5 knockout mice present abnormally high branched dendrites in cerebellar neurons and reduced LTP [64, 67]. These animals present motor deficits and diminished innate fear levels in response to aversive stimuli [67, 68]. These data support a role of TRPC5 in nervous system development and a possible involvement in intellectual disability.

12.5 TRPC3, TRPC7, and Bipolar Disorder

Bipolar disorder (BD), also known as maniac-depressive disorder, is a mental illness characterized by unusual shifts in mood, swinging from periods of maniac episodes, when the individual experiences an overexcited state, being abnormally happy, energetic or irritable, and periods of depressive mood, in which the individual presents sadness and poor interest and pleasure in most activities. As with most of the neurological disorders previously discussed, BD is a disorder with heterogeneous etiology, in which genetic and environmental factors play a role [40]. Many genes have already been associated with BD based on exome studies, case-control cohorts, or genetic studies of families in which BD segregates throughout generations [39, 77]. Evidences from functional studies have also brought support for involvement of some genes in BD, as are the cases of *trpc3* and *trpc7*. Reports of abnormalities in basal and agonist-stimulated Ca^{2+} concentrations in different cell types of BD individuals suggest that Ca^{2+} homeostasis plays a significant role in BD pathophysiology [18, 20], especially considering the importance of Ca^{2+} signaling for neuroplasticity [26]. Moreover, a series of studies using lithium treatment (one of the most widely used drug treatment for BD) show that this chemical affects Ca^{2+} signaling in neurons and in non-excitable cells such as platelets, lymphocytes, and glial cells, providing another line of evidence supported for the involvement of Ca^{2+} signaling in BD [13, 17, 81, 85]. The evidences that Ca^{2+} signaling abnormalities in non-excitable cells from BD individuals prompt scientists to investigate mechanisms involved in Ca^{2+} influx in these types of cells, particularly in the glial cells [82]. Among the molecular mechanisms responsible for Ca^{2+} dynamics regulation in non-excitable cells, store-operated calcium entry is proposed to be involved in BD pathogenesis [32, 84], which directs the attention to channels from TRP family. TRPC7 expression was reduced in a subgroup of BD patients [86]. Interestingly, this reduction was inversely correlated to Ca^{2+} basal levels in B lymphoblasts cell lines (BLCLs) from BD subjects. Andreopoulos et al. [6] report that chronic lithium treatment significantly reduces TRPC3 protein levels in BLCLs of BD individuals. Decreased levels of TRPC3 proteins are also reported in cerebral cortex of rats chronically treated with lithium [88]. Interestingly, neither of the studies detects decreases in TRPC3 mRNA levels. A faster Ca^{2+} influx is observed in BLCLs of BD individuals upon lysophosphatidic acid (LPA) stimulation [63]. In this model, the authors suggested that LPA-mediated Ca^{2+} influx is probably mediated by TRPC3, since LPA is structurally homologue to 1-oleoyl-2-acetyl-sn-glycerol (OAG), a known activator of TRPC3/6/7, but only TRPC3 is expressed in BLCLs [69], suggesting a role of TRPC3 in BD pathogenesis. Oxidative stress has also been implicated in BD [5, 60, 83]. Based on the fact that TRPC channels can be sensitive to redox state of the cells, Roedding and collaborators [70] investigate whether TRPC3 expression and function could be affected by oxidative stress in BLCLs from BD individuals and controls. The authors found that chronic mitochondrial-generated oxidative stress reduces TRPC3 protein levels as well as the Ca^{2+} influx through these channels, though this reduction is equally found in controls and BD individuals. A similar reduction is detected in primary rat cortical neurons exposed to stressor conditions [71]. Why no differences between subjects and controls were found regarding TRPC3 expression and function response to stressor conditions tested and why these conditions induced a decrease in protein levels (which is an opposite consequence expected considering the effects of lithium in similar cellular models) are questions that deserve further investigation.

12.6 TRPC3 and Williams–Beuren Syndrome

Williams–Beuren syndrome (WBS) is a multisystem disorder caused by a deletion on chromosome 7, specifically at 7q11.23 region. WBS patients present characteristic facial features (such as wide mouth, flattened nasal bridge, and widely spaced teeth), connective tissue problems, and cardiovascular malformations. They also present neurodevelopmental problems that include moderate mental retardation, difficulty with visual-spatial tasks, in contrast to strong verbal skills and a friendly personality, demonstrating an extreme interest in other people. Patients with WBS can also have hypercalcemia, which is most frequently present during childhood [4, 43]. The region deleted in WBS harbors 26–28 genes. Their relative contribution to the phenotype is still poorly understood, except for elastin gene, which is related to the cardiovascular problems presented by the patients [21]. One of the genes in 7q11.23 is *gtf3I* (general transcription factor IIi). This gene codes for the transcription factor TFII-I, which is thought to be one of the main causative factors for cognitive dysfunctions in WBS based on the definition of a shared region always present between patients with different sizes of the deletion [7]. Although TFII-I is primarily a transcription factor, it may also have a role outside the nucleus, as it is evidenced by its presence in cytosol, for example, of B lymphocytes [61] and in dendritic trees of Purkinje cells [16]. In human B lymphocytes, it has been described that TFII-I suppresses accumulation of TRPC3 in cell surface, thus, decreasing Ca^{2+} entry into the cells [9]. Based on this observation, Letavernier and colleagues [45] investigate whether TRPC3 expression is altered in a WBS individual reported to have hypercalcemia. By using immunohistochemistry, they found an increased protein expression of TRPC3 in lymphocytes and epithelial intestinal cells from this patient compared to control samples. Although TRPC3 mRNA levels seems not to be altered in neuronal cells derived from iPSC of WBS individuals [1, 41, 44], we cannot disregard that TFII-I might be acting at the protein level as it is seen in B lymphocytes [9].

12.7 TRPC Channels, Hyperforin, and Depression

Depression is a mental disorder which etiology and pathology remain largely unknown, but it is hypothesized to be associated with reduced function of monoamine chemicals, such as serotonin and norepinephrine (reviewed by Dale et al. [15]). In fact, the most successful approaches to treat major depressive disorder (MDD) target the monoamine system by either the selective serotonin reuptake inhibitors (SSRI) or by serotonin and norepinephrine reuptake inhibitors (SNRI) increasing monoamine transmission [15]. However, only about 50% of patients diagnosed with MDD evolve into clinical remission under these treatments [74]. In this regard, hyperforin, one of the main bioactive compounds of the medicinal plant Saint John's wort (SJW), has shown to present unique antidepressant effect and to be significantly more effective than those in the first line of antidepressant drugs [22, 25, 78]. In contrast to conventional drugs, hyperforin acts as a nonselective neurotransmitter reuptake inhibitor, blocking the uptake of serotonin, norepinephrine, dopamine, and other neurotransmitters [89]. Part of hyperforin's antidepressant effects has been attributed to its property to activate TRPC6 channel. TRPC6 channels are permeable to sodium in the presynaptic membrane and its activation contributes to decreasing the sodium gradient that drives the neurotransmitters reuptake through the transporters, contributing to increasing the levels of the monoamine neurotransmitters in the synaptic cleft [46, 47, 80].

Depression is also commonly associated with decreased levels of BDNF in the hippocampus [38, 75]. As BDNF is a neurotrophic factor fundamental to the modulation of dendritic architecture and synapse [37], it is also hypothesized that depression is associated with loss of hippocampal synapses and dendritic spines contributing to dysfunction of synaptic plasticity [19, 29, 66]. As previously mentioned, synaptogenic properties of BDNF is thought to be in part mediated by calcium transients evoked by TRPC channels [2, 3]. Similarly, the potential effectiveness of hyperforin in the treatment of depression may also be attributed to the calcium influx through TRPC6 channel, which can contribute to the modulation of dendritic spine density and morphology [47].

Although hyperforin has shown to present exciting antidepressant properties and now to be commonly prescribed for the treatment of mild to moderate depression [52], studies in vivo demonstrating the potential effectiveness of this compound are still lacking and require further investigation. In summary, although there are no genetic findings directly correlating TRPC channels with depression, the studies mentioned above show how these channels (in particular, TRPC6) modulate etiopathological mechanisms involved in this disorder and how these channels represent potential therapeutic targets, not only for depression but also for other neurological diseases with dysfunction of these channels.

12.8 Conclusions and Future Perspectives

Genetic and functional evidences have indicated the participation of TRPC channels in the pathophysiology of several mental disorders (as summarized in Fig. 12.1). Further investigation of such relationships would be of great value since these proteins are ion channels and, thus, can be directly targeted to regulate their activity, which may lead to an effective therapeutic effect.

As illustrated by the case of *trpc6* disruption in an ASD individual, the use of cell reprogramming to obtain neuronal cells derived from pluripotent stem cells from individuals with neurological disorders is a powerful tool to explore possible roles of TRPC channels in the brain. Furthermore, considering that TRPC channels are one of the major classes of channels responsible for ion influx in non-excitable cells, their involvement in the pathophysiology of neurological disorders should be explored not only in neurons but also neuronal progenitor cells and glial cells. For instance, it would be interesting to validate the results found by the expression and functional studies conducted in BLCLs of patients with BD in neuronal cells derived from these patients using such approach. If these stud-

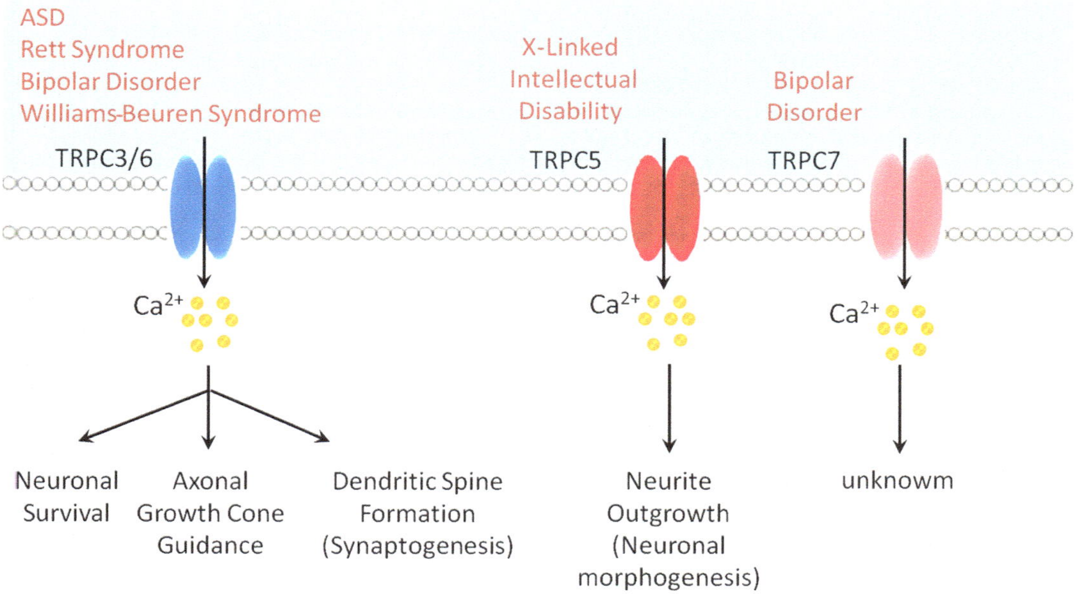

Fig. 12.1 TRPC channels are mediators of sensory signals with marked effects on neuronal functions. TRPC3 proteins can form heteromeric interactions with TRPC6 and regulate neuronal survival, axonal growth cone guidance, and dendritic spine formation. Disturbances in the functioning of these proteins and in the cation influx through the channels they form have been implicated in the pathophysiology of several neurological diseases, such as ASD, Rett syndrome, and Williams–Beuren syndrome. Similarly, X-linked intellectual disability pathophysiology has been attributed to disturbances in the calcium influx elicited through TRPC5 channels and altered neurite outgrowth. Finally, altered calcium influx, possibly through TRPC3 or TRPC7, has been associated with Bipolar disorder

ies point to a participation of TRPC channels in the etiology of the disease, the same system can also be used to screen for drugs that can modulate TRPC channel activity in target cell types in an attempt to rescue cellular function and, possibly, contributing to the treatment of the diseases.

Acknowledgments This work was supported by grants from the California Institute for Regenerative Medicine (CIRM) TR2-01814 and TR4-06747; the National Institutes of Health through the P01 NICHD033113; NIH Director's New Innovator Award Program 1-DP2-OD006495-01, R01MH094753, R01MH103134, and U19MH107367; and a NARSAD Independent Investigator.

References

1. Adamo A, Atashpaz S, Germain PL, Zanella M, D'Agostino G, Albertin V, Chenoweth J, Micale L, Fusco C, Unger C, Augello B, Palumbo O, Hamilton B, Carella M, Donti E, Pruneri G, Selicorni A, Biamino E, Prontera P, McKay R, Merla G, Testa G (2015) 7q11.23 dosage-dependent dysregulation in human pluripotent stem cells affects transcriptional programs in disease-relevant lineages. Nat Genet 47(2):132–141. doi:10.1038/ng.3169
2. Amaral MD, Pozzo-Miller L (2007a) TRPC3 channels are necessary for brain-derived neurotrophic factor to activate a nonselective cationic current and to induce dendritic spine formation. J Neurosci 27(19):5179–5189. doi:10.1523/JNEUROSCI.5499-06.2007
3. Amaral MD, Pozzo-Miller L (2007b) BDNF induces calcium elevations associated with IBDNF, a nonselective cationic current mediated by TRPC channels. J Neurophysiol 98(4):2476–2482. doi:10.1152/jn.00797.2007
4. Amenta S, Sofocleous C, Kolialexi A, Thomaidis L, Giouroukos S, Karavitakis E, Mavrou A, Kitsiou S, Kanavakis E, Fryssira H (2005) Clinical manifestations and molecular investigation of 50 patients with Williams syndrome in the Greek population. Pediatr Res 57(6):789–795. doi:10.1203/01.PDR.0000157675.06850.68
5. Andreazza AC, Shao L, Wang J-F, Young LT (2010) Mitochondrial complex I activity and oxidative damage to mitochondrial proteins in the prefrontal cortex of patients with bipolar disorder. Arch Gen Psychiatry 67(4):360–368. doi:10.1001/archgenpsychiatry.2010.22
6. Andreopoulos S, Wasserman M, Woo K, Li PP, Warsh JJ (2004) Chronic lithium treatment of B lymphoblasts from bipolar disorder patients reduces transient receptor potential channel 3 levels. Pharm J 4(6):365–373. doi:10.1038/sj.tpj.6500266
7. Antonell A, Del Campo M, Magano LF, Kaufmann L, de la Iglesia JM, Gallastegui F, Flores R, Schweigmann U, Fauth C, Kotzot D, Pérez-Jurado LA (2010) Partial 7q11.23 deletions further implicate GTF2I and GTF2IRD1 as the main genes responsible for the Williams-Beuren syndrome neurocognitive profile. J Med Genet 47(5):312–320. doi:10.1136/jmg.2009.071712
8. Beis D, Schwarting RKW, Dietrich A (2011) Evidence for a supportive role of classical transient receptor potential 6 (TRPC6) in the exploration behavior of mice. Physiol Behav 102(2):245–250. doi:10.1016/j.physbeh.2010.11.002
9. Caraveo G, van Rossum DB, Patterson RL, Snyder SH, Desiderio S (2006) Action of TFII-I outside the nucleus as an inhibitor of agonist-induced calcium entry. Science 314(5796):122–125. doi:10.1126/science.1127815
10. Chahrour M, Jung SY, Shaw C, Zhou X, Wong ST, Qin J, Zoghbi HY (2008) MeCP2, a key contributor to neurological disease, activates and represses transcription. Science 320(5880):1224–1229. doi:10.1126/science.1153252
11. Chen HC, Wang CH, Shih CP, Chueh SH, Liu SF, Chen HK, Lin YC (2015) TRPC1 is required for survival and proliferation of cochlear spiral ganglion stem/progenitor cells. Int J Pediatr Otorhinolaryngol 79(12):2290–2294. doi:10.1016/j.ijporl.2015.10.027
12. Chen WG, Chang Q, Lin Y, Meissner A, West AE, Griffith EC, Jaenisch R, Greenberg ME (2003) Depression of BDNF transcription involves calcium-dependent phosphorylation of MeCP2. Science 302(5646):885–889. doi:10.1126/science.1086446
13. Chen Y, Hertz L (1996) Inhibition of noradrenaline stimulated increase in [Ca2+]i in cultured astrocytes by chronic treatment with a therapeutically relevant lithium concentration. Brain Res 711(1–2):245–248. doi:10.1016/0006-8993(95)01199-4
14. Chiurazzi P, Schwartz CE, Getz J, Neri G (2008) XLMRgenes: update 2007. Eur J Hum Genet 16(4):422–434. doi:10.1038/sj.ejhg.5201994
15. Dale E, Bang-Andersen B, Sánchez C (2015) Emerging mechanisms and treatments for depression beyond SSRIs and SNRIs. Biochem Pharmacol 95(2):81–97. doi:10.1016/j.bcp.2015.03.011
16. Danoff SK, Taylor HE, Blackshaw S, Desiderio S (2004) TFII-I, a candidate gene for Williams syndrome cognitive profile: parallels between regional expression in mouse brain and human phenotype. Neuroscience 123(4):931–938. doi:10.1016/j.neuroscience.2003.08.038
17. Dubovsky SL, Lee C, Christiano J, Murphy J (1991) Lithium decreases platelet intracellular calcium concentration in bipolar patients. Lithium 2:167–174
18. Dubovsky SL, Murphy J, Thomas M, Rademacher J (1992) Abnormal intracellular calcium ion concentration in platelets and lymphocytes of bipolar patients. Am J Psychiatry 149(1):118–120. http://dx.doi.org/10.1176/ajp.149.1.118

19. Duman RS (2002) Pathopshysiology of depression: the concept of synaptic plasticity. Eur Psychiatry 17(Suppl 3):306–310. doi:10.1016/S0924-9338(02)00654-5
20. Emamghoreishi M, Schlichter L, Li PP, Parikh S, Sen J, Kamble A, Warsh JJ (1997) High intracellular calcium concentrations in transformed lymphoblasts from subjects with bipolar I disorder. Am J Psychiatry 154(7):976–982. http://dx.doi.org/10.1176/ajp.154.7.976
21. Ewart AK, Morris CA, Atkinson D, Jin W, Sternes K, Spallone P, Stock AD, Leppert M, Keating MT (1993) Hemizygosity at the elastin locus in a developmental disorder, Williams syndrome. Nat Genet 5(1):11–16. doi:10.1038/ng0993-11
22. Fava M, Alpert J, Nierenberg AA, Mischoulon D (2005) A double-blind, randomized trial of St. John's wort, fluoxetine, and placebo in major depressive disorder. J Clin Psychopharmacol 25(5):441–447. doi:10.1097/01.jcp.0000178416.60426.29
23. Fiorio Pla A, Maric D, Brazer SC, Giacobini P, Liu X, Chang YH, Ambudkar IS, Barker JL (2005) Canonical transient receptor potential 1 plays a role in basic fibroblast growth factor (bFGF)/FGF receptor-1-induced Ca2+ entry and embryonic rat neural stem cell proliferation. J Neurosci 25(10):2687–2701. doi:10.1523/JNEUROSCI.0951-04.2005
24. Fortin DA, Srivastava T, Dwarakanath D, Pierre P, Nygaard S, Derkach VA, Soderling TR (2012) Brain-derived neurotrophic factor activation of CaM-Kinase kinase via transient receptor potential canonical channels induces the translation and synaptic incorporation of GluA1-containing calcium-permeable AMPA receptors. J Neurosci 32(24):8127–8137. doi:10.1523/JNEUROSCI.6034-11.2012
25. Gastpar M, Singer A, Zeller K (2005) Efficacy and tolerability of hypericum extract STW3 in long-term treatment with a once-daily dosage in comparison with sertraline. Pharmacopsychiatry 38(2):78–86. doi:10.1055/s-2005-837807
26. Ghosh A, Greenberg ME (1995) Calcium signaling in neurons: molecular mechanisms and cellular consequences. Science 268(5208):239–247. doi:10.1126/science.7716515
27. Greka A, Navarro B, Oancea E, Duggan A, Clapham DE (2003) TRPC5 is a regulator of hippocampal neurite length and growth cone morphology. Nat Neurosci 6(8):837–845. doi:10.1038/nn1092
28. Griesi-Oliveira K, Acab A, Gupta AR, Sunaga DY, Chailangkarn T, Nicol X, Nunez Y, Walker MF, Murdoch JD, Sanders SJ, Fernandez TV, Ji W, Lifton RP, Vadasz E, Dietrich A, Pradhan D, Song H, Ming G, Gu X, Haddad G, Marchetto MCN, Spitzer N, Passos-Bueno MR, State MW, Muotri AR (2015) Modeling non-syndromic autism and the impact of TRPC6 disruption in human neurons. Mol Psychiatry 20(11):1350–1365. doi:10.1038/mp.2014.141
29. Hajszan T, Dow A, Warner-Schmidt JL, Szigeti-Buck K, Sallam NL, Parducz A, Leranth C, Duman RS (2009) Remodeling of hippocampal spine synapses in the rat learned helplessness model of depression. Biol Psychiatry 65(5):392–400. doi:10.1016/j.biopsych.2008.09.031
30. He Z, Jia C, Feng S, Zhou K, Tai Y, Bai X, Wang Y (2012) TRPC5 channel is the mediator of neurotrophin-3 in regulating dendritic growth via CaMKIIa in rat hippocampal neurons. J Neurosci 32(27):9383–9395. doi:10.1523/JNEUROSCI.6363-11.2012
31. Heo DK, Chung WY, Park HW, Yuan JP, Lee MG, Kim JY (2012) Opposite regulatory effects of TRPC1 and TRPC5 on neurite outgrowth in PC12 cells. Cell Signal 24(4):899–906. doi:10.1016/j.cellsig.2011.12.011
32. Hough C, Lu S-J, Davis CL, Chuang D-M, Post RM (1999) Elevated basal and thapsigargin-stimulated intracellular calcium of platelets and lymphocytes from bipolar affective disorder patients measured by a fluorometric microassay. Biol Psychiatry 46(2):247–255. doi:10.1016/S0006-3223(98)00308-4
33. Jeon JP, Roh SE, Wie J, Kim J, Kim H, Lee KP, Yang D, Jeon JH, Cho NH, Kim IG, Kang DE, Kim HJ, So I (2013) Activation of TRPC4β by Gαi subunit increases Ca2+ selectivity and controls neurite morphogenesis in cultured hippocampal neuron. Cell Calcium 54(4):307–319. doi:10.1016/j.ceca.2013.07.006
34. Jia Y, Zhou J, Tai Y, Wang Y (2007) TRPC channels promote cerebellar granule neuron survival. Nat Neurosci 10(5):559–567. doi:10.1038/nn1870
35. Kaczmarek JS, Riccio A, Clapham DE (2012) Calpain cleaves and activates the TRPC5 channel to participate in semaphorin 3A-induced neuronal growth cone collapse. Proc Natl Acad Sci U S A 109(20):7888–7892. doi:10.1073/pnas.1205869109
36. Katz DM (2014) Brain-derived neurotrophic factor and Rett syndrome. Handb Exp Pharmacol 220:481–495. doi:10.1007/978-3-642-45106-5_18
37. Kellner Y, Gödecke N, Dierkes T, Thieme N, Zagrebelsky M, Korte M (2014) The BDNF effects on dendritic spines of mature hippocampal neurons depend on neuronal activity. Front Synaptic Neurosci 6(5):1–17. doi:10.3389/fnsyn.2014.00005
38. Kerman IA (2012) New insights into BDNS signaling: relevance to major depression and antidepressant action. Am J Psychiatry 169(11):1137–1140. doi:10.1176/appi.ajp.2012.12081053
39. Kerner B (2014) Genetics of bipolar disorder. Appl Clin Genet 7:33–42. doi:10.2147/TACG.S39297
40. Kerner B (2015) Toward a deeper understanding of the genetics of bipolar disorder. Front Psychiatry 6:105. doi:10.3389/fpsyt.2015.00105
41. Khattak S, Brimble E, Zhang W, Zaslavsky K, Strong E, Ross PJ, Hendry J, Mital S, Salter MW, Osborne LR, Ellis J (2015) Human induced pluripotent stem cell derived neurons as a model for Williams-Beuren syndrome. Mol Brain 8(1):77. doi:10.1186/s13041-015-0168-0

42. Kim SJ (2013) TRPC3 channel underlies cerebellar long-term depression. Cerebellum 12(3):334–337. doi:10.1007/s12311-013-0455-1
43. Kruse K, Pankau R, Gosch A, Wohlfahrt K (1992) Calcium metabolism in Williams-Beuren syndrome. J Pediatr 121(6):902–907. doi:10.1016/S0022-3476(05)80336-1
44. Lalli MA, Jang J, Park JC, Wang Y, Guzman E, Zhou H, Audouard M, Bridges D, Tovar KR, Papuc SM, Tutulan-Cunita AC, Huang Y, Budisteanu M, Arghir A, Kosik KS (2016) Haploinsufficiency of BAZ1B contributes to Williams syndrome through transcriptional dysregulation of neurodevelopmental pathways. Hum Mol Genet 25(7):1294–1306. doi:10.1093/hmg/ddw010
45. Letavernier E, Rodenas A, Guerrot D, Haymann J-P (2012) Williams-Beuren syndrome hypercalcemia: is TRPC3 a novel mediator in calcium homeostasis? Pediatrics 129(6):e1626–e1630. doi:10.1542/peds.2011-2507
46. Leuner K, Kazanski V, Müller M, Essin K, Henke B, Gollasch M, Harteneck C, Müller WE (2007) Hyperforin – a key constituent of St. John's wort specifically activates TRPC6 channels. FASEB J 21(14):4101–4111. doi:10.1096/fj.07-8110com
47. Leuner K, Li W, Amaral MD, Rudolph S, Calfa G, Schuwald AM, Harteneck C, Inoue T, Pozzo-Miller L (2013) Hyperforin modulates dendritic spine morphology in hippocampal pyramidal neurons by activating Ca(2+)-permeable TRPC6 channels. Hippocampus 23(1):40–52. doi:10.1002/hipo.22052
48. Li H, Xu XS, Montell C (1999) Activation of a TRPC3-dependent cation current through the neurotrophin BDNF. Neuron 24(1):261–273. doi:10.1016/S0896-6273(00)80838-7
49. Li M, Chen C, Zhou Z, Xu S, Yu Z (2012) A TRPC1-mediated increase in store-operated Ca2+ entry is required for the proliferation of adult hippocampal neural progenitor cells. Cell Calcium 51(6):486–496. doi:10.1016/j.ceca.2012.04.014
50. Li W, Calfa G, Larimore J, Pozzo-Miller L (2012) Activity-dependent BDNF release and TRPC signaling is impaired in hippocampal neurons of Mecp2 mutant mice. Proc Natl Acad Sci U S A 109(42):17087–17092. doi:10.1073/pnas.1205271109
51. Li Y, Jia Y-C, Cui K, Li N, Zheng Z-Y, Wang Y-Z, Yuan X-B (2005) Essential role of TRPC channels in the guidance of nerve growth cones by brain-derived neurotrophic factor. Nature 434(7035):894–898. doi:10.1038/nature03477
52. Linde K, Mulrow CD, Berner M, Egger M (2005) St John's wort for depression. Cochrane Database Syst Rev 2:CD000448. doi:10.1002/14651858.CD000448.pub2
53. Marchetto MCN, Carromeu C, Acab A, Yu D, Yeo GW, Mu Y, Chen G, Gage FH, Muotri AR (2010) A model for neural development and treatment of Rett syndrome using human induced pluripotent stem cells. Cell 143(4):527–539. doi:10.1016/j.cell.2010.10.016
54. Marchetto MC, Belinson H, Tian Y, Freitas BC, Fu C, Vadodaria KC, Beltrao-Braga PC, Trujillo CA, Mendes AP, Padmanabhan K, Nunez Y, Ou J, Ghosh H, Wright R, Brennand KJ, Pierce K, Eichenfield L, Pramparo T, Eyler LT, Barnes CC, Courchesne E, Geschwind DH, Gage FH, Wynshaw-Boris A, Muotri AR (2016) Altered proliferation and networks in neural cells derived from idiopathic autistic individuals. Mol Psychiatry. doi:10.1038/mp.2016.95
55. Martinowich K, Hattori D, Wu H, Fouse S, He F, Hu Y, Fan G, Sun YE (2003) DNA methylation-related chromatin remodeling in activity-dependent BDNF gene regulation. Science 302(5646):890–893. doi:10.1126/science.1090842
56. Maulik PK, Mascarenhas MN, Mathers CD, Dua T, Saxena S (2011) Prevalence of intellectual disability: a meta-analysis of population-based studies. Res Dev Disabil 32(2):419–436. doi:10.1016/j.ridd.2010.12.018
57. Mignon-Ravix C, Cacciagli P, Choucair N, Popovici C, Missirian C, Milh M, Mégarbané A, Busa T, Julia S, Girard N, Badens C, Sigaudy S, Philip N, Villard L (2014) Intragenic rearrangements in X-linked intellectual deficiency: results of a-CGH in a series of 54 patients and identification of TRPC5 and KLHL15 as potential XLID genes. Am J Med Genet A 164A(8):1991–1997. doi:10.1002/ajmg.a.36602
58. Nageshappa S, Carromeu C, Trujillo CA, Mesci P, Espuny-Camacho I, Pasciuto E, Vanderhaeghen P, Verfaillie CM, Raitano S, Kumar A, Carvalho CM, Bagni C, Ramocki MB, Araujo BH, Torres LB, Lupski JR, Van Esch H, Muotri AR (2016) Altered neuronal network and rescue in a human MECP2 duplication model. Mol Psychiatry 21(2):178–188. doi:10.1038/mp.2015.128
59. Neuner SM, Wilmott LA, Hope KA, Hoffmann B, Chong JA, Abramowitz J, Birnbaumer L, O'Connell KM, Tryba AK, Greene AS, Savio Chan C, Kaczorowski CC (2015) TRPC3 channels critically regulate hippocampal excitability and contextual fear memory. Behav Brain Res 281:69–77. doi:10.1016/j.bbr.2014.12.018
60. Ng F, Berk M, Dean O, Bush AI (2008) Oxidative stress in psychiatric disorders: evidence base and therapeutic implications. Int J Neuropsychopharmacol 11(6):851–876. doi:10.1017/S1461145707008401
61. Novina CD, Kumar S, Bajpai U, Cheriyath V, Zhang K, Pillai S, Wortis HH, Roy AL (1999) Regulation of nuclear localization and transcriptional activity of TFII-I by Bruton's tyrosine kinase. Mol Cell Biol 19(7):5014–5024. doi:10.1128/MCB.19.7.5014
62. Paez PM, Fulton D, Spreuer V, Handley V, Campagnoni AT (2011) Modulation of canonical transient receptor potential channel 1 in the proliferation of oligodendrocyte precursor cells by the golli products of the myelin basic protein gene. J Neurosci 31(10):3625–3637. doi:10.1523/JNEUROSCI.4424-10.2011

63. Perova T, Wasserman MJ, Li PP, Warsh JJ (2008) Hyperactive intracellular calcium dynamics in B lymphoblasts from patients with bipolar I disorder. Int J Neuropsychopharmacol 11(2):185–196. doi: http://dx.doi.org/10.1017/S1461145707007973
64. Phelan KD, Shwe UT, Abramowitz J, Wu H, Rhee SW, Howell MD, Gottschall PE, Freichel M, Flockerzi V, Birnbaumer L, Zheng F (2013) Canonical transient receptor channel 5 (TRPC5) and TRPC1/4 contribute to seizure and excitotoxicity by distinct cellular mechanisms. Mol Pharmacol 83(2):429–438. doi:10.1124/mol.112.082271
65. Piton A, Redin C, Mandel JL (2013) XLID-causing mutations and associated genes challenged in light of data from large-scale human exome sequencing. Am J Hum Genet 93(2):368–383. doi:10.1016/j.ajhg.2013.06.013
66. Pittenger C, Duman RS (2008) Stress, depression, and neuroplasticity: a convergence of mechanisms. Neuropsychopharmacology 33(1):88–109. doi:10.1038/sj.npp.1301574
67. Puram SV, Riccio A, Koirala S, Ikeuchi Y, Kim AH, Corfas G, Bonni A (2011) A TRPC5-regulated calcium signaling pathway controls dendrite patterning in the mammalian brain. Genes Dev 25(24):2659–2673. doi:10.1101/gad.174060.111
68. Riccio A, Li Y, Moon J, Kim KS, Smith KS, Rudolph U, Gapon S, Yao GL, Tsvetkov E, Rodig SJ, Van't Veer A, Meloni EG, Carlezon WA, Bolshakov VY, Clapham DE (2009) Essential role for TRPC5 in amygdala function and fear-related behavior. Cell 137(4):761–772. doi:10.1016/j.cell.2009.03.039
69. Roedding AS, Li PP, Warsh JJ (2006) Characterization of the transient receptor potential channels mediating lysophosphatidic acid-stimulated calcium mobilization in B lymphoblasts. Life Sci 80(2):89–97. doi:10.1016/j.lfs.2006.08.021
70. Roedding AS, Gao AF, Au-Yeung W, Scarcelli T, Li PP, Warsh JJ (2012) Effect of oxidative stress on TRPM2 and TRPC3 channels in B lymphoblast cells in bipolar disorder. Bipolar Disord 14(2):151–161. doi:10.1111/j.1399-5618.2012.01003.x
71. Roedding AS, Tong SY, Au-Yeung W, Li PP, Warsh JJ (2013) Chronic oxidative stress modulates TRPC3 and TRPM2 channel expression and function in rat primary cortical neurons: relevance to the pathophysiology of bipolar disorder. Brain Res 1517:16–27. doi:10.1016/j.brainres.2013.04.025
72. Ropers HH (2010) Genetics of early onset cognitive impairment. Annu Rev Genomics Hum Genet 11:161–187. doi:10.1146/annurev-genom-082509-141640
73. Rosenberg SS, Spitzer NC (2011) Calcium signaling in neuronal development. Cold Spring Harb Perspect Biol 3(10):a004259. doi:10.1101/cshperspect.a004259
74. Rush AJ, Trivedi MH, Wisniewski SR, Nierenberg AA, Stewart JW, Warden D, Niederehe G, Thase ME, Lavori PW, Lebowitz BD, McGrath PJ, Rosenbaum JF, Sackeim HA, Kupfer DJ, Luther J, Fava M (2006) Acute and longer-term outcomes in depressed outpatients requiring one or several treatment steps: a STAR*D report. Am J Psychiatry 163(11):1905–1917
75. Sen S, Duman R, Sanacora G (2008) Serum brain-derived neurotrophic factor, depression, and antidepressant medications: meta-analysis and implications. Biol Psychiatry 64(6):527–532. doi:10.1016/j.biopsych.2008.05.005
76. Shin HY, Hong YH, Jang SS, Chae HG, Paek SL, Moon HE, Kim DG, Kim J, Paek SH, Kim SJ (2010) A role of canonical transient receptor potential 5 channel in neuronal differentiation from A2B5 neural progenitor cells. PLoS ONE 5(5):e10359. doi:10.1371/journal.pone.0010359
77. Shinozaki G, Potash JB (2014) New developments in the genetics of bipolar Disorder. Curr Psychiatry Rep 16(11):493. doi:10.1007/s11920-014-0493-5
78. Szegedi A, Kohnen R, Dienel A, Kieser M (2005) Acute treatment of moderate to severe depression with hypericum extract WS5570 (St John's wort): randomized controlled double blind non-inferiority trial versus paroxetine. BMJ 330(7490):503–507. doi: http://dx.doi.org/10.1136/bmj.38356.655266.82
79. Tai Y, Feng S, Ge R, Du W, Zhang X, He Z, Wang Y (2008) TRPC6 channels promote dendritic growth via the CaMKIV-CREB pathway. J Cell Sci 121:2301–2307. doi:10.1242/jcs.026906
80. Treiber K, Singer A, Henke B, Müller WE (2005) Hyperforin activates nonselective cation channels (NSCCs). Br J Pharmacol 145(1):75–83. doi:10.1038/sj.bjp.0706155
81. Varney MA, Godfrey PP, Drummond AH, Watson SP (1992) Chronic lithium treatment inhibits basal and agonist-stimulated responses in rat cerebral cortex and GH3 pituitary cells. Mol Pharmacol 42(4):671–678
82. Verkhratsky A, Orkand RK, Kettenmann H (1998) Glial calcium: homeostasis and signaling function. Physiol Rev 78(1):99–141
83. Wang J-F, Shao L, Sun X, Young LT (2009) Increased oxidative stress in the anterior cingulate cortex of subjects with bipolar disorder and schizophrenia. Bipolar Disord 11(5):523–529. doi:10.1111/j.1399-5618.2009.00717.x
84. Wasserman MJ1, Corson TW, Sibony D, Cooke RG, Parikh SV, Pennefather PS, Li PP, Warsh JJ (2004) Chronic lithium treatment attentuates intracellular calcium mobilization. Neuropsychopharmacology 29(4):759–769. doi:10.1038/sj.npp.1300400
85. Yamaji T, Kagaya A, Uchitomi Y, Yokota N, Yamawaki S (1997) Chronic treatment with antidepressants, verapamil, or lithium inhibits the serotonin-induced intracellular calcium response in individual C6 rat glioma cells. Life Sci 60(11):817–823. doi:10.1016/S0024-3205(97)00010-6
86. Yoon I-S, Li PP, Siu K-P, Kennedy JL, Macciardi F, Cooke RG, Parikh SV, Warsh JJ (2001) Altered

TRPC7 gene expression in bipolar-I disorder. Biol Psychiatry 50(8):620–626. doi:10.1016/S0006-3223(01)01077-0

87. Yuan JP, Kiselyov K, Shin DM, Chen J, Shcheynikov N, Kang SH, Dehoff MH, Schwarz MK, Seeburg PH, Muallem S, Worley PF (2003) Homer binds TRPC family channels and is required for gating of TRPC1 by IP3 receptors. Cell 114(6):777–789. doi:10.1016/S0092-8674(03)00716-5

88. Zaeri S, Farjadian S, Emamghoreishi M (2015) Decreased levels of canonical transient receptor potential channel 3 protein in the rat cerebral cortex after chronic treatment with lithium or valproate. Res Pharm Sci 10(5):397–406

89. Zanoli P (2004) Role of hyperforin in the pharmacological activities of St. John's Wort. CNS Drug Rev 10(3):203–218. doi:10.1111/j.1527-3458.2004.tb00022.x

90. Zhou J, Du W, Zhou K, Tai Y, Yao H, Jia Y, Ding Y, Wang Y (2008) Critical role of TRPC6 channels in the formation of excitatory synapses. Nat Neurosci 11(7):741–743. doi:10.1038/nn.2127

TRPC Channels and Cell Proliferation

13

Cheng Zhan and Yu Shi

Abstract

TRPCs have been demonstrated to be widely expressed in different cancers. In recent years, a number of studies closely investigated the roles of TRPCs in cancer cells. Most of the results show that both mRNA and protein levels of TRPCs significantly increase in cancer tissues compared with healthy controls. TRPCs regulate Ca^{2+} homeostasis, contribute to cell cycle regulation and the expression/activation of Ca^{2+}-related factors, and thus play critical roles in the proliferation of cancer cells. Therefore, TRPCs could act as potential drug targets for cancer diagnosis and therapy.

Keywords

Calcium • Cancer • Cell cycle • Proliferation • TRPCs

13.1 Introduction

Cancer is the leading cause of death. The social burden of cancer is expected to grow year by year, with an estimated 14.1 million new cancer cases and 8.2 million cancer deaths occurred in 2012 throughout the world [35]. The most fundamental character of cancer cells is their ability of sustained proliferation [18]. About decades ago, several studies proposed that Ca^{2+}could regulate cell proliferation, and since then it has been intensively investigated in various types of cancers [3]. Now it is well accepted that Ca^{2+} plays a key role throughout the cell cycle and is critical in early G1 phase, the G1/S and G2/M transitions [7]. Ca^{2+}can regulate the expression of transcription factors (e.g., MYC, CREB, NF-κB, and NFAT) and stimulate multiple signaling pathways (e.g., Ca^{2+}/CaM/CaMK, and Ca^{2+}/calcineurin) [9, 24, 25]. Through these factors, Ca^{2+} coordinates expression of cell cycle regulators, such as cyclin D1, cyclin-dependent kinase 4 (CDK4), and retinoblastoma (RB1), to ensure that the cell enters cell cycle and passes through the boundary of checkpoints [24]. Meanwhile,

C. Zhan • Y. Shi (✉)
Department of Thoracic Surgery, Zhongshan Hospital, Fudan University, Shanghai, People's Republic of China
e-mail: shi.yu@zs-hospital.sh.cn

Y. Wang (ed.), *Transient Receptor Potential Canonical Channels and Brain Diseases*, Advances in Experimental Medicine and Biology 976, DOI 10.1007/978-94-024-1088-4_13

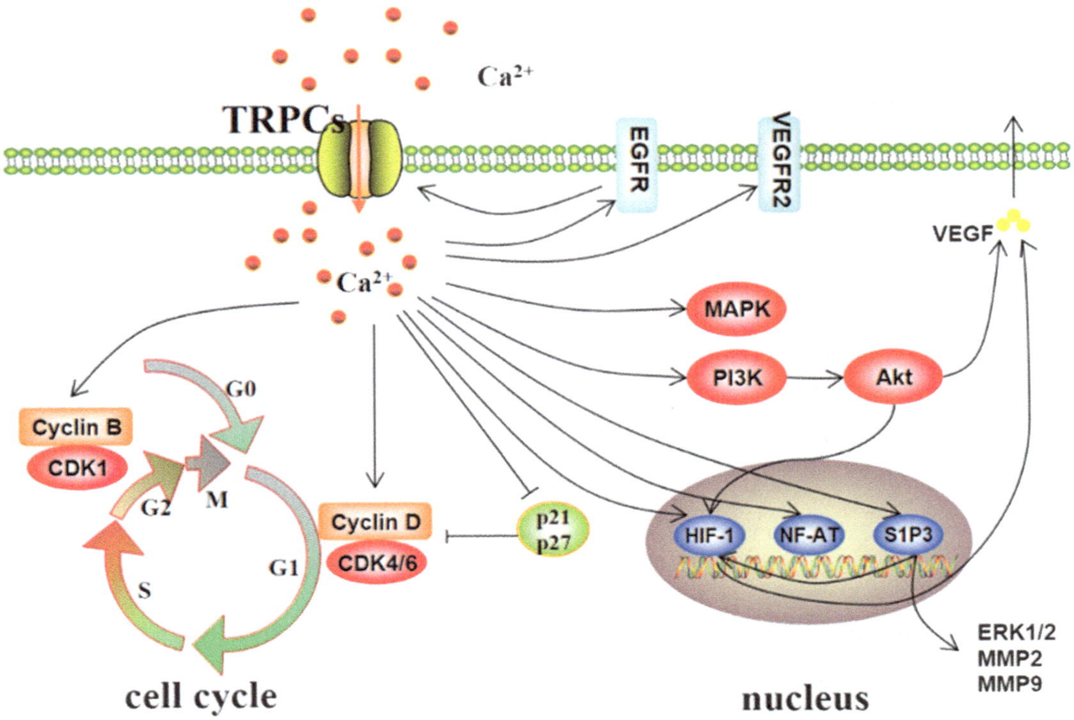

Fig. 13.1 Roles of TRPCs in cancer cell proliferation. TRPCs play important roles in cell proliferation via regulating the intracellular Ca^{2+} concentration, the cell cycle, and the expression/activation of Ca^{2+}-related factors

Ca^{2+} signaling is also involved in the process of cell growth, survival, and death [26].

As a family of nonselective cation channels permanent to Ca^{2+}, TRPCs are widely expressed in different types of cancer cells [27, 29]. Many of the TRPC members are demonstrated to maintain intracellular Ca^{2+} homeostasis and play important roles in cell cycle and Ca^{2+}-related factors and pathways (Fig. 13.1). Moreover, blockade of TRPCs leads to significant inhibition of tumor cell proliferation. Therefore, TRPCs could be potential targets in the treatment of cancers. So far, studies on exploring the role of TRPCs in cancer proliferation mainly focus on TRPC1 and TRPC6, while TRPC2 and TRPC7 have not been investigated.

13.2 TRPC1

Studies have shown that the reduced expression or activity of TRPC1 can result in inhibition of cell proliferation in various cancers both in vitro and in vivo. It has been reported that the cell proliferation is greatly suppressed when TRPC1 is knocked down or loss-of-function mutated in ML-1 cells, a thyroid cancer cell line, and cell proliferation is recovered when the expression of TRPC1 is introduced in TRPC1 knockdown cells [1]. Additionally, inhibition of TRPC1 activity by pharmacological agents or inhibition of its expression by shRNA both lead to suppression of cell proliferation and incomplete cell division, finally resulting in multinucleated cells commonly found in patient biopsies [5]. In a flank tumor model, tumor size is markedly decreased when TRPC1 expression was disrupted by doxycycline-induced shRNA [5]. It has been reported that TRPC1 is overexpressed in human breast ductal adenocarcinoma compared to the adjacent normal tissue [10], which is strongly correlated with proliferative parameters (histo-prognosis grade, Ki67 proliferation index, and tumor size). Downregulation of TRPC1 [14] impairs the calcium-sensing receptor (CaSR)-

stimulated proliferation of breast cancer MCF-7 cells. Selli, C et al. found that when TRPC1 is downregulated by siRNA in human hepatocellular carcinoma cell line Huh7, the cell proliferation is significantly suppressed, and the carcinoma cells have a longer doubling time [28]. Tajeddine, N et al. showed that TRPC1 knockdown by siRNA in non-small cell lung carcinoma cell lines results in dramatic decrease in cell proliferation [33]. Jiang, H N et al. reported that the proliferation of A549 cells is inhibited by the neutralizing antibody T1E3 which specifically targets TRPC1, while proliferation is increased in cells overexpressed with TRPC1 [19]. Zeng, B et al. reported inhibition of TRPC1 reduces the cell proliferation, while overexpression of TRPC1 or a spliced isoform of TRPC1 with exon 9 deletion increase the cancer cell colony growth in human ovarian cancer cells [38].

Most studies have showed that Ca^{2+} entry is attenuated when TRPC1 is knocked down in cancer cells, resulting in the alteration of Ca^{2+}-related cell cycle and signaling pathways. In TRPC1 knockdown ML-1 cells, there was a significant increase of cell population in the G1 phase along with a significant decrease in the S and G2 phases of the cell cycle. Moreover, the expression of several cell cycle regulatory proteins is altered, such as an increase in levels of cyclin-dependent kinase (CDK) inhibitors p21 and p27, while a reduction in levels of cyclin D2, cyclin D3, and CDK6 [1]. Tajeddine, N et al. showed that TRPC1 knockdown in non-small cell lung carcinoma cell lines results in G0/G1 cell cycle arrest and leads tothe reduced expression of cyclin D1 and cyclin D3 [33]. These results suggest the Ca^{2+} influx via TRPC1 is crucial for the cell cycle progress.

Furthermore, the pro-migratory sphingosine-1-phosphate 3 (S1P3) and vascular endothelial growth factor receptor 2 (VEGFR2) are significantly downregulated in the TRPC1 knockdown cells due to the attenuated calcium entry [1]. Hypoxia-inducible factor-1 (HIF-1), extracellular regulated protein kinases 1/2 (ERK1/2), matrix metalloproteinase 2 (MMP2), as well as MMP9, factors known critical for cell proliferation and migration, have been found to be regulated by S1P3, and knockdown of TRPC1 results in decreased expression or inactivation of these proteins, leading to inhibition of cell proliferation [4, 20, 21]. TRPC1 is also an important component of a Ca^{2+}-dependent amplification of epidermal growth factor (EGF)-dependent cell proliferation. Depletion of TRPC1 is associated with a decreased phosphorylation and activation of epidermal growth factor receptor (EGFR). Additionally, a disruption of phosphatidylinositol 3-kinase (PI3K)/AKT serine/threonine kinase (Akt) and mitogen-activated protein kinase (MAPK) a downstream pathway is also found when TPRC1 is suppressed [14]. Activation of EGFR by EGF induces Ca^{2+} entry through TRPC1, and Ca^{2+} entry further activates EGFR and regulates TRPC1. In contrast, inhibition of EGFR reduces the expression of TRPC1 and suppresses the cell proliferation [14]. Therefore, TRPC1 can act as a helper to the factors that induce cell proliferation. Moreover, activated TRPC1 can in turn mediate the signaling to induce the factor expression. In this context, TRPC1 can be regarded as a novel potential therapeutic target for the cancer cells.

13.3 TRPC3

There are only a few studies which focus on the role of TRPC3 in cancer cell proliferation. Zeng, B et al. reported that transfection with siRNA or application of specific blocking antibodies against TRPC3 inhibits the cell proliferation in ovarian cancer, while overexpression of TRPC3 increases the cancer cell colony growth [38]. Yang, S L et al. found that TRPC3 protein levels in ovarian cancer specimens are greatly increased compared with those in normal specimens [37]. Downregulating TRPC3 expression in SKOV3 cells leads to a reduction of proliferation, suppression in EGF-induced Ca^{2+} influx, dephosphorylation of CDK1 and calcium-/calmodulin-dependent protein kinase II alpha (CamKIIA), and prolonged progression through M phase of these cells. In contrast, reducing the expression of TRPC3 suppresses the tumor formation in nude mice [37]. Tao, X et al. also reported that TRPC3 plays important roles in follicle-stimulating hormone (FSH)-stimulated cell proliferation in ovarian cancers. The expression of TRPC3 is increased after treatment with FSH in HEY and ES-2 cells [34]. Knockdown of TRPC3 blocks the FSH-induced facilitation of Ca^{2+} influx, abrogates the activation of Akt phosphorylation, and leads to decreased expression of downstream effectors including survivin, HIF-1a, and VEGF [34]. These results suggest that TRPC3 could be a potential therapeutic target in ovarian cancer.

Another study reported that the histone variant macroH2A represses the expression of TRPC3 in bladder cancer cells and attenuates TRPC3-mediated Ca^{2+}-dependent proliferation responses. MacroH2A1 recruits histone deacetylase 1 (HDAC1) and HDAC2, inhibits the histone acetylation, and impairs the transcription of TRPC3, which finally results in a decreased cell growth [22].

13.4 TRPC4

There is only one work reporting the function of TRPC4 in cell proliferation. Zeng et al. reported that both downregulation of TRPC4 by siRNA and blockage of TRPC4 by specific antibodies inhibit cell proliferation, while overexpression of TRPC4 increases the cancer cell colony growth [38].

13.5 TRPC5

TRPC5 and P-glycoprotein (P-gp) expression are largely upregulated in the adriamycin-resistant breast cancer cell line MCF-7/ADM. Inhibition of TRPC5 activity or reduction in its expression causes a remarkable reduction in P-gp protein expression as well as inhibition of cell growth. And the adriamycin resistance in MCF-7/ADM cells is restored [23]. NFATc3, an isoform of Ca^{2+}-dependent transcription factor NFAT (nuclear factor of activated T cells), regulates the transcription of P-gp. Activation of TRPC5 induces a time-dependent NFATc3 translocation from cytoplasm to nucleus [23]. Inhibition of TRPC5 causes a reduction of VEGF release from MCF-7/ADM cells, decreases the expression of HIF-1a, and inhibits the HIF-1a translocation from cytosol to nucleus [39]. These results suggest a role of the TRPC5-NFATc3-P-gp and TRPC5-HIF-1-VEGF signaling pathways in cell proliferation and drug resistance in breast cancer cells.

13.6 TRPC6

Recently, the important role of TRPC6 in cell proliferation has been intensively studied, and most of the studies attribute its function to the alternated Ca^{2+}-related cell cycle. Aydar, E et al. reported silencing of TRPC6 by siRNAs in MDA-MB-231 cells results in inhibition of cell growth [2]. The expression of TRPC6 is lower in isolated hepatocytes from healthy patients compared with tumor samples. The rate of cell proliferation in TRPC6-overexpressed cells is 80% faster than untransfected cells, and their SOCE amplitude was 160% higher, whereas proliferation rate is 50% lower and SOCE amplitude is 85% lower in TRPC6-knockdown cells compared with controls [13]. The epithelial growth factor and hepatocyte growth factor increase the expression level of TRPC6, and cyclin D1 expression is decreased when TRPC6 is knocked down [13]. The expression of TRPC6 is greatly upregulated in human gastric cancer epithelial cells compared with that in normal gastric epithelial cell. Treatment with TRPC channel inhibitor ofgastric cancer cells inhibits Ca^{2+} elevation evoked by histamine, arrests cell cycle in G2/M phase, suppresses cell growth, and inhibits the formation of gastric tumors in nude mice [6]. Expressing a dominant-negative form of TRPC6 also arrests cell cycle in G2/M phase and inhibits cell growth [6]. TRPC6 is overexpressed in human glioma cells and is essential for glioma development by regulation of G2/M phase transition. Inhibition of TRPC6 activity or expression attenuates the increased intracellular Ca^{2+} by platelet-derived growth factor, inhibits cell growth and clonogenic ability, induces cell cycle arrest at the G2/M phase, and enhances the antiproliferative effect of ionizing radiation [11].

Moreover, TRPC6 regulates CDK1 activation and the expression of cell division cycle 25 homolog C, both of which are crucial in the cell cycle arrest. Inhibition of TRPC6 activity also reduces tumor volume and increases survival of mice [11].

TRPC6 is also a critical factor that contributes to VEGF-evoked Ca^{2+} current in human umbilical vein endothelial cells (HUVECs). Inhibition of TRPC6 arrests HUVECs at G2/M phase and suppresses VEGF-induced HUVEC proliferation [15]. Guilbert et al. reported that TRPC6 mRNA and protein levels are elevated in breast carcinoma specimens compared to normal breast tissues [16]. The cell proliferation is increased in the A549 cells overexpressing TRPC6 [19]. The expression of TRPC6 is remarkably increased in human esophageal cancer specimens compared to normal human esophageal tissues by immunohistochemistry staining [30]. Blockade of TRPC6 channels inhibits the elevation of intracellular Ca^{2+} concentration and the activation of CDK1 in human esophageal cancer cells, then arrests the cell cycle at G2/M phase, and suppresses the cell growth both in vitro and in vivo [30]. Blockade of TRPC6 channels also suppresses basal cell proliferation and partially inhibits HGF-induced cell proliferation via prolonging the transition through G2/M phase in renal cell carcinoma (RCC) [31]. Furthermore, the TRPC6 protein is detected in 73.3% RCC samples, with a much higher level than in normal renal samples (30.0%), and the expression of TRPC6 is markedly associated with RCC histological grades [31]. In addition, Song et al. found that lovastatin inhibits the cell proliferation in human B lymphoma cells, but not in the cells when TRPC6 is knocked down [32]. It is proposed that intracellular oxidative stress due to exogenous cholesterol would be responsible for malignant proliferation via TRPC6-mediated elevation of intracellular Ca^{2+}. Together, these results strongly suggest that Ca2+ influx via TRPC6 is critical for different kinds of cells to pass the G2/M phase in the cell cycle. Therefore, blocking this channel activity will be a novel strategy for radiotherapy as the cells arrested at G2/M phase are sensitive to radiation. Indeed, application of HGF significantly increases the expression of TRPC6 and the calcium influx, while downregulation of TRPC6 inhibits the calcium influx. Consistently, the prostate cells are arrested at the G2/M phase, and HGF-induced cell proliferation is also suppressed [36]. Moreover, Zeng et al. reported targeting TRPC6 with siRNA inhibits the cell proliferation, while overexpressing TRPC increases the cancer cell colony growth in human ovarian cancer [38].

Kim et al. also reported that in bladder cancer cells macroH2A, the expression of TRPC6 is downregulated through the inhibition of histone acetylation [22]. Pharmacological inhibition of Notch signaling greatly abrogates hypoxia-induced TRPC6 expression, and TRPC6-mediated Ca^{2+} entry increases NFAT activation and promotes NFAT-dependent cell proliferation [8].

13.7 TRPC Inhibitors and Blocking Antibodies

SKF96365 (1-[2-[3-(4-Methoxyphenyl)propoxy]-2-(4-methoxyphenyl)ethyl]-1*H*-imidazole hydrochloride) and 2-APB (2-aminoethoxydiphenylborate) are the two most commonly used inhibitors to the nonselective cation TRPCs. Both SKF96365 and 2-APB cause the inhibition of cell proliferation in cancers. For example, 2-APB significantly reduces the S phase of the cell cycle and the cell proliferation in MCF-7 cells, and SKF96365 inhibits the cell growth of a glioma cell line D54MG and arrests it at G2/M phase of the cell cycle with cytokinesis defects.

Current pharmacological inhibitors do not provide enough selectivity to discriminate TRPC individualmember. Given the nature of specificity of antibodies, targeting the third extracellular region (E3) has been demonstrated to specifically block some TRPC channels. So far, T1E3 and T5E3, polyclonal antibodies targeting directly to TRPC1 and TRPC5, respectively, have been used in some researches. However, their impacts on cell proliferation remain to be explored [12, 17].

13.8 Summary and Perspectives

Cancer is a leading cause of mortality with an increase in morbidity worldwidely. The important role of TRPC members in cell proliferation in human cancers makes them as potential targets for cancer diagnosis and therapy.

Acknowledgment This study was supported by the National Natural Science Foundation of China (Grant Nos. 81401875, 81472225) (http://www.nsfc.gov.cn/) and Natural Science Foundation of Shanghai, China (Grant No. 14ZR1406000) (http://www.stcsm.gov.cn/).

References

1. Asghar MY, Magnusson M, Kemppainen K, Sukumaran P, Lof C, Pulli I, Kalhori V, Tornquist K (2015) Transient Receptor Potential Canonical 1 (TRPC1) channels as regulators of sphingolipid and VEGF receptor expression: implications for thyroid cancer cell migration and proliferation. J Biol Chem 290(26):16116–16131. doi:10.1074/jbc.M115.643668
2. Aydar E, Yeo S, Djamgoz M, Palmer C (2009) Abnormal expression, localization and interaction of canonical transient receptor potential ion channels in human breast cancer cell lines and tissues: a potential target for breast cancer diagnosis and therapy. Cancer Cell Int 9:23. doi:10.1186/1475-2867-9-23
3. Balk SD (1971) Calcium as a regulator of the proliferation of normal, but not of transformed, chicken fibroblasts in a plasma-containing medium. Proc Natl Acad Sci U S A 68(2):271–275
4. Bergelin N, Lof C, Balthasar S, Kalhori V, Tornquist K (2010) S1P1 and VEGFR-2 form a signaling complex with extracellularly regulated kinase 1/2 and protein kinase C-alpha regulating ML-1 thyroid carcinoma cell migration. Endocrinology 151(7):2994–3005. doi:10.1210/en.2009-1387
5. Bomben VC, Sontheimer H (2010) Disruption of transient receptor potential canonical channel 1 causes incomplete cytokinesis and slows the growth of human malignant gliomas. Glia 58(10):1145–1156. doi:10.1002/glia.20994
6. Cai R, Ding X, Zhou K, Shi Y, Ge R, Ren G, Jin Y, Wang Y (2009) Blockade of TRPC6 channels induced G2/M phase arrest and suppressed growth in human gastric cancer cells. Int J Cancer 125(10):2281–2287. doi:10.1002/ijc.24551
7. Capiod T (2011) Cell proliferation, calcium influx and calcium channels. Biochimie 93(12):2075–2079. doi:10.1016/j.biochi.2011.07.015
8. Chigurupati S, Venkataraman R, Barrera D, Naganathan A, Madan M, Paul L, Pattisapu JV, Kyriazis GA, Sugaya K, Bushnev S, Lathia JD, Rich JN, Chan SL (2010) Receptor channel TRPC6 is a key mediator of Notch-driven glioblastoma growth and invasiveness. Cancer Res 70(1):418–427. doi:10.1158/0008-5472.CAN-09-2654
9. Deliot N, Constantin B (2015) Plasma membrane calcium channels in cancer: alterations and consequences for cell proliferation and migration. Biochim Biophys Acta 1848(10 Pt B):2512–2522. doi:10.1016/j.bbamem.2015.06.009
10. Dhennin-Duthille I, Gautier M, Faouzi M, Guilbert A, Brevet M, Vaudry D, Ahidouch A, Sevestre H, Ouadid-Ahidouch H (2011) High expression of transient receptor potential channels in human breast cancer epithelial cells and tissues: correlation with pathological parameters. Cell Physiol Biochem 28(5):813–822. doi:10.1159/000335795
11. Ding X, He Z, Zhou K, Cheng J, Yao H, Lu D, Cai R, Jin Y, Dong B, Xu Y, Wang Y (2010) Essential role of TRPC6 channels in G2/M phase transition and development of human glioma. J Natl Cancer Inst 102(14):1052–1068. doi:10.1093/jnci/djq217
12. Dong Y, Pan Q, Jiang L, Chen Z, Zhang F, Liu Y, Xing H, Shi M, Li J, Li X, Zhu Y, Chen Y, Bruce IC, Jin J, Ma X (2014) Tumor endothelial expression of P-glycoprotein upon microvesicular transfer of TrpC5 derived from adriamycin-resistant breast cancer cells. Biochem Biophys Res Commun 446(1):85–90. doi:10.1016/j.bbrc.2014.02.076
13. El BC, Bidaux G, Enfissi A, Delcourt P, Prevarskaya N, Capiod T (2008) Capacitative calcium entry and transient receptor potential canonical 6 expression control human hepatoma cell proliferation. Hepatology 47(6):2068–2077. doi:10.1002/hep.22263
14. El HY, Ahidouch A, Lehen'Kyi V, Hague F, Gouilleux F, Mentaverri R, Kamel S, Lassoued K, Brule G, Ouadid-Ahidouch H (2009) Extracellular signal-regulated kinases 1 and 2 and TRPC1 channels are required for calcium-sensing receptor-stimulated MCF-7 breast cancer cell proliferation. Cell Physiol Biochem 23(4–6):335–346. doi:10.1159/000218179
15. Ge R, Tai Y, Sun Y, Zhou K, Yang S, Cheng T, Zou Q, Shen F, Wang Y (2009) Critical role of TRPC6 channels in VEGF-mediated angiogenesis. Cancer Lett 283(1):43–51. doi:10.1016/j.canlet.2009.03.023
16. Guilbert A, Dhennin-Duthille I, Hiani YE, Haren N, Khorsi H, Sevestre H, Ahidouch A, Ouadid-Ahidouch H (2008) Expression of TRPC6 channels in human epithelial breast cancer cells. BMC Cancer 8:125. doi:10.1186/1471-2407-8-125
17. Hai L, Kawarabayashi Y, Imai Y, Honda A, Inoue R (2011) Counteracting effect of TRPC1-associated Ca2+ influx on TNF-alpha-induced COX-2-dependent prostaglandin E2 production in human colonic myofibroblasts. Am J Physiol Gastrointest Liver Physiol 301(2):G356–G367. doi:10.1152/ajpgi.00354.2010
18. Hanahan D, Weinberg RA (2011) Hallmarks of cancer: the next generation. Cell 144(5):646–674. doi:10.1016/j.cell.2011.02.013

19. Jiang HN, Zeng B, Zhang Y, Daskoulidou N, Fan H, Qu JM, Xu SZ (2013) Involvement of TRPC channels in lung cancer cell differentiation and the correlation analysis in human non-small cell lung cancer. PLoS One 8(6):e67637. doi:10.1371/journal.pone.0067637
20. Kalhori V, Kemppainen K, Asghar MY, Bergelin N, Jaakkola P, Tornquist K (2013) Sphingosine-1-phosphate as a regulator of hypoxia-induced factor-1alpha in thyroid follicular carcinoma cells. Plos One 8(6):e66189. doi:10.1371/journal.pone.0066189
21. Kalhori V, Tornquist K (2015) MMP2 and MMP9 participate in S1P-induced invasion of follicular ML-1 thyroid cancer cells. Mol Cell Endocrinol 404:113–122. doi:10.1016/j.mce.2015.01.037
22. Kim JM, Heo K, Choi J, Kim K, An W (2013) The histone variant MacroH2A regulates Ca(2+) influx through TRPC3 and TRPC6 channels. Oncogenesis 2:e77. doi:10.1038/oncsis.2013.40
23. Ma X, Cai Y, He D, Zou C, Zhang P, Lo CY, Xu Z, Chan FL, Yu S, Chen Y, Zhu R, Lei J, Jin J, Yao X (2012) Transient receptor potential channel TRPC5 is essential for P-glycoprotein induction in drug-resistant cancer cells. Proc Natl Acad Sci U S A 109(40):16282–16287. doi:10.1073/pnas.1202989109
24. Mellstrom B, Savignac M, Gomez-Villafuertes R, Naranjo JR (2008) Ca2+-operated transcriptional networks: molecular mechanisms and in vivo models. Physiol Rev 88(2):421–449. doi:10.1152/physrev.00041.2005
25. Pinto MC, Kihara AH, Goulart VA, Tonelli FM, Gomes KN, Ulrich H, Resende RR (2015) Calcium signaling and cell proliferation. Cell Signal 27(11):2139–2149. doi:10.1016/j.cellsig.2015.08.006
26. Roderick HL, Cook SJ (2008) Ca2+ signalling checkpoints in cancer: remodelling Ca2+ for cancer cell proliferation and survival. Nat Rev Cancer 8(5):361–375. doi:10.1038/nrc2374
27. Santoni G, Farfariello V (2011) TRP channels and cancer: new targets for diagnosis and chemotherapy. Endocr Metab Immune Disord Drug Targets 11(1):54–67
28. Selli C, Erac Y, Kosova B, Erdal ES, Tosun M (2015) Silencing of TRPC1 regulates store-operated calcium entry and proliferation in Huh7 hepatocellular carcinoma cells. Biochem Pharmacol 71:194–200. doi:10.1016/j.biopha.2015.02.024
29. Shapovalov G, Lehen'Kyi V, Skryma R, Prevarskaya N (2011) TRP channels in cell survival and cell death in normal and transformed cells. Cell Calcium 50(3):295–302. doi:10.1016/j.ceca.2011.05.006
30. Shi Y, Ding X, He ZH, Zhou KC, Wang Q, Wang YZ (2009) Critical role of TRPC6 channels in G2 phase transition and the development of human oesophageal cancer. Gut 58(11):1443–1450. doi:10.1136/gut.2009.181735
31. Song J, Wang Y, Li X, Shen Y, Yin M, Guo Y, Diao L, Liu Y, Yue D (2013) Critical role of TRPC6 channels in the development of human renal cell carcinoma. Mol Biol Rep 40(8):5115–5122. doi:10.1007/s11033-013-2613-4
32. Song X, Liu BC, Lu XY, Yang LL, Zhai YJ, Eaton AF, Thai TL, Eaton DC, Ma HP, Shen BZ (2014) Lovastatin inhibits human B lymphoma cell proliferation by reducing intracellular ROS and TRPC6 expression. Biochim Biophys Acta 1843(5):894–901. doi:10.1016/j.bbamcr.2014.02.002
33. Tajeddine N, Gailly P (2012) TRPC1 protein channel is major regulator of epidermal growth factor receptor signaling. J Biol Chem 287(20):16146–16157. doi:10.1074/jbc.M112.340034
34. Tao X, Zhao N, Jin H, Zhang Z, Liu Y, Wu J, Bast RJ, Yu Y, Feng Y (2013) FSH enhances the proliferation of ovarian cancer cells by activating transient receptor potential channel C3. Endocr Relat Cancer 20(3):415–429. doi:10.1530/ERC-12-0005
35. Torre LA, Bray F, Siegel RL, Ferlay J, Lortet-Tieulent J, Jemal A (2015) Global cancer statistics, 2012. CA Cancer J Clin 65(2):87–108. doi:10.3322/caac.21262
36. Wang Y, Yue D, Li K, Liu YL, Ren CS, Wang P (2010) The role of TRPC6 in HGF-induced cell proliferation of human prostate cancer DU145 and PC3 cells. Asian J Andrologia 12(6):841–852. doi:10.1038/aja.2010.85
37. Yang SL, Cao Q, Zhou KC, Feng YJ, Wang YZ (2009) Transient receptor potential channel C3 contributes to the progression of human ovarian cancer. Oncogene 28(10):1320–1328. doi:10.1038/onc.2008.475
38. Zeng B, Yuan C, Yang X, Atkin SL, Xu SZ (2013) TRPC channels and their splice variants are essential for promoting human ovarian cancer cell proliferation and tumorigenesis. Curr Cancer Drug Targets 13(1):103–116
39. Zhu Y, Pan Q, Meng H, Jiang Y, Mao A, Wang T, Hua D, Yao X, Jin J, Ma X (2015) Enhancement of vascular endothelial growth factor release in long-term drug-treated breast cancer via transient receptor potential channel 5-Ca(2+)-hypoxia-inducible factor 1alpha pathway. Pharmacol Res 93:36–42. doi:10.1016/j.phrs.2014.12.006

TRPC Channels and Glioma 14

Shanshan Li and Xia Ding

Abstract

Glioma is the most common type of brain tumors and malignant glioma is extremely lethal, with patients' 5-year survival rate less than 10%. Treatment of gliomas poses remarkable clinical challenges, not only because of their particular localization but also because glioma cells possess several malignant biological features, including highly proliferative, highly invasive, highly angiogenic, and highly metabolic aberrant. All these features make gliomas highly recurrent and drug resistant. Finding new and effective molecular drug targets for glioma is an urgent and critical task for both basic and clinical research. Recent studies have proposed a type of non-voltage-gated calcium channels, namely, canonical transient receptor potential (TRPC) channels, to be newly emerged potential drug targets for glioma. They are heavily involved in the proliferation, migration, invasion, angiogenesis, and metabolism of glioma cells. Abundant evidence from both cell models and preclinical mouse models has demonstrated that inhibition of TRPC channels shows promising anti-glioma effect. In this chapter, we will give a comprehensive review on the current progress in the studies on TRPC channels and glioma and discuss their potential clinical implication in glioma therapy.

Keywords

Glioma • Drug targets • TRPC

S. Li (✉)
Department of Molecular and Cellular Biology,
Baylor College of Medicine,
1 Baylor Plaza, Houston 77030, TX, USA
e-mail: shanshan.li@bcm.edu

X. Ding (✉)
Mouse Cancer Genetics Program,
National Cancer Institute, NIH,
Frederick, MD 21702, USA
e-mail: xia.ding@nih.gov

Y. Wang (ed.), *Transient Receptor Potential Canonical Channels and Brain Diseases*,
Advances in Experimental Medicine and Biology 976, DOI 10.1007/978-94-024-1088-4_14

14.1 Introduction

Glioma is the most common tumor in the brain. It constitutes ~30% of all brain tumors and 80% of all malignant brain tumors [16]. Glioma has a yearly incidence of 3–5 per 100,000 population, and it occurs in all age groups but is most prevalent in adults over the age of 45. Studies on the tumor cell origin suggest that glioma could originate from neural stem cells, transit amplifying cells, neural/glial progenitors, astrocytes, or oligodendrocytes [54]. Based on their histological characters, glioma is classified into four main types: ependymomas, astrocytomas, oligodendrogliomas, and mixed gliomas [1]. Glioma is further classified according to their pathologic features (WHO grade I–IV), and glioblastoma multiforme (GBM) is the most aggressive glioma that accounts for more than 50% of gliomas [35]. Despite great progresses have been made in conventional therapeutic approaches, the overall 5-year survival rate of GBM is less than 5% and even worse for the elderly [12]. Therefore, it is fundamentally important to explore new molecular targets, and combination of those new treatments with conventional therapeutic approaches may improve the effects for patients with GBM.

Gliomagenesis and development are complex processes, which are only partially understood. Maintenance of intracellular Ca^{2+} homeostasis is essential for a large number of cellular processes [3]. The Ca^{2+} exerts biphasic effects on cellular growth. For example, a modest increase in intracellular Ca^{2+} concentration ($[Ca^{2+}]_i$) could promote cell proliferation, whereas high $[Ca^{2+}]_i$ would result in elevated mitochondrial Ca^{2+} levels and eventually the release of pro-apoptotic factors, leading to cell death [47]. To regulate the intracellular Ca^{2+} levels, cells have evolved multiple mechanisms, mainly by regulating the function of Ca^{2+} channels. Important players in this regulation are members of the TRP superfamily of ion channels (TRPC, TRPV, TRPM, TRPML, TRPP, TRPA, and TRPN). Cancer cells have abnormally regulated Ca^{2+} homeostasis [13], resulting in their abnormal biological behavior. Dysregulation of TRPC channels is one of the culprits leading to these malignancies.

14.2 Role of TRPC Channels in Glioma

Channels formed by the TRPC proteins are a subfamily of the TRP proteins, which are widely expressed in mammalian cells. It has been reported that the TRPC family consists of six members in human (TRPC1 and TRPC3-7) and seven in rodents (TRPC1-7) [55]. Diverse studies have suggested that TRPC channels are likely to play a role in glioma growth and development at different levels by regulating cellular metabolism, controlling cell proliferation, promoting angiogenesis, and triggering the migration and invasion during glioma progression (Fig. 14.1).

In each square are the represented members of the TRPC family, which are involved in the main processes driving glioma progression.

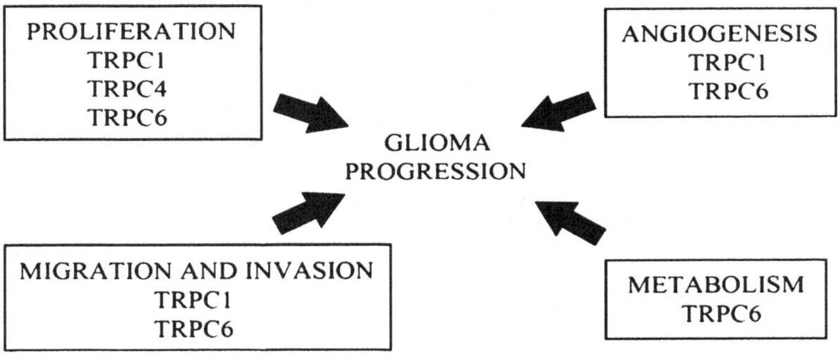

Fig. 14.1 TRPC and glioma progression

We are going to summarize the role of TRPC channels in glioma to deepen our understanding of glioma biology and help to find new and effective drug targets.

14.3 Role of TRPC Channels in Cell Proliferation in Glioma

The Ca^{2+} signaling plays a vital role in cell proliferation and is required at multiple stages of cell cycle [25]. The signaling involves receptors, channels, transducers, Ca^{2+} effectors, Ca^{2+}-sensitive enzymes, Ca^{2+} pumps, and Ca^{2+} exchangers [11]. These Ca^{2+} signaling proteins contribute to the proliferative capacity of cells.

Growth control of cancer cells has been studied extensively over the past decades, and some members of the TRPC family have been identified to affect the proliferation of many types of cancer cells [38]. Studies have demonstrated the expression of five TRPC channel proteins (TRPC1, 3, 4, 5, and 6) in glioma cell lines and patient-derived tissues [5]. Ca^{2+} influx through these channels activates intracellular Ca^{2+} effectors and Ca^{2+}-sensitive enzymes, which in turn mediate cellular activities necessary for cell cycle progression and proliferation [49].

Glioma, particularly the GBM, proliferates extensively and cells often undergo incomplete cell divisions, resulting in multinucleated cells. The recent study in D54MG glioma cells indicates that pharmacological or short hairpin RNA (shRNA)-mediated inhibition of TRPC1 channels, which are involved in agonist-induced Ca^{2+} influx and reloading of intracellular Ca^{2+} stores, leads to reduced cell proliferation and higher percentages of multinucleated cells. Moreover, decreased TRPC1 protein expression correlates with an increased percentage of multinucleated cells in GBM patient biopsies and impairs in vivo tumor growth in a xenograft tumor model. These results suggest that TRPC1 plays a critical role in glioma cell division by regulating Ca^{2+} signaling during cytokinesis [4].

In addition, a role for TRPC4 channels in glioma cell proliferation has been suggested. TRPC activation has been demonstrated as downstream signal of the epidermal growth factor receptor (EGFR) stimulation [44]. EGFR is the major growth factor receptor activated in malignant gliomas, as mutated or amplified EGFR is often observed in malignant gliomas and is associated with increased cell proliferation [7]. In Cos-7 cells, EGFR activation leads to TRPC4 phosphorylation and channel insertion into the plasma membrane [44]. Furthermore, knockdown of TRPC4 in human corneal epithelial cells suppresses EGF-induced cell proliferation, again linking proliferation to TRPC channels [62].

Another TRPC channel member, TRPC6, also plays a role in the control of glioma cell cycle and proliferation. Studies have shown that the mRNA and protein levels of TRPC6, among the TRPC family, are specifically enhanced in human glioma tissues compared to normal brain tissues. The increased expression of TRPC6 is associated with glioma malignancy grades. Moreover, functional TRPC6 channels are present in U251MG, U87MG, and T98G glioma cells. In these cells, inhibition of TRPC6 activity or expression by using a dominant-negative form of TRPC6 (DNC6) [21] or shRNA, respectively, causes a decrease in Ca^{2+} influx stimulated by platelet-derived growth factor (PDGF), which suppresses cell growth, induces cell cycle arrest at the G2/M phase, and enhances the anti-proliferation effect of irradiation. Further analysis shows that inhibition of TRPC6 suppresses the activation of cyclin-dependent kinase 1 (CDK1) via downregulation of cell division cycle protein 25C (CDC25C) expression, which is responsible for the cell cycle arrest. More importantly, inhibition of TRPC6 activity also significantly reduces tumor volumes in nude mice subcutaneous xenograft model and increases mean survival in nude mice intracranial xenograft model [11]. Besides, another study shows that TRPC6 channels could promote the proliferation and malignant growth of glioma during hypoxia by activating the calcineurin-nuclear factor of activated T-cell (NFAT) pathway [10]. Taken together, these findings strongly imply the reliance of glioma cell proliferation on TRPC6 channels.

14.4 Role of TRPC Channels in Angiogenesis in Glioma

Angiogenesis is recognized as a key event in glioma progression [2]. Glioma cells stimulate the growth of new blood vessels to support their energy demands, a process known as neovascularization. Neovascularization in the brain tumors is positively correlated with the biological aggressiveness, degree of malignancy, and clinical recurrence and negatively correlated with the postoperative survival of patients with glioma [48]. GBM is the most common form of malignant brain tumor, and the growth of these tumors is highly angiogenesis dependent, meanwhile higher-grade malignant astrocytomas have a higher degree of vascularity [57]. Tumor angiogenesis is not only resulted from adaptation to hypoxia in response to the increasing tumor mass but also resulted from genetic mutations that activate gene transcription for angiogenesis [31]. Ca^{2+} entry through plasma membrane affects angiogenesis, and several reports indicate that TRPC channels are activated during glioma angiogenesis.

Vascular endothelial growth factor (VEGF) is the most potent angiogenic factor implicated in tumor angiogenesis, whose expression is activated by hypoxia [34]. In U87MG cells, Ca^{2+} influx through TRPC channels plays a critical role in hypoxia-induced VEGF gene expression. Importantly, silencing of TRPC1, not other TRPC members, largely suppresses the upregulation of VEGF expression by hypoxia, suggesting the involvement of TRPC1 channels in glioma angiogenesis [57] Recently, it is reported that in zebra fish, knockdown of TRPC1 severely disrupts angiogenic sprouting of intersegmental vessels (ISVs), which is attributable to the impairment of filopodia extension, migration, and proliferation of ISV tip cells. Furthermore, TRPC1 acts synergistically with VEGF-A in controlling ISV growth and seems to be downstream of VEGF-A in controlling angiogenesis. Therefore, TRPC1 is essential for angiogenesis in vivo [63].

Other TRPC channels have also been found to be involved in glioma angiogenesis. Ca^{2+} entry through TRPC6 increases endothelial permeability and promotes angiogenesis [26]. TRPC6 can be activated by VEGF. Overexpression of a dominant-negative TRPC6 construct in human microvascular endothelial cells (HMVECs) suppresses the VEGF-mediated increase in intracellular Ca^{2+} levels, migration, sprouting, and proliferation. In contrast, overexpression of a wild-type TRPC6 construct promotes the proliferation and migration of HMVECs [19]. Additionally, inhibition of TRPC6 in human umbilical vein endothelial cells (HUVECs) by pharmacological or genetic approaches arrests cells at the G2/M phase and suppresses the VEGF-induced proliferation and tube formation, which are key steps in angiogenesis. Furthermore, inhibition of TRPC6 abolishes VEGF-induced angiogenesis in the chicken embryo chorioallantoic membrane (CAM) [15]. The above reports indicate that VEGF activates TRPC6 channels to regulate the angiogenesis. Conversely, TRPC6 activation in endothelial cells (ECs) may stimulate transcription and release of angiogenic growth factors such as VEGF, which stimulate angiogenesis. For example, hypoxia induces Notch1 activation, increases TRPC6 expression, and thus elevates $[Ca^{2+}]_i$ that is coupled to the activation of the calcineurin-NFAT pathway, resulting in glioma angiogenesis [10]. It remains to be determined whether Notch pathway directly or indirectly regulates TRPC6 expression [17, 52].

14.5 Role of TRPC Channels in Cell Migration and Invasion in Glioma

GBM is extremely invasive and thus the clinical prognosis for patients is desperate. Numerous studies have focused on understanding the molecular mechanisms of glioma cell invasion, and Ca^{2+} signaling has been shown to play a role in it [28].

TRPC1-mediated migration and chemotaxis have been demonstrated in different cell types such as myoblasts [36], renal epithelial cells [14], and nervous cells [58]. Recently, it has been reported that TRPC1 regulates glioma chemotaxis induced by EGF and the localization of TRPC1 in lipid rafts is essential for the function [6]. In response to EGF, TRPC1 is enriched in the leading edge of D54MG cells and co-localized with lipid raft proteins. Pharmacological or shRNA-mediated inhibition of TRPC1 channels abolishes EGF-induced cell migration, without affecting the motility of un-stimulated cells. Moreover, disruption of lipid rafts not only decreases chemotaxis but also decreases store-operated Ca^{2+} entry and impairs TRPC currents. These results indicate that TRPC1 association with lipid rafts is essential for glioma chemotaxis in response to specific stimuli.

As mentioned above, TRPC6 expression is markedly enhanced during hypoxia in a manner that is dependent on Notch activation [10]. Notch signaling is reported to mediate hypoxia-induced tumor migration and invasion [50]. TRPC6 is also required for the development of this aggressive phenotype because knockdown of TRPC6 decreases glioma invasion.

The last step of invasion requires cytoskeletal rearrangements and formation of lamellipodia and filopodia, where the Rho family of GTPases plays a critical role. The role of TRPC6 in Rho activation and actin cytoskeletal rearrangements has been indicated in several studies [51]. Therefore, the TRPC6-mediated Ca^{2+} influx likely contributes to glioma invasion by promoting actin-myosin interactions and the cell-substratum adhesion assembly and disassembly that are important for migration [39, 50].

14.6 Role of TRPC Channels in Cellular Metabolism of Glioma

Cellular metabolism influences survival and death decisions. In recent years, an emerging theme in tumor biology is that metabolic regulation is closely linked to tumor progression [42]. Interest has been renewed as it has become clear that many cancer-related pathways have significant impacts on tumor metabolism and that many tumors become dependent on specific metabolic processes.

It has been reported that all major tumor suppressors and oncogenes have intimate connections with metabolic pathways [29]. Genetic alterations (affecting HIF-1, AMPK, p53, Myc, and PI_3K) drive the metabolic inputs into multiple pathways that not only supply cellular energy (i.e., ATP) but also provide macromolecular precursors (i.e., ribose and acetyl-CoA) as well as the reducing power for biosynthetic processes and redox regulation (NADPH). More and more studies reveal that altered cellular metabolism could be one of the major routes by which oncogenes promote tumor formation and progression [23]. By far, there is no direct evidence existing about the role of Ca^{2+} in tumor metabolism, although being a key player in tumor progression. However, it should be mentioned that some key molecules (i.e., HIF-1, AMPK, p53, Myc) involved in tumor metabolism are sensitive to Ca^{2+} [8]. It is therefore possible that Ca^{2+} affects tumor metabolism by regulating these metabolic regulators.

A recent study confirmed this hypothesis [32]. Insulin-like growth factor 1 (IGF-1) is specifically released from human glioma cells during hypoxia. TRPC6, the specifically upregulated TRPC member in human glioma tissues, is then rapidly activated by IGF-1 receptor (IGF-1R)/phospholipase C (PLC)/IP_3 receptor (IP_3R) pathway. IGF-1 released in hypoxia stimulates IGF-1R, leading to the activation of PLC, which hydrolyzes phosphatidylinositol 4,5-bisphosphate (PIP_2) into diacylglycerol (DAG) and inositol 1,4,5-trisphosphate (IP_3). IP_3 induces Ca^{2+} release from internal stores, which in turn activates TRPC6 to mediate Ca^{2+} influx. TRPC6 activation decreases α-ketoglutarate (α-KG) levels and inhibits HIF prolyl hydroxylase (PHD) activities, leading to HIF-1α accumulation. HIF-1 plays a critical role in tumor cell glucose metabolism via regulating the expression of the metabolic related genes [61]. And TRPC6 is shown to enhance glucose uptake through HIF-1α. Inhibition of TRPC6 by using DNC6 dramatically suppresses hypoxia-induced glucose transporter 1 (GLUT1, also known as SLC2A1) expression. Expressing exogenous HIF-1 indeed reverses DNC6 inhibition of GLUT1 expression. Importantly, expressing HIF-1 reversed DNC6 suppression on glucose uptake in hypoxia. Further, treatment of the cells with 1-oleoyl-2-acetyl-sn-glycerol (OAG), the membrane-permeable DAG analogue known to induce Ca^{2+} entry through TRPC6, increases GLUT1 mRNA and protein levels in hypoxia, which is significantly decreased by expressing DNC6 [10]. It should be mentioned that two other well-known genes involved in energy metabolism, lactate dehydrogenase A (*LDHA*) and lactate dehydrogenase B (*LDHB*), are not affected by TRPC6. Collectively, these results imply that activation of TRPC6 promotes glucose metabolism in hypoxia, yet however does not stimulate glycolysis to produce energy. Therefore, TRPC6 controls the metabolite levels to regulate the rapid hydroxylation and stability of HIF-1α and affect the consequent glucose metabolism during hypoxia. In this context, The Ca^{2+} influx via TRPC6 can act as an important metabolic regulator (Fig. 14.2). In addition, the inhibitory effect of DNC6 on GLUT1 expression may partially explain its suppression of glioma cell development [10, 11]. Therefore, studying the possible role of TRPC in the regulation of glioma metabolism likely has profound clinical significance.

TRPC6-mediated regulation of the metabolite levels promotes glucose uptake through an increase in HIF-1α stability in human glioma cells under hypoxia [32].

Based on the above findings, TRPC channels could be potential drug targets in glioma treatment. SKF96365, known as a selective inhibitor of receptor-mediated Ca^{2+} entry and voltage-gated Ca^{2+} entry, can inhibit many other types of Ca^{2+} channels besides TRPC channels, leading to strong nonspecific effects [40]. In the past decades, great effort was made to explore the agents specifically targeting Ca^{2+} channels including TRPC channels (Table 14.1). Recently, a potent and selective TRPC4/5 antagonist, ML204, is identified. Its selectivity is far superior to other pharmacological blockers currently used in TRPC research [41]. However, such agents for TRPC1 or TRPC6 channels have not yet been reported. Therefore, in order to facilitate the clinical treatment of glioma, the development of specific blockers targeting TRPC1 and TRPC6 channels is in urgent need.

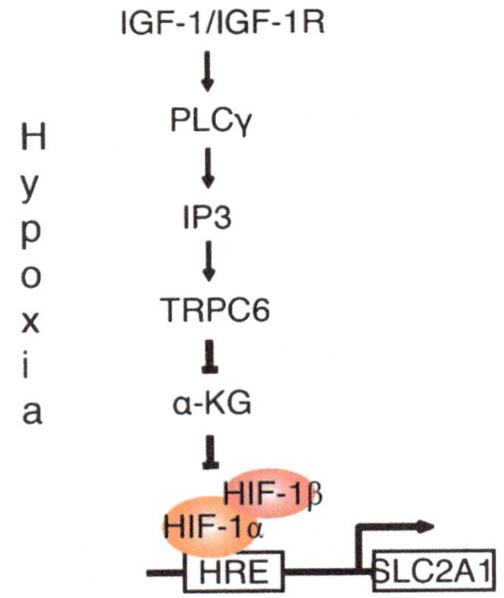

Fig. 14.2 TRPC6 and glioma metabolism

Table 14.1 Pharmacological antagonists of TRPC channels

TRPCs	Channel blockers
TRPC1	La^{3+}, Gd^{3+}, 2-APB, SKF96365, GsMTx-4 [53, 64]
TRPC3	Gd^{3+}, BTP2, La^{3+}, 2-APB, SKF96365, ACAA, Ni^{2+}, KB-R7943, Pyr3, [18, 20, 27, 33]
TRPC4	Niflumic acid, ML204, 2-APB, La^{3+}, SKF96365 [41, 56]
TRPC5	La^{3+}, KB-R7943, progesterone, ML204, bromoenol lactone, 2-APB, Mg^{2+}, chlorpromazine, BTP2, SKF96365, flufenamic acid, GsMTx-4 [9, 24, 30, 37, 41, 43, 59]
TRPC6	Gd^{3+}, La^{3+}, SKF96365, amiloride, Cd^{2+}, extracellular H^+, 2-APB, ACAA, GsMTx-4, KB-R7943, ML9 [22]
TRPC7	La^{3+}, SKF96365, amiloride, 2-APB [46]

14.7 Conclusion and Prospective

In this chapter, we have summarized the important role of TRPC channels in glioma growth and development and pointed out that TRPC channels are potential therapeutic targets for glioma. Although the link between TRPC and glioma becomes clearer, there are areas still relatively less explored. Until now, almost all the evidence comes from loss-of-function experiments. One interesting thing for the future is that whether

TRPC overexpression indeed contributes to gliomagenesis, which will deepen our understanding of TRPC in glioma biology. Another aspect of glioma biology where TRPC is going to be critical but has not been explored is the tumor epigenetics. "Epigenetics" is critical for gene regulation, where Ca^{2+} also has important roles. An increasing number of studies showed that the epigenetic differences could influence tumorigenesis [45, 60]. Further studies on the epigenetic differences resulting from TRPC dysfunction and thus the gliomagenesis are required. It is hoped that extensive knowledge about TRPC would offer effective therapeutic targets and specific chemotherapeutic agents and enable the use of TRPC channels as invaluable markers of diagnosis and prognosis in patients.

References

1. Behin A, Hoang-Xuan K et al (2003) Primary brain tumours in adults. Lancet 361(9354):323–331
2. Bello L, Giussani C et al (2004) Angiogenesis and invasion in gliomas. Cancer Treat Res 117:263–284
3. Berridge MJ, Bootman MD et al (2003) Calcium signalling: dynamics, homeostasis and remodelling. Nat Rev Mol Cell Biol 4(7):517–529
4. Bomben VC, Sontheimer H (2010) Disruption of transient receptor potential canonical channel 1 causes incomplete cytokinesis and slows the growth of human malignant gliomas. Glia 58(10):1145–1156
5. Bomben VC, Sontheimer HW (2008) Inhibition of transient receptor potential canonical channels impairs cytokinesis in human malignant gliomas. Cell Prolif 41(1):98–121
6. Bomben VC, Turner KL et al (2011) Transient receptor potential canonical channels are essential for chemotactic migration of human malignant gliomas. J Cell Physiol 226(7):1879–1888
7. Bryant JA, Finn RS et al (2004) EGF activates intracellular and intercellular calcium signaling by distinct pathways in tumor cells. Cancer Biol Ther 3(12):1243–1249
8. Cairns RA, Harris IS et al (2011) Regulation of cancer cell metabolism. Nat Rev Cancer 11(2):85–95
9. Chakraborty S, Berwick ZC et al (2011) Bromoenol lactone inhibits voltage-gated Ca2+ and transient receptor potential canonical channels. J Pharmacol Exp Ther 339(2):329–340
10. Chigurupati S, Venkataraman R et al (2010) Receptor channel TRPC6 is a key mediator of Notch-driven glioblastoma growth and invasiveness. Cancer Res 70(1):418–427
11. Ding X, He Z et al (2010) Essential role of TRPC6 channels in G2/M phase transition and development of human glioma. J Natl Cancer Inst 102(14):1052–1068
12. Dolecek TA, Propp JM et al (2012) CBTRUS statistical report: primary brain and central nervous system tumors diagnosed in the United States in 2005–2009. Neuro-Oncology 14(Suppl 5):v1–49
13. El Boustany C, Bidaux G et al (2008) Capacitative calcium entry and transient receptor potential canonical 6 expression control human hepatoma cell proliferation. Hepatology 47(6):2068–2077
14. Fabian A, Fortmann T et al (2008) TRPC1 channels regulate directionality of migrating cells. Pflugers Arch 457(2):475–484
15. Ge R, Tai Y et al (2009) Critical role of TRPC6 channels in VEGF-mediated angiogenesis. Cancer Lett 283(1):43–51
16. Goodenberger ML, Jenkins RB (2012) Genetics of adult glioma. Cancer Genet 205(12):613–621
17. Gustafsson MV, Zheng X et al (2005) Hypoxia requires notch signaling to maintain the undifferentiated cell state. Dev Cell 9(5):617–628
18. Halaszovich CR, Zitt C et al (2000) Inhibition of TRP3 channels by lanthanides. Block from the cytosolic side of the plasma membrane. J Biol Chem 275(48):37423–37428
19. Hamdollah Zadeh MA, Glass CA et al (2008) VEGF-mediated elevated intracellular calcium and angiogenesis in human microvascular endothelial cells in vitro are inhibited by dominant negative TRPC6. Microcirculation 15(7):605–614
20. Hellwig N, Albrecht N et al (2005) Homo- and heteromeric assembly of TRPV channel subunits. J Cell Sci 118(Pt 5):917–928
21. Hofmann T, Schaefer M et al (2002) Subunit composition of mammalian transient receptor potential channels in living cells. Proc Natl Acad Sci U S A 99(11):7461–7466
22. Inoue R, Okada T et al (2001) The transient receptor potential protein homologue TRP6 is the essential component of vascular alpha(1)-adrenoceptor-activated Ca(2+)-permeable cation channel. Circ Res 88(3):325–332
23. Jones NP, Schulze A (2012) Targeting cancer metabolism – aiming at a tumour's sweet-spot. Drug Discov Today 17(5–6):232–241
24. Jung S, Muhle A et al (2003) Lanthanides potentiate TRPC5 currents by an action at extracellular sites close to the pore mouth. J Biol Chem 278(6):3562–3571
25. Kahl CR, Means AR (2003) Regulation of cell cycle progression by calcium/calmodulin-dependent pathways. Endocr Rev 24(6):719–736
26. Kini V, Chavez A et al (2010) A new role for PTEN in regulating transient receptor potential canonical channel 6-mediated Ca2+ entry, endothelial permeability, and angiogenesis. J Biol Chem 285(43):33082–33091

27. Kiyonaka S, Kato K et al (2009) Selective and direct inhibition of TRPC3 channels underlies biological activities of a pyrazole compound. Proc Natl Acad Sci U S A 106(13):5400–5405
28. Komuro H, Kumada T (2005) Ca2+ transients control CNS neuronal migration. Cell Calcium 37(5):387–393
29. Koppenol WH, Bounds PL et al (2011) Otto Warburg's contributions to current concepts of cancer metabolism. Nat Rev Cancer 11(5):325–337
30. Kraft R (2007) The Na+/Ca2+ exchange inhibitor KB-R7943 potently blocks TRPC channels. Biochem Biophys Res Commun 361(1):230–236
31. Krock BL, Skuli N et al (2011) Hypoxia-induced angiogenesis: good and evil. Genes Cancer 2(12):1117–1133
32. Li S, Wang J et al (2015) Crucial role of TRPC6 in maintaining the stability of HIF-1alpha in glioma cells under hypoxia. J Cell Sci 128(17):3317–3329
33. Lievremont JP, Bird GS et al (2005) Mechanism of inhibition of TRPC cation channels by 2-aminoethoxydiphenylborane. Mol Pharmacol 68(3):758–762
34. Liu Y, Cox SR et al (1995) Hypoxia regulates vascular endothelial growth factor gene expression in endothelial cells. Identification of a 5′ enhancer. Circ Res 77(3):638–643
35. Louis DN, Ohgaki H et al (2007) The 2007 WHO classification of tumours of the central nervous system. Acta Neuropathol 114(2):97–109
36. Louis M, Zanou N et al (2008) TRPC1 regulates skeletal myoblast migration and differentiation. J Cell Sci 121(Pt 23):3951–3959
37. Majeed Y, Amer MS et al (2011) Stereo-selective inhibition of transient receptor potential TRPC5 cation channels by neuroactive steroids. Br J Pharmacol 162(7):1509–1520
38. Malarkey EB, Ni Y et al (2008) Ca2+ entry through TRPC1 channels contributes to intracellular Ca2+ dynamics and consequent glutamate release from rat astrocytes. Glia 56(8):821–835
39. Mareel M, Leroy A (2003) Clinical, cellular, and molecular aspects of cancer invasion. Physiol Rev 83(2):337–376
40. Merritt JE, Armstrong WP et al (1990) SK&F 96365, a novel inhibitor of receptor-mediated calcium entry. Biochem J 271(2):515–522
41. Miller M, Shi J et al (2011) Identification of ML204, a novel potent antagonist that selectively modulates native TRPC4/C5 ion channels. J Biol Chem 286(38):33436–33446
42. Munoz-Pinedo C, El Mjiyad N et al (2012) Cancer metabolism: current perspectives and future directions. Cell Death Dis 3:e248
43. Obukhov AG, Nowycky MC (2005) A cytosolic residue mediates Mg2+ block and regulates inward current amplitude of a transient receptor potential channel. J Neurosci 25(5):1234–1239
44. Odell AF, Scott JL et al (2005) Epidermal growth factor induces tyrosine phosphorylation, membrane insertion, and activation of transient receptor potential channel 4. J Biol Chem 280(45):37974–37987
45. Oermann EK, Wu J et al (2012) Alterations of metabolic genes and metabolites in cancer. Semin Cell Dev Biol 23(4):370–380
46. Okada T, Inoue R et al (1999) Molecular and functional characterization of a novel mouse transient receptor potential protein homologue TRP7. Ca(2+)-permeable cation channel that is constitutively activated and enhanced by stimulation of G protein-coupled receptor. J Biol Chem 274(39): 27359–27370
47. Orrenius S, Nicotera P (1994) The calcium ion and cell death. J Neural Transm Suppl 43:1–11
48. Radner H, Blumcke I et al (2002) The new WHO classification of tumors of the nervous system 2000. Pathology and genetics. Pathologe 23(4):260–283
49. Roderick HL, Cook SJ (2008) Ca2+ signalling checkpoints in cancer: remodelling Ca2+ for cancer cell proliferation and survival. Nat Rev Cancer 8(5):361–375
50. Sahlgren C, Gustafsson MV et al (2008) Notch signaling mediates hypoxia-induced tumor cell migration and invasion. Proc Natl Acad Sci U S A 105(17): 6392–6397
51. Singh I, Knezevic N et al (2007) Galphaq-TRPC6-mediated Ca2+ entry induces RhoA activation and resultant endothelial cell shape change in response to thrombin. J Biol Chem 282(11):7833–7843
52. Song LL, Peng Y et al (2008) Notch-1 associates with IKKalpha and regulates IKK activity in cervical cancer cells. Oncogene 27(44):5833–5844
53. Strübing C, Krapivinsky G et al (2001) TRPC1 and TRPC5 form a novel cation channel in mammalian brain. Neuron 29(3):645–655
54. Van Meir EG, Hadjipanayis CG et al (2010) Exciting new advances in neuro-oncology: the avenue to a cure for malignant glioma. CA Cancer J Clin 60(3):166–193
55. Venkatachalam K, Montell C (2007) TRP channels. Annu Rev Biochem 76:387–417
56. Walker RL, Koh SD et al (2002) TRPC4 currents have properties similar to the pacemaker current in interstitial cells of Cajal. Am J Phys Cell Physiol 283(6):C1637–C1645
57. Wang B, Li W et al (2009) Hypoxia up-regulates vascular endothelial growth factor in U-87 MG cells: involvement of TRPC1. Neurosci Lett 459(3):132–136
58. Wang GX, Poo MM (2005) Requirement of TRPC channels in netrin-1-induced chemotropic turning of nerve growth cones. Nature 434(7035):898–904
59. Xu SZ, Zeng F et al (2005) Block of TRPC5 channels by 2-aminoethoxydiphenyl borate: a differential, extracellular and voltage-dependent effect. Br J Pharmacol 145(4):405–414

60. Xu W, Yang H et al (2011) Oncometabolite 2-hydroxyglutarate is a competitive inhibitor of alpha-ketoglutarate-dependent dioxygenases. Cancer Cell 19(1):17–30
61. Yang F, Zhang H et al (2014) Reciprocal regulation of HIF-1alpha and lincRNA-p21 modulates the Warburg effect. Mol Cell 53(1):88–100
62. Yang H, Mergler S et al (2005) TRPC4 knockdown suppresses epidermal growth factor-induced store-operated channel activation and growth in human corneal epithelial cells. J Biol Chem 280(37):32230–32237
63. Yu P-C, Gu S-Y et al (2010) TRPC1 is essential for in vivo angiogenesis in zebrafish. Circ Res 106(7):1221–1232
64. Zitt C, Zobel A et al (1996) Cloning and functional expression of a human Ca2+-permeable cation channel activated by calcium store depletion. Neuron 16(6):1189–1196

Printed by Printforce, the Netherlands